W9-CNV-238

Hepatocyte and Kupffer Cell Interactions

Hepatocyte and Kupffer Cell Interactions

Edited by

Timothy R. Billiar
Samuel P. Harbison Assistant Professor of Surgery
University of Pittsburgh, Pittsburgh, Pennsylvania

and

Ronald D. Curran
Research Assistant Professor and Chief Resident
Department of Surgery, University of Pittsburgh
Pittsburgh, Pennsylvania

CRC Press
Boca Raton Ann Arbor London Tokyo

Library of Congress Cataloging-in-Publication Data

Hepatocyte and Kupffer Cell interactions / edited by Timothy R. Billiar and
Ronald D. Curran.
p. cm.
Includes bibliographical references and index.
ISBN 0-8493-6109-5
1. Liver—Pathophysiology. 2. Liver cells. 3. Kupffer cells.
I. Billiar, Timothy R. II. Curran, Ronald D.
RC846.9.H45 1992
616.3′6207—dc20 92-4151
CIP

This book represents information obtained from authentic and highly regarded sources. Reprinted material is quoted with permission, and sources are indicated. A wide variety of references are listed. Every reasonable effort has been made to give reliable data and information, but the author and the publisher cannot assume responsibility for the validity of all materials or for the consequences of their use.

Direct all inquiries to CRC Press, Inc., 2000 Corporate Blvd., N.W., Boca Raton, Florida, 33431.

International Standard Book Number 0-8493-6109-5

Library of Congress Card Number 92-4151

Printed in the United States of America 1 2 3 4 5 6 7 8 9 0

Printed on acid-free paper

PREFACE

Compared to the accumulated data on the functions and responses of individual liver cell populations, specifically hepatocytes and Kupffer cells, relatively little is known about the interactions between these cells. The intimate association of Kupffer cells and hepatocytes makes it unlikely that these cells act independently of each other. Recent evidence from a number of laboratories suggests that interactions between Kupffer cells and hepatocytes play roles in normal liver physiology, the normal response of the liver in disease states, and perhaps even hepatic dysfunction or injury. In a very general sense, the association between these cell types can be viewed in one of two ways: (1) this cell-cell interface represents the anatomic connection of two separate organ systems—the Kupffer cell (representing the reticuloendothelial system with its scavenging, antimicrobial, and antitumor functions) and the hepatocyte (which performs the hepatic functions of metabolism, biosynthesis, and detoxification); and (2) the Kupffer cell-hepatocyte combination can be looked upon as a functional unit with ongoing bidirectional communication and interdependent control mechanisms. Our bias is toward the second possibility, although many specific functions remain quite isolated to each cell type. Much of what is known about this complex interaction is a result of experimental studies in disease models.

This volume presents detailed literature reviews as well as current hypotheses on the role of Kupffer cell and hepatocyte interactions in a number of disease states. Sections on nonseptic disease states and septic disease states follow an introductory section on the anatomic and physiologic relationships of Kupffer cells and hepatocytes. Keeping in mind that the study of cell interaction in the liver is relatively new, we hope that this book will serve as an up-to-date summary, useful both as a general reference and a guide for additional investigation.

T.R. Billiar
R.D. Curran

THE EDITORS

Timothy R. Billiar, M.D., is the Samuel P. Harbison Assistant Professor of Surgery at the University of Pittsburgh, Pittsburgh, PA. He is the first recipient of an endowed surgical assistant professorship in the United States.

Dr. Billiar graduated summa cum laude in 1979 from Doane College in Crete, NE with a bachelor's degree in Natural Sciences, and obtained his M.D. degree from the University of Chicago Pritzker School of Medicine in 1983. His specialty training in general surgery and his research fellowship were performed at both the University of Minnesota, Minneapolis, and at the University of Pittsburgh.

Dr. Billiar is a member of the American Society for the Advancement of Science, the Society for Leukocyte Biology, the Surgical Infection Society, the Association for Academic Surgery, and the American Federation for Clinical Research.

Among other awards, he has received two National Research Service Awards from the National Institutes of Health, the American College of Surgeons Scholarship, and the Stuart Pharmaceutical Scholarship in Surgical Infectious Disease. He is the first recipient of the George H.A. Clowes Memorial Scholarship of the American College of Surgeons and was named the Outstanding Young Alumnus of Doane College in 1988.

Dr. Billiar has given over 30 invited and guest presentations at research meetings and universities, and he has published nearly 100 articles. His research interests include cell interactions and metabolic changes within the liver during infection and sepsis, and he continues to perform NIH-funded research in these areas.

Ronald D. Curran, M.D., is currently a Research Assistant Professor and Chief Resident in the Department of Surgery at the University of Pittsburgh. In 1993, he will begin a fellowship in Cardiovascular and Thoracic Surgery at Northwestern University in Chicago, IL.

Dr. Curran graduated in 1980 from DePauw University, Greencastle, IN with a bachelor's degree in Chemistry and Zoology. He obtained his M.D. degree in 1984 from Rush University, Chicago, IL and then began his post-graduate training in general surgery at the University of Minnesota. In 1987, Dr. Curran transferred to the University of Pittsburgh to pursue his research interests in the laboratory of Dr. Richard L. Simmons and to complete his training in general surgery.

Dr. Curran is a candidate member of the American College of Surgeons and a resident member of the American Medical Association. He is a member of the American Association for the Advancement of Science, the Association for Academic Surgery, and the honorary society Sigma Xi.

In medical school, Dr. Curran was awarded a Rush University Herrick research fellowship and, more recently during his surgical training, has been the recipient of research awards from the Surgical Infection Society, the Society of Critical Care Medicine, and the American College of Surgeons. He has received funding from the American Cancer Society and the pharmaceutical industry. His research has resulted in numerous publications and presentation. He currently is investigating the role of nitric oxide in modulating cardiac physiology and pathophysiology.

CONTRIBUTORS

P. Stephen Almond, M.D.
General Surgery Resident
University of Minnesota
Minneapolis, Minnesota

Timothy R. Billiar, M.D.
Samuel P. Harbison Assistant Professor of Surgery
Department of Surgery
University of Pittsburgh
Pittsburgh, Pennsylvania

Luc Bouwens, Ph.D.
Senior Researcher
Department of Cell Biology and Histology
Free University Brussels
Brussels, Belgium

Ginny L. Bumgardner, M.D.
Liver Transplant Fellow
Liver Transplant Division
University of California
San Francisco, California

Stefan A. Cohen, Ph.D.
Research Associate Professor of Medicine
Department of Medicine
State University of New York
Buffalo, New York

Ronald D. Curran, M.D.
Research Assistant Professor
Chief Resident
Department of Surgery
University of Pittsburgh
Pittsburgh, Pennsylvania

Jean-Pierre Gut, M.D., Ph.D.
Maître de Conférences
Laboratoire de Virologie
Faculté de Médecine
Strasbourg, France

Brian G. Harbrecht, M.D.
Department of Surgery
University of Louisville
Louisville, Kentucky

Michael L. Heaney, M.D.
Surgical Research Fellow
Department of Surgery
Hennepin County Medical Center
Minneapolis, Minnesota

Andre Kirn, M.D.
Professor of Virology
Director
INSERM Unit 74
Strasbourg, France

Paul H. Kispert, M.D.
Assistant Professor
Department of Surgery
University of Pittsburgh
Pittsburgh, Pennsylvania

Debra L. Laskin, Ph.D.
Associate Professor
Department of Pharmacology and Toxicology
Rutgers University
Piscataway, New Jersey

Thomas W. Lysz, Ph.D.
Associate Professor
Department of Surgery
UMDNJ
New Jersey Medical School
Newark, New Jersey
and
Research Pharmacologist
Department of Surgery
Veterans Administration Hospital
East Orange, New Jersey

George Machiedo, M.D.
Professor and Vice Chairman
Department of Surgery
UMDNJ
New Jersey Medical School
and
Director of Clinical Surgery
Department of Surgery
University Hospital
Newark, New Jersey

Andreas K. Nussler, Ph.D.
Research Assistant Professor
Department of Surgery
University of Pittsburgh
Pittsburgh, Pennsylvania

Theresa C. Peterson, Ph.D.
Assistant Professor
Departments of Medicine and Pharmacology
Dalhousie University
Halifax, Nova Scotia, Canada

Shie-Pon Tzung, M.D., Ph.D.
Medical Resident
Department of Medicine
Duke University Medical Center
Durham, North Carolina

Denis Vetter, M.D.
Professor
Department of Hepato-Gastroenterology
Hôpitaux Universitaires de Strasbourg
Strasbourg, France

Michael A. West, M.D., Ph.D.
Clinical Assistant Professor
Department of Surgery
University of Minnesota
and
Director of Critical Care Research
Department of Surgery
Hennepin County Medical Center
Minneapolis, Minnesota

S.N. Wickramasinghe, Sc.D., Ph.D., M.B.B.S., F.R.C.P., F.R.C.Path.
Professor of Haematology and Chairman of the Division of Pathology Sciences
St. Mary's Hospital Medical School
Imperial College of Science, Technology and Medicine
University of London
London, England

Eddie Wisse, Ph.D.
Professor
Free University Brussels
Medical School
Department of Cell Biology and Histology
Brussels, Belgium

ACKNOWLEDGMENTS

The editors are deeply indebted to Richard L. Simmons for his guidance and support. He recognized the potential importance of Kupffer cell and hepatocyte interactions many years ago, and a number of investigators, including ourselves, have benefited greatly from his ideas and encouragement.

We are also extremely grateful to Rebecca Pfeifer and Karen Holleran for their excellent secretarial assistance.

TABLE OF CONTENTS

Section I

The Normal Liver

Chapter 1

THE ORIGIN OF KUPFFER CELLS AND THEIR ANATOMIC RELATIONSHIP TO HEPATOCYTES

L. Bouwens and E. Wisse

TABLE OF CONTENTS

ISBN 0-8493-6109-5

I. THE DISCOVERY OF KUPFFER CELLS

Kupffer cells represent the largest population of macrophages in the mammalian body. They have a unique, strategic position in view of their intravascular location in the hepatic capillary bed. A considerable amount of information concerning the structural and functional aspects of Kupffer cells and other sinusoidal cells has been gained during the past 20 years. Methods have become available to study these cells both *in vivo* and *in vitro,* and these methods have been reviewed recently by Wake et al.[1] Nevertheless, the origin of Kupffer cells and their relation with other cell types has formed a controversial debate spanning many years. The current concept has evolved over more than a century and Kupffer cells have been called not only macrophages, but also histiocytes or phagocytes, with adjectives describing their position (hepatic, fixed, sinusoidal, littoral), shape (stellate, *Sternzellen*), or function (phagocytic, reticuloendothelial). This nomenclature reflects the considerable uncertainty that has existed about the identity of these cells. Still in many studies of the liver, nonparenchymal hepatic cells are described as Kupffer cells based on their localization in the sinusoidal region or on their size, although endothelial and fat-storing (Ito) cells are present as well and even outnumber the Kupffer cells.

It all started with the discovery of the *Sternzellen* by von Kupffer in 1876. The *Sternzellen* (i.e., the German word for "star cells") were extensively described in the liver of various mammals by using a special gold chloride staining method, and they were reported to be localized in the perisinusoidal connective tissue (or the space of Disse). Later, von Kupffer changed his opinion after observing that sinusoidal endothelial cells ingested India ink after its intravenous injection in rabbits. These endocytotic cells were identified with the *Sternzellen,* which led the author to conclude that the latter were part of the sinusoidal endothelium. This misunderstanding was noted by Wake,[2] but it contributed to the concept of the "reticuloendothelial system", or RES, which was first proposed in 1924 by Aschoff.[3] The RES grouped different cell types into a "family" of cells called reticuloendothelial cells, which have in common the ability to ingest colloidal dyes and which were thought to have a common origin. In the case of the liver (one of the major reticuloendothelial organs), Aschoff referred in his paper to the sinusoidal cells as phagocytic endothelial cells which store fat and vitamin A. Now it is known that this description represents an amalgam of the characteristics of the three main sinusoidal cell types, i.e., Kupffer cells, endothelial cells, and fat-storing cells. Still, according to the RES concept, the hepatic endothelial cells were considered to be the immature differentiation stage of the Kupffer cells, since the latter are more active in the uptake of colloidal material from the circulation. Monocytes were then considered as a further maturation step, i.e., as macrophages "bud off" to form the circulating elements. It must be stressed that histology handbooks published in the 1960s were still referring to this endothelial origin of Kupffer cells, whereas other tissue macrophages were considered derivatives from monocytes (or the monocyte phagocyte system; see below).

In the 1970s, new light was shed on the identity of hepatic sinusoidal cells by the combined application of the following histological techniques: vascular (portal) perfusion fixation to remove the sinusoidal blood and better preserve the sinusoidal microarchitecture; cytochemical staining for endogenous peroxidase, a highly specific characteristic of Kupffer cells and other resident macrophages in the rat; intravenous injection of particulate material, such as latex beads, zymosan particles, or Thorotrast, which is phagocytosed by Kupffer cells; and electron microscopy.[2,4-9]

By these combined techniques it was possible to characterize, unequivocally, four different cell types as resident inhabitants of the sinusoidal wall, i.e., the endothelial cells, Kupffer cells, pit cells, and fat-storing cells. These were shown to represent different cell types without transitional forms, even under conditions reported to enhance the frequency of such transitions, such as RES stimulation. The morphological/ultrastructural characteristics of these sinusoidal cell types *in vivo* and *in vitro* were recently reviewed.[10-12]

II. MICROSCOPIC ANATOMY OF THE KUPFFER CELL

A. POSITION AND INTERCELLULAR CONTACTS

In the rat liver, Kupffer cells have been extensively investigated by electron microscopy.[4-8] Kupffer cells have a highly variable shape and position in the sinusoids, in contrast to endothelial and fat-storing cells. This aspect suggests that Kupffer cells may be more mobile than the other constituents of the sinusoidal wall. Mostly Kupffer cells are found upon or embedded within the endothelium. In these cases, the Kupffer cell cytoplasm is bulging into the sinusoidal lumen (Figure 1). Sometimes, and especially under experimental or pathological conditions, they are observed within the perisinusoidal space of Disse, i.e., in close contact with fat-storing cells and hepatocytes. In the case of hepatocytes, however, small Kupffer cells usually are involved (Figure 2).

Contact between Kupffer and endothelial cells can be formed by simple apposition of cell membranes with some electron-dense material in between. No desmosomes or other specialized intercellular contacts have been observed between Kupffer cells and endothelial cells. The sinusoidal endothelium under physiological conditions is considered to be closed with only small fenestrations ($\approx$0.1 μm diameter), allowing for the translocation of fluids between the blood and the hepatocyte. These endothelial fenestrations sometimes contain microvilli from the Kupffer cell which invade the space of Disse. In these places, the Kupffer cell extensions intermingle with the microvilli of the hepatocyte, which may represent a form of anchorage or communication

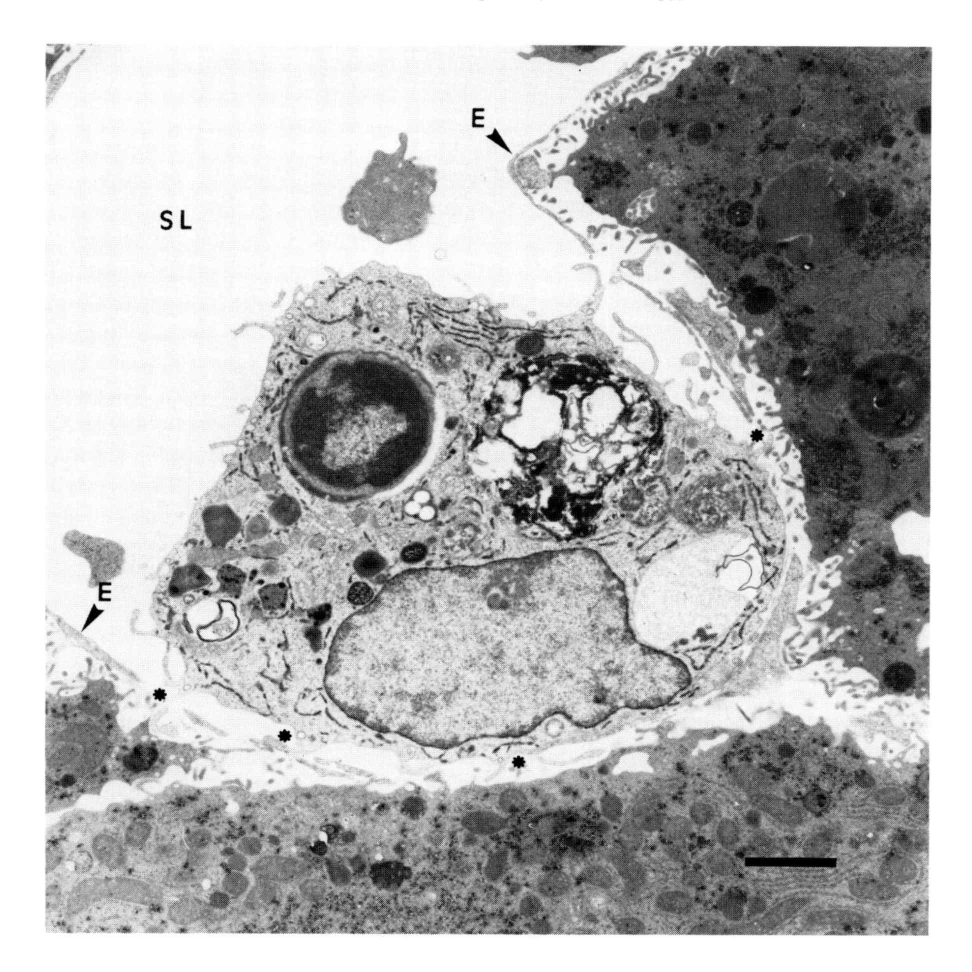

FIGURE 1. Electron micrograph (EM) of a Kupffer cell stained for endogenous peroxidase. The cell contains an ingested lymphocyte and another, partially degraded cell. Lysosomes are abundant. Note the microvilli that penetrate through the endothelium and make contact with the microvilli of the adjacent hepatocytes (asterisks). SL, Sinusoidal lumen. E, Endothelial lining. Bar = 2 μm.

(Figure 1). However, larger gaps in the endothelium must exist at those sites where larger insertions of the Kupffer cell cytoplasm are present. In any case, the Kupffer cells have a large part of their cell surface in contact with the endothelium, which indicates that they are spread out and well attached to the sinusoidal wall. This differentiates them from marginated leukocytes, such as monocytes, which are abundant during experimental inflammations. The latter have a spherical or oval shape with much less cell surface in contact with the sinusoidal wall (Figure 3).[13] In summary, then, Kupffer cells are

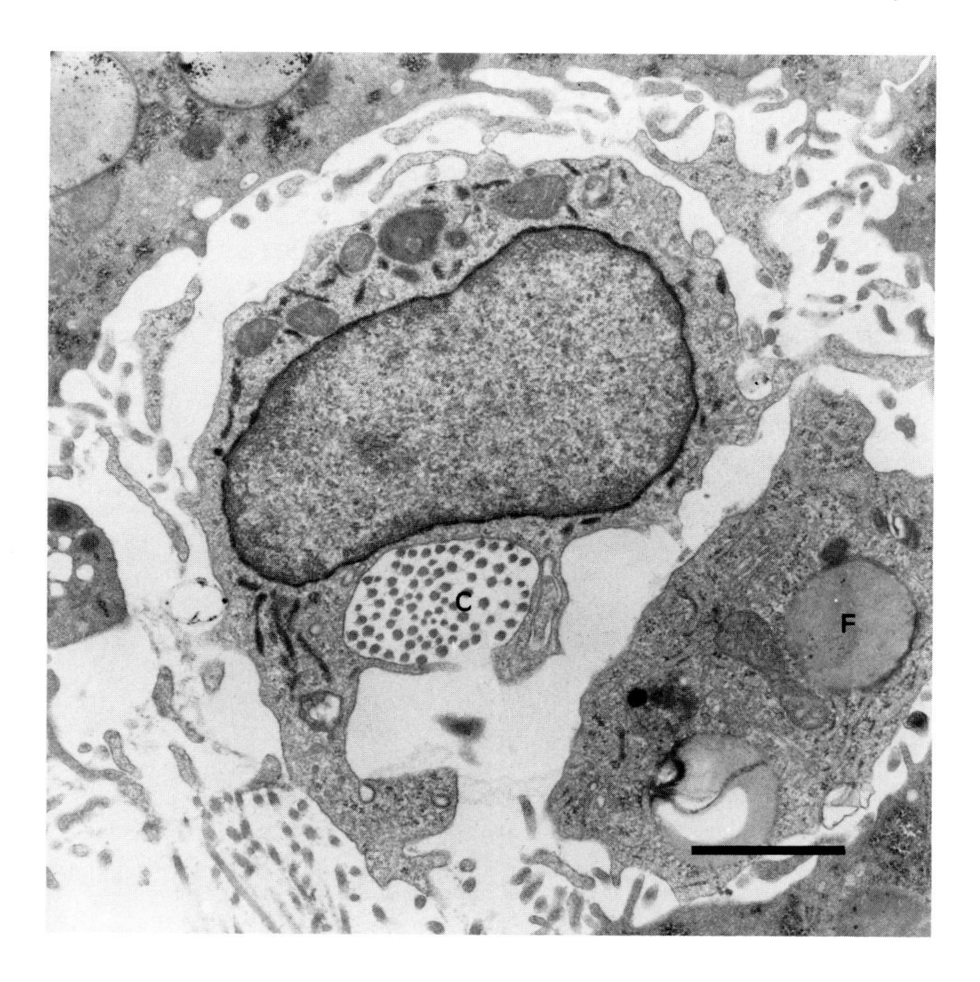

FIGURE 2. EM of a (peroxidase-positive) Kupffer cell present in the space of Disse. The cell surrounds a bundle of collagen fibers (C). F, Adjacent fat-storing (Ito) cell. Bar = 2 μm.

located mostly in the sinusoidal lumen and appear to be well fixed to the sinusoidal wall (this does not exclude locomotion along it). They have direct intercellular contact with the endothelial cells, pit cells, blood cells, and, via endothelial fenestrations or gaps, with the fat-storing cells and hepatocytes.

In scanning electron microscopy, one has a better three-dimensional impression of the shape and attachment structures of Kupffer cells. Besides microvilli, thinner and longer cell processes are seen to extend from the Kupffer cell surface to the endothelial lining, and have been compared to the guy ropes of a tent.[1]

During inflammatory conditions induced by the intravenous injection of biological response modifiers, lymphocytes and especially pit cells are typ-

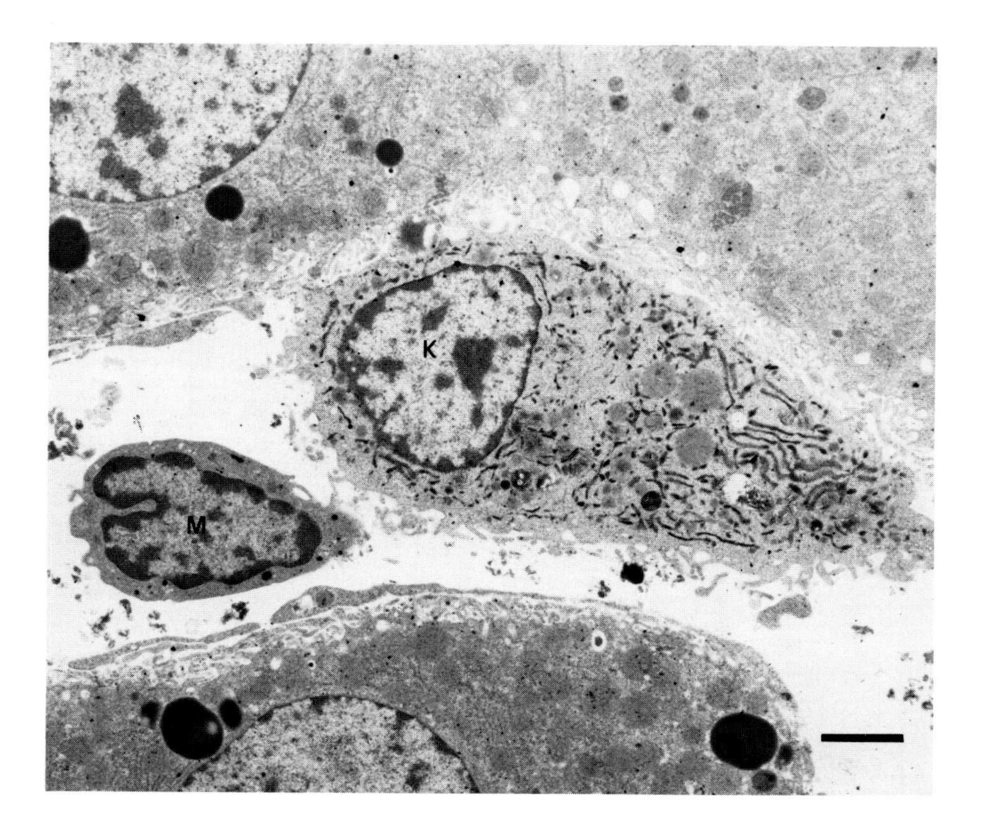

FIGURE 3. EM of a Kupffer cell (K) and a monocyte (M) in a sinusoid. Monocytes have peroxidase activity in granules, but not in the endoplasmic reticulum. Bar = 2 μm.

ically seen in close contact with Kupffer cells, although extensive vascular perfusion has washed out most erythrocytes and granulocytes in the sinusoids (Figure 4).[14] In the case of particle injections, the presence of platelets surrounding the Kupffer cells shortly after the injection is consistently observed (Figure 5). These cells, pit cells, lymphocytes, and platelets may be attracted or bound to the Kupffer cells, but are not phagocytosed.

In experimental pathological conditions such as after injection of yeast cell wall derivatives (zymosan, glycan), Kupffer cells have been observed to accumulate in granulomas.[8,14] We have also frequently observed Kupffer cells in close contact with tumor cells at the periphery of experimental colon carcinoma metastases.[15]

B. HETEROGENEITY AND LOBULAR DISTRIBUTION

In the normal physiological situation, Kupffer cells are heterogeneous in their volume, phagocytic activity, and in some enzymatic activities. Larger

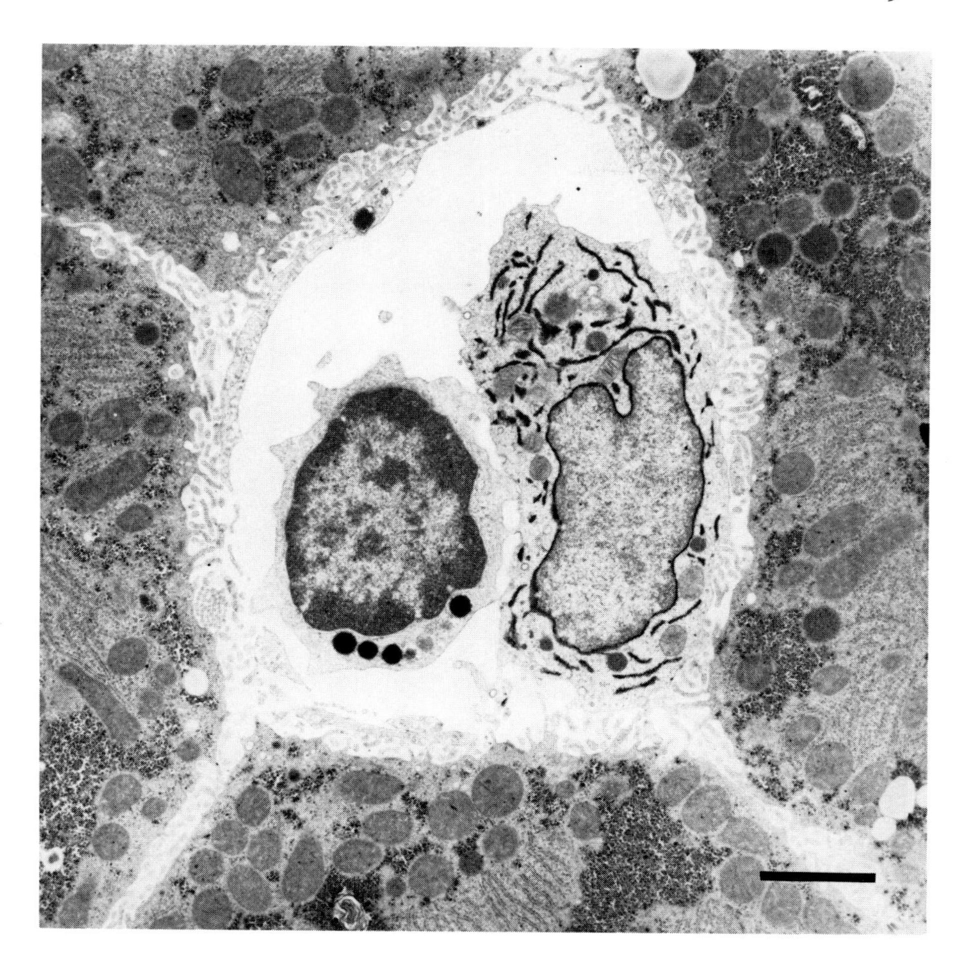

FIGURE 4. EM showing contact between a Kupffer cell (with peroxidase activity in the endoplasmic reticulum) and a pit cell or large granular lymphocyte. Bar = 2 μm.

and more active cells are predominantly located in the periphery of the liver lobule, i.e., close to the portal tracts.[16] The smaller Kupffer cells are predominantly located in the central zone of the liver lobule (close to the central veins) and are much less active in the phagocytosis of latex particles. The latter is true even when the particles are injected in a retrograde fashion via the hepatic vein or are administered at overloading doses.[16,17] As a result, we have found that only a maximum of 64% of all Kupffer cells in the rat liver could be labeled *in situ* with phagocytosed latex particles.[17]

C. FINE STRUCTURE

Kupffer cells are well characterized in electron microscopy.[5,7,8] Their shape is variable from thin and elongated to rounded and bulky or star shaped.

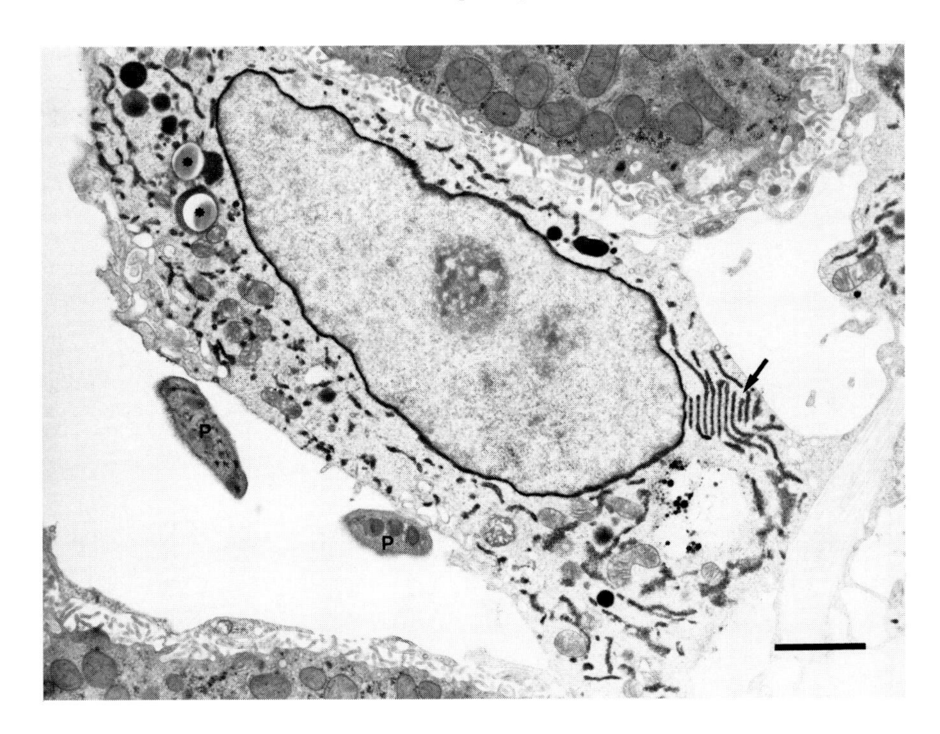

FIGURE 5. EM of a Kupffer cell that has phagocytosed two latex beads of 0.8 μm diameter (asterisk). Two platelets are seen close to the Kupffer cell (P). The arrow points to the annulate lamellae (with peroxidase activity). Bar = 2 μm.

Their cytoplasmic volume is considerable, which makes them the largest sinusoidal hepatic cells. The Kupffer cell cytoplasm contains a large number of lysosomes, phagosomes, and vacuoles varying widely in size, density, and shape, which is typical for tissue macrophages. Two main types of such inclusions have been identified.[7] The first type comprises electron-lucent vacuoles of variable diameter composed of a limiting membrane which is often paralleled by a granular internal layer corresponding to the surface "fuzzy coat" (see below). These vacuoles probably represent recent internalization of plasma membrane and extracellular fluid. The second type concerns dense bodies (lysosomes) which are electron dense and have different sizes and shapes. Both types can contain dense inclusions with a globular or myelinlike form. Whereas the dense bodies or lysosomes are often arranged in groups lying near the nucleus and the cytocenter, the vacuoles tend to be located in the periphery of the cell or close to the plasma membrane. Large vacuoles containing phagocytosed material such as leukocytes or cell debris are sometimes observed (Figure 1). These vacuoles can be surrounded by dense bodies that are seen to fuse with it.

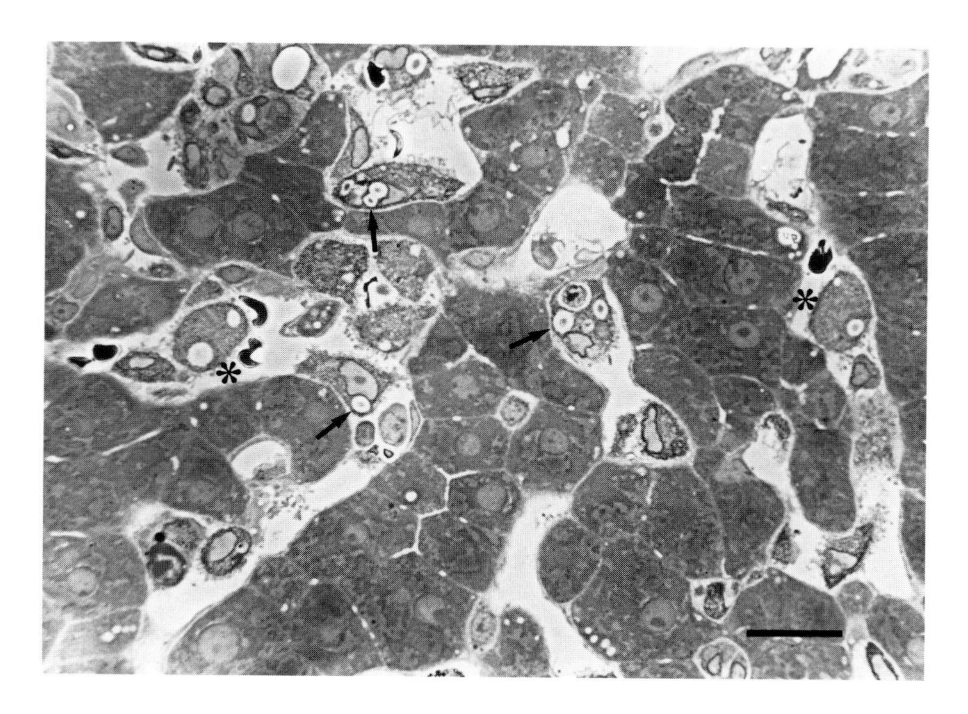

FIGURE 6. Light micrograph of a semithin liver section, 3 d after the intravenous injection of zymosan (yeast cell) particles. Peroxidase reaction product is found in the cytoplasm (endoplasmic reticulum) and nuclear envelope of the Kupffer cells. The latter frequently contain one or more zymosan particles (arrows). Two mitoses are seen (asterisks). Bar = 2 μm.

Golgi apparatuses are regularly seen in groups close to the cytocenter or in the juxtanuclear cytoplasm. The cytocenter is formed by a pair of centrioles from which microtubules radiate in all directions. From the Golgi apparatuses, irregularly shaped vesicles as well as bristle-coated vesicles are seen to pinch off.

Rough endoplasmic reticulum (RER) is abundant in Kupffer cells. Both short profiles and parallel stacks of long cisternae are found. This RER and the nuclear envelope contain peroxidase enzymatic activity that can be demonstrated cytochemically both for light and electron microscopy (Figures 1 through 6).[5,7,8] In the rat liver, Kupffer cells are the only cell type which have this peroxidase staining pattern. Peroxidase activity is also present in the annulate lamellae, which are continuous with the RER and are composed of a stack of short parallel cisternae with porelike constrictions (Figures 6 and 7). Annulate lamellae are regularly encountered in Kupffer cells, but are found in only very few other cell types (e.g., oocytes). Mitochondria, clusters of free ribosomes, microtubules, and filaments are dispersed in the cytoplasm. Free fat droplets and glycogen particles are normally not found in Kupffer cells.

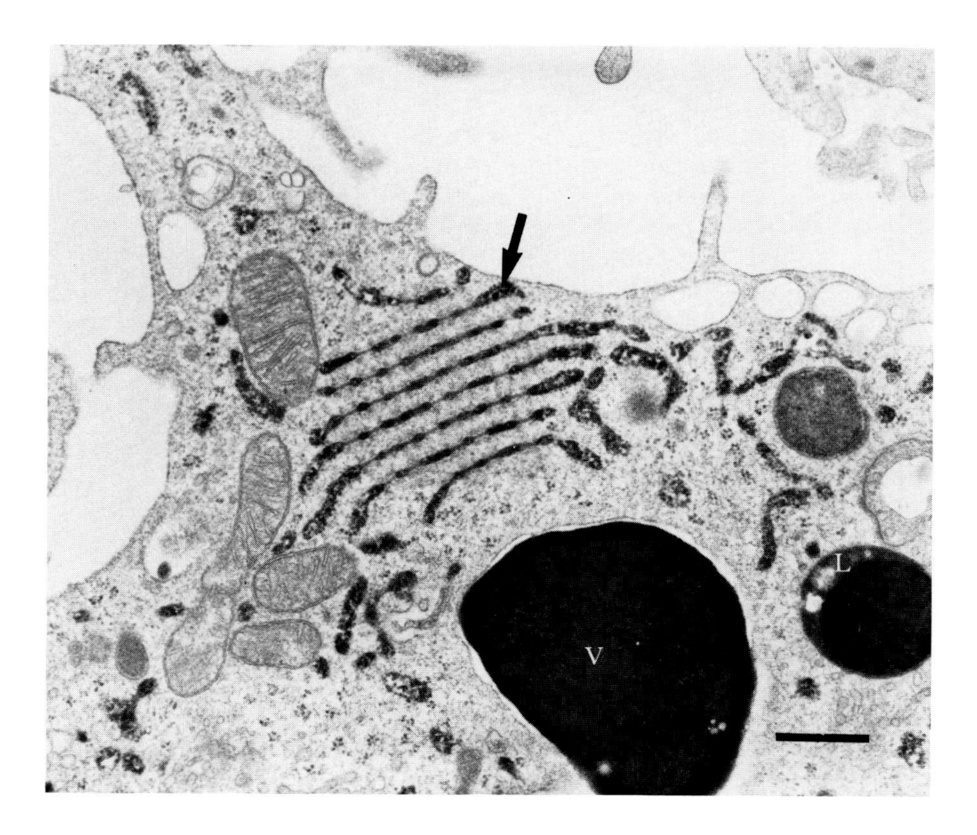

FIGURE 7. EM detail showing annulate lamellae with peroxidase activity (arrow). The large, very dense material represents a phagocytosed erythrocyte. One large vacuole (V) is seen and a smaller one is fused with a lysosome (L). Bar = 0.5 μm.

Kupffer cells have one nucleus with an oval shape and a characteristically finely dispersed euchromatin. There is only a very thin perinuclear rim of heterochromatin. The nucleus contains one or two characteristically fluffy nucleoli.

The plasma membrane of Kupffer cells is ruffled and extends microvilli, filopodia, lamellipodia, and pseudopodia. These processes are composed of organelle-free, but microfilamentous hyaloplasm. Bristle-coated micropinocytotic vesicles are found close to the plasma membrane. Another endocytotic structure is represented by the "wormlike structures", highly characteristic of Kupffer cells (although not found in all Kupffer cells; Figure 8). These are invaginations of the cell membrane and have an average width of 140 nm. They can penetrate deeply into the cytoplasm. The wormlike structures have a central dense granular layer which corresponds to the surface "fuzzy coat". The fuzzy coat of the cell has a width of 70 nm and covers the Kupffer cell surface or is found in the wormlike invagination and some vacuoles within

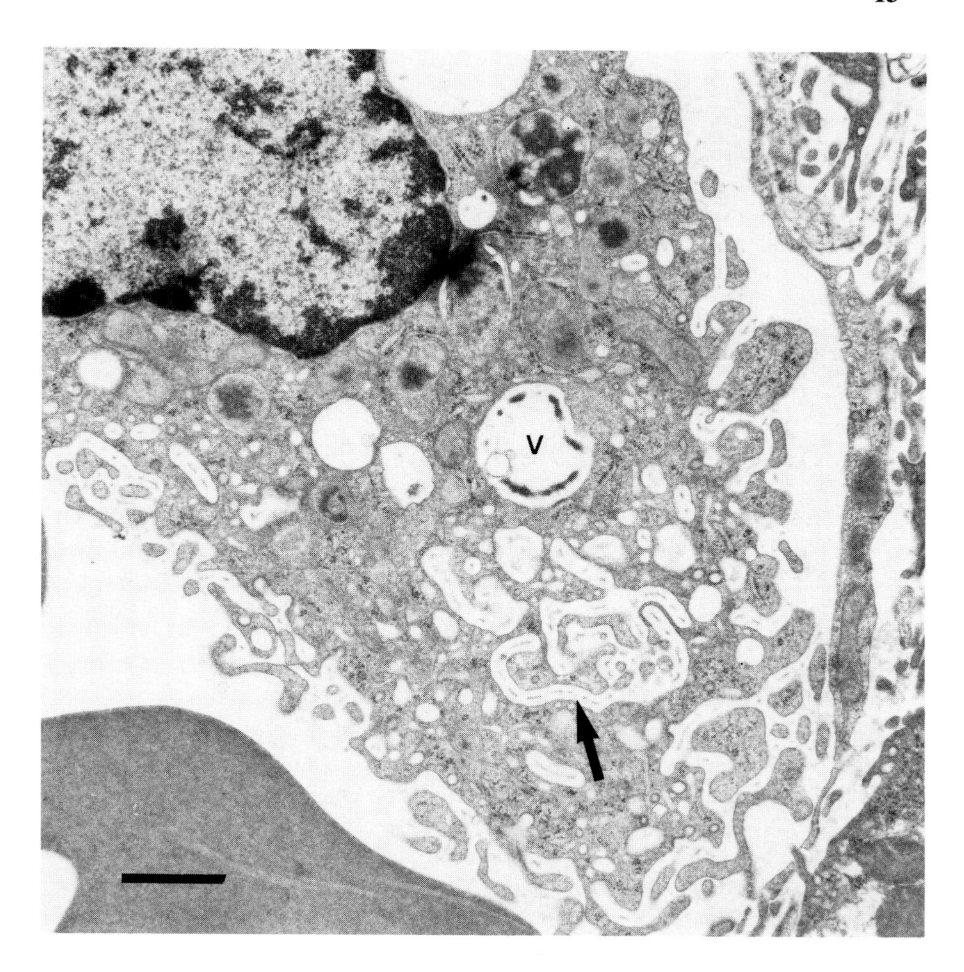

FIGURE 8. EM detail showing wormlike structures (arrow). The fuzzy surface coat is also found in these structures as well as in a vacuole (V). Bar = 1 μm.

the cytoplasm. This cell coat is rich in sugar groups and is not preserved by glutaraldehyde perfusion fixation (a protein fixative). To visualize the fuzzy coat, either osmium-fixed frozen sections must be used or it must be stabilized by the injection of colloidal particles or erythrocyte ghosts.[1,7] Although the exact nature of this cell coat is not known, it is probably involved in the attachment of particles to be phagocytosed. Fuzzy coat vacuoles that can be observed in the cytoplasm (see above) must represent a kind of macropinocytosis.

In summary, Kupffer cells have a variable shape, ruffled cell surface, and several ultrastructural characteristics that differentiate it from other cell types, e.g., the annulate lamellae, wormlike structures, and fuzzy coat. Kupffer cells possess several morphologically recognizable mechanisms to accomplish endocytosis and which have been shown to contain or transport exogenous materials. These mechanisms include bristle-coated micropinocytosis, macropinocytotic vacuoles, wormlike invagination, and phagocytosis via pseudopod fusion.[7,8] Interestingly, these structures can be seen in normal Kupffer cells from untreated healthy animals (including germfree animals), indicating that these endocytotic mechanisms are constantly operating *in vivo*. The abundance and relative volume of the lysosomal apparatus in Kupffer cells clearly indicate a considerable capacity for the subsequent digestion of internalized material. The presence of very well-developed RER and Golgi complexes suggests also that normal Kupffer cells are actively involved in protein secretion.

III. ORIGIN AND POPULATION KINETICS

A. INTRODUCTION

In many experimental and pathological conditions, an increase in the number (hyperplasia) of Kupffer cells is known to occur. The RES concept stated that this increase resulted from an increased proliferation and differentiation of reticuloendothelial cells, including the sinusoidal endothelial cells. In the same period that the endothelial origin of Kupffer cells was refuted, a new hypothesis was put forward according to which all tissue macrophages have the circulating monocyte as a common precursor. Within this concept, called the mononuclear phagocyte system (MPS), Van Furth et al.[19] stated that resident macrophages would represent "end cells" which are incapable of mitotic division (local proliferation) and are continuously renewed by an influx of blood monocytes. Still, according to this concept, all macrophage types would belong to a single lineage. Although most textbooks and scientific publications accept this MPS concept, a considerable controversy has arisen and still continues concerning the origin of macrophages. Some investigators have put in doubt whether or not there is a single macrophage lineage, but rather several lineages, each with its own precursor in the bone marrow.[20,21] Others have reported conflicting, but sound data which demonstrate that mature resident macrophages can perform local proliferation even in the total absence of circulating monocytes.[22] They propose that under inflammatory conditions monocytes migrate to sites of inflammation where they differentiate into "exudate" macrophages. The latter would remain distinct, however, from resident macrophages. We will try to briefly review the major reports on this controversial issue and will restrict ourselves to those reports that deal with the hepatic macrophage.

B. BONE MARROW-DERIVED MACROPHAGES IN THE LIVER

A number of reports have been presented since 1969 in which either marrow or liver transplantation was used to determine the origin of macrophages in the liver. Shand and Bell[23] transplanted rat bone marrow cells into irradiated mice and subsequently found macrophages of the donor type in the mouse liver as evidenced by immunological markers. Similar results were reported by Gale et al.,[24] who studied the liver macrophages from irradiated human patients receiving a bone marrow graft. After sex chromosome staining, they found macrophages in the liver of the recipient that were of the other sex when male-female transplants had been carried out. In human liver transplants, macrophages of the recipient type were reported to have invaded the sinusoids of the donor liver.[25] Others used kinetic experiments with tritiated thymidine and autoradiography to demonstrate that under certain conditions macrophages can be recruited from the marrow to several tissues or body compartments, including the liver.[26-28]

C. SELF-RENEWAL OF KUPFFER CELLS BY LOCAL PROLIFERATION

Other investigators reported extensively on the local proliferative capacity of resident macrophages or Kupffer cells. Interesting data were obtained in the early fetal rat liver where, from the 11th day of gestation on, phagocytosing and dividing macrophages were observed which had the characteristic peroxidase pattern of mature Kupffer cells. At this ontogenic stage, no monocytic cells were observed since they arise only from the 17th day of gestation on.[29-32] In adult life, proliferating Kupffer cells were demonstrated either by the observation of metaphases or by autoradiographic detection of cells that had incorporated tritiated thymidine. Dividing Kupffer cells were reported at low frequency during the normal steady state.[8,17,33] A much higher frequency of locally proliferating Kupffer cells is observed after partial hepatectomy and in inflammatory conditions (Figures 6, 9, and 10).[8,33-36] *In vitro,* growth and even clonal expansion of Kupffer cells has been reported.[37,38] Recently, Yamada et al.[39] investigated the population kinetics of Kupffer cells in severely monocytopenic mice. A prolonged monocytopenia was obtained by local bone marrow irradiation using the ^{89}Sr isotope combined with or without splenectomy. Kupffer cells were characterized unequivocally by immunohistochemistry with the anti-mouse macrophage monoclonal antibody F4/80, electron microscopy, and endogenous peroxidase cytochemistry. Local proliferation in the liver was analyzed by cell counting and by tritiated thymidine autoradiography. The authors could conclude from their experiments that, in their model, Kupffer cells were self-renewing through local cell division and could even increase in number and form granulomas during inflammation.

The increasing amount of data that cannot be reconciled with the MPS concept has led to considerable criticism and to the proposal that Kupffer cells represent a discrete, self-renewing cell population.[40]

D. DUAL ORIGIN OF KUPFFER CELLS

The reports cited thus far give convincing evidence both for a monocyte origin of Kupffer cells under conditions such as inflammation, and for a self-

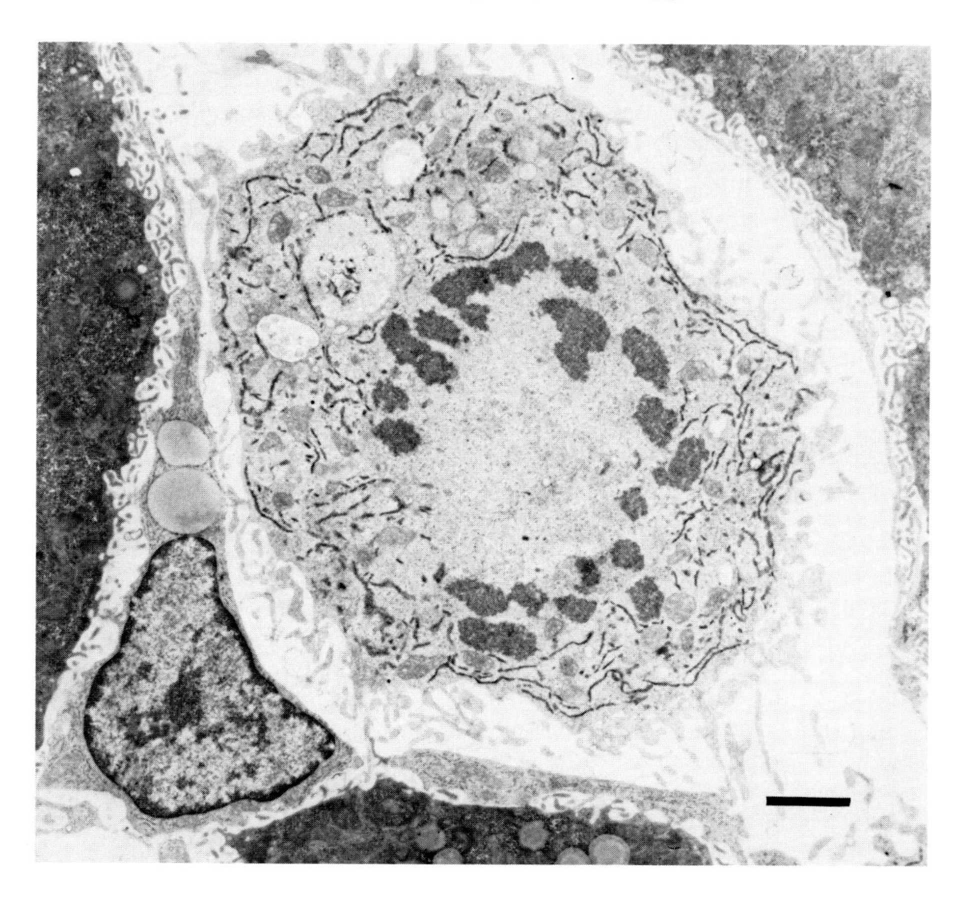

FIGURE 9. EM of a Kupffer cell mitosis (metaphase) with peroxidase activity. Bar = 2 μm.

renewal of Kupffer cells via mitosis. Most authors, however, were not able to reconcile these data and were trying to prove either the mitotic capacity or the extrahepatic origin of Kupffer cells. Only a very limited number of studies pointed out that the kinetics of the Kupffer cell population is regulated by two different mechanisms: (1) local proliferation of mature cells and (2) extrahepatic recruitment of macrophage precursors formed in the bone marrow. This means that the Kupffer cells have a dual origin.[26,41,42] They can proliferate locally when receiving the proper stimulus, such as partial hepatectomy, and although they are normally self-renewing they can be recruited from the marrow, as happens under conditions of immunological stress, after irradiation, or due to severe damage to the resident population.[23-27] Bouwens et al.[34,35] quantified the numerical increase and the local mitotic production of Kupffer cells after stimulation of the macrophages with the yeast component zymosan. After a single intravenous injection of the

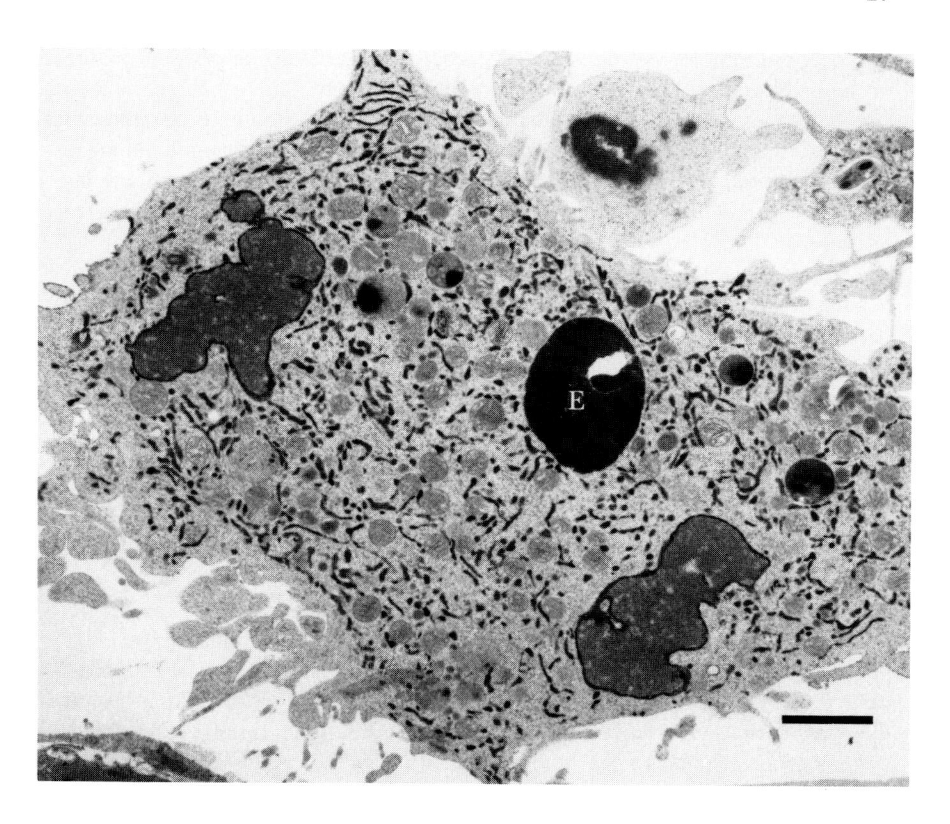

FIGURE 10. EM of a Kupffer cell mitosis (anaphase) with peroxidase activity and an ingested erythrocyte (E). Bar = 2 μm.

inflammatory agent in rats, there was a fourfold increase in Kupffer cell number within 5 d postinjection. During this hyperplasia, mitotic activity of Kupffer cells was measured with the metaphase arrest technique (stathmokinetic analysis) and tritiated thymidine incorporation experiments combined with autoradiography. The results of these kinetic experiments demonstrated that the local production of new Kupffer cells by mitosis contributed to 60 to 80% of the observed numerical increase. Experiments with partial irradiation (8.5 Gy) prior to zymosan stimulus were carried out to selectively inhibit either the local proliferation of Kupffer cells (only the liver region was irradiated; the rest of the body was shielded) or the extrahepatic recruitment (total irradiation with shielded liver). With these experiments, Bouwens et al.[43] confirmed that between 60 and 80% of the Kupffer cell expansion following zymosan stimulation resulted from local proliferation of mature Kupffer cells. The remaining 20 to 40% of the growth was attributed to recruitment of macrophage precursors from the marrow. In these studies, the Kupffer cells were unequivocally identified by their peroxidase staining, by electron microscopy, and by the ingestion of particles. The mature or resident nature of the proliferating Kupffer cells was demonstrated by labeling the cells with latex particles prior to the application of the growth stimulus (latex did not influence the Kupffer cell kinetics). In this way, the distinction could be made between recently recruited macrophages which were devoid of latex particles and Kupffer cells that were already present in the liver before the application of the growth stimulus because they contained latex in their cytoplasm.[34] We have proposed that the relative importance of local proliferation and extrahepatic recruitment (≈70/30% in the case of zymosan as a growth stimulus) varies and depends on the nature of the growth stimulus. In the case of liposome-encapsulated muramyl dipeptide as a growth stimulus, Daemen et al.[44] recently found that minimally 35% of the observed increase in Kupffer cells was due to the local proliferation of resident cells expressing a monoclonal marker of mature macrophages which is not present on monocytes (ED2). Maximally 65% of the increase was accounted for by the immigration of immature monocyte/macrophages (recognized by another marker, ED1) and by the immigration of macrophages with the mature (ED2) marker. In addition, the recently recruited macrophages further contributed to the total expansion by local mitotic activity in the liver.

Clearly, the kinetics of the hepatic macrophage population are more complex than initially stated in the MPS concept: mature Kupffer cells retain the capacity of cell division and may contribute significantly to the expansion of their population under inflammatory and other conditions. Immigrating monocytes also proliferate in the liver during inflammation, and leave the organ via the central veins.[13] Also of note, the exact nature of the Kupffer cell precursor (present in the bone marrow) has not been established yet.[1,13] In addition to monocytes, macrophages with a more mature phenotype have been observed immigrating into the liver.[44] Thus far, there is no direct evidence demonstrating that monocytes can differentiate in the liver into Kupffer cells.

In the steady state, or normal physiological condition, we found that latex-labeled Kupffer cells have a very long residence time in the rat liver (more than 32 months).[17] From cytokinetic analysis of thymidine-labeled cells, the residence time of Kupffer cells in the liver has been calculated to average between 3 and 17 months.[17] Under these conditions, the mitotic activity of Kupffer cells is very low.

Kupffer cells are normally a self-renewing population with long residence time (slow turnover). Yet, under the influence of inflammatory or extreme environmental stimuli, macrophages can immigrate within the liver and become resident, or leave the liver at a later stage. It has been reported that after unusual experimental stress, macrophages leave the liver via the blood circulation and reach the lungs, where they are eliminated. Immigration of carbon-loaded Kupffer cells to the portal interstitium and thereafter to the regional lymph nodes has also been reported.[46]

It is evident that the origin and differentiation of macrophages are still incompletely understood. The mechanisms regulating the complex and flexible system of macrophage diversity await further investigation.

ACKNOWLEDGMENT

The authors would like to thank C. Derom for printing the micrographs.

REFERENCES

1. **Wake, K., Decker, K., Kirn, A., Knook, D. L., McCuskey, R. S., Bouwens, L., and Wisse, E.,** Cell biology and kinetics of Kupffer cells in the liver, *Int. Rev. Cytol.*, 118, 173, 1989.
2. **Wake, K.,** Perisinusoidal stellate cells (fat-storing cells, interstitial cells, lipocytes), their related structure in and around the liver sinusoids, and vitamin A-storing cells in extrahepatic organs, *Int. Rev. Cytol.*, 66, 303, 1980.
3. **Aschoff, L.,** Das Reticuloendotheliale System, *Ergeb. Inn. Med. Kinderheilkd.*, 26, 1, 1924.
4. **Wisse, E.,** An electron microscopic study of the fenestrated endothelial lining of rat liver sinusoids, *J. Ultrastruct. Res.*, 31, 125, 1970.
5. **Widmann, J. J., Cotran, R. S., and Fahimi, H. D.,** Mononuclear phagocytes (Kupffer cells) and endothelial cells. Identification of two functional cell types in rat liver sinusoids by endogenous peroxidase activity, *J. Cell Biol.*, 52, 159, 1972.
6. **Wisse, E.,** An ultrastructural characterization of the endothelial cell in the rat liver sinusoid under normal and various experimental conditions, as a contribution to the distinction between endothelial and Kupffer cells, *J. Ultrastruct. Res.*, 38, 528, 1972.
7. **Wisse, E.,** Observations on the fine structure and peroxidase cytochemistry of normal rat liver Kupffer cells, *J. Ultrastruct. Res.*, 46, 393, 1974.
8. **Wisse, E.,** Kupffer cell reactions in rat liver under various conditions as observed in the electron microscope, *J. Ultrastruct. Res.*, 46, 499, 1974.
9. **Wisse, E., Van't Noordende, J. M., Van der Meulen, J., and Daems, W. T.,** The pit cell: description of a new type of cell occurring in rat liver sinusoids and peripheral blood, *Cell Tissue Res.*, 173, 423, 1976.
10. **McCuskey, R. S. and McCuskey, P. A.,** Fine structure and function of Kupffer cells, *J. Electron Microsc. Technol.*, 14, 237, 1990.
11. **Geerts, A., Bouwens, L., and Wisse, E.,** Ultrastructure and function of hepatic fat-storing and pit cells, *J. Electron Microsc. Technol.*, 14, 247, 1990.
12. **Deleeuw, A. M., Brouwer, A., and Knook, D. L.,** Sinusoidal endothelial cells of the liver — fine structure and function in relation to age, *J. Electron Microsc. Technol.*, 14, 218, 1990.
13. **Bouwens, L. and Wisse, E.,** Proliferation, kinetics, and fate of monocytes in rat liver during a zymosan-induced inflammation, *J. Leuk. Biol.*, 37, 531, 1985.

14. **Bouwens, L. and Wisse, E.,** Tissue localization and kinetics of pit cells or large granular lymphocytes in the liver of rats treated with biological response modifiers, *Hepatology,* 8, 46, 1988.
15. **Bouwens, L., Charles, K., Van Dalen, P., and Wisse, E.,** Sinusoidal cell reactions associated with colon cancer metastases in rat liver, in *Cells of the Hepatic Sinusoid,* Vol. 2, E. Wisse, D. L. Knook, and K. Decker, Eds., Kupffer Cell Foundation, Rijswijk, The Netherlands, 1989, 237.
16. **Sleyster, E. D. and Knook, D. L.,** Relation between localization and function of rat liver Kupffer cells, *Lab. Invest.,* 47, 484, 1982.
17. **Bouwens, L., Baekeland, M., De Zanger, R., and Wisse, E.,** Quantitation, tissue distribution and proliferation kinetics of Kupffer cells in normal rat liver, *Hepatology,* 6, 718, 1986.
18. **Kaneda, K. and Wake, K.,** Distribution and morphological characteristics of the pit cell in the liver of the rat, *Cell Tissue Res.,* 233, 485, 1983.
19. **Van Furth, R., Cohn, Z. A., Hirsch, J. G., Humphrey, J. H., Spector, W. G., and Langevoort, H. L.,** The MPS: a new classification of macrophages, monocytes and their precursor cells, *Bull. W.H.O.,* 46, 845, 1972.
20. **Bursuker, I. and Goldman, R.,** On the origin of macrophage heterogeneity: a hypothesis, *J. Reticuloend. Soc.,* 33, 207, 1983.
21. **Daems, W. T. and De Bakker, J. M.,** Do resident macrophages proliferate?, *Immunobiology,* 161, 204, 1982.
22. **Volkman, A., Chang, N. C., Strausbauch, P. H., and Morahan, P. S.,** Differential effects of chronic monocyte depletion on macrophage populations, *Lab. Invest.,* 49, 291, 1983.
23. **Shand, F. L. and Bell, E. B.,** Studies on the distribution of macrophages derived from rat bone marrow cells in xenogeneic radiation chimeras, *Immunology,* 22, 549, 1972.
24. **Gale, R. P., Sparkes, R. S., and Golde, D. W.,** Bone marrow origin of hepatic macrophages (Kupffer cells) in humans, *Science,* 201, 937, 1978.
25. **Portmann, B., Schindler, A. M., Murray-Lion, I. M., and Williams, R.,** Histological sexing of a reticulum cell sarcoma arising after liver transplantation, *Gastroenterology,* 70, 82, 1976.
26. **North, R. J.,** The relative importance of blood monocytes and fixed macrophages to the expression of cell-mediated immunity to infection, *J. Exp. Med.,* 132, 521, 1970.
27. **Diesselhoff-Den Dulk, M. M. C., Crofton, R. W., and Van Furth, R. V.,** Origin and kinetics of Kupffer cells during an acute inflammatory response, *Immunology,* 37, 7, 1979.
28. **Crofton, R. W., Diesselhoff-Den Dulk, M. M. C., and Van Furth, R.,** The origin, kinetics, and characteristics of the Kupffer cell in the normal steady state, *J. Exp. Med.,* 148, 1, 1978.
29. **Naito, M. and Wisse, E.,** Observations on the fine structure and cytochemistry of sinusoidal cells in fetal and neonatal rat liver, in *Kupffer Cells and Other Liver Sinusoidal Cells,* D. L. Knook and E. Wisse, Eds., Elsevier Biomedical, Amsterdam, 1977, 497.
30. **Deimann, W. and Fahimi, H. D.,** Peroxidase and ultrastructure of resident macrophages in fetal rat liver. A developmental study, *Dev. Biol.,* 66, 43, 1978.

31. **Pino, R. M. and Bankston, P. W.,** The development of the sinusoids of fetal rat liver: localization of endogenous peroxidase in fetal Kupffer cells, *J. Histochem. Cytochem.*, 27, 643, 1979.
32. **Bankston, P. W. and Pino, R. M.,** The development of the sinusoids of fetal rat liver: morphology of endothelial cells, Kupffer cells, and the transmural migration of blood cells into the sinusoids, *Am. J. Anat.*, 159, 1, 1980.
33. **Widmann, J. J. and Fahimi, H. D.,** Proliferation of mononuclear phagocytes (Kupffer cells) and endothelial cells in regenerating liver, *Am. J. Anat.*, 80, 349, 1975.
34. **Bouwens, L., Baekeland, M., and Wisse, E.,** Importance of local proliferation in the expanding Kupffer cell population of rat liver after zymosan stimulation and partial hepatectomy, *Hepatology*, 4, 213, 1984.
35. **Bouwens, L., Baekeland, M., and Wisse, E.,** Cytokinetic analysis of the expanding Kupffer cell population in rat liver, *Cell Tissue Kinet.*, 19, 217, 1986.
36. **North, R. J.,** The mitotic potential of fixed phagocytes in the liver as revealed during the development of cellular immunity, *J. Exp. Med.*, 130, 315, 1969.
37. **Clark, J. M. and Pateman, J. A.,** Long-term culture of chinese hamster Kupffer cell lines isolated by a primary cloning step, *Exp. Cell Res.*, 112, 207, 1978.
38. **Decker, T., Kiderlen, A. F., and Lohmann-Matthes, M. L.,** Liver macrophages (Kupffer cells) as cytotoxic effector cells in extracellular and intracellular cytotoxicity, *Infect. Immun.*, 50, 358, 1985.
39. **Yamada, M., Naito, M., and Takahashi, K.,** Kupffer cell proliferation and glycan-induced granuloma formation in mice depleted of blood monocytes by strontium-89, *J. Leuk. Biol.*, 47, 195, 1990.
40. **Volkman, A.,** The unsteady state of the Kupffer cell, in *Kupffer Cells and Other Sinusoidal Cells*, E. Wisse and D. L. Knook, Eds., Elsevier Biomedical, Amsterdam, 1977, 459.
41. **Warr, G. W. and Sljivic, V. S.,** Origin and division of liver macrophages during stimulation of the mononuclear phagocyte system, *Cell Tissue Kinet.*, 7, 559, 1974.
42. **Sljivic, V. S. and Warr, G. W.,** Role of cellular proliferation in the stimulation of MPS phagocytic activity, *Br. J. Exp. Pathol.*, 56, 314, 1975.
43. **Bouwens, L., Knook, D. L., and Wisse, E.,** Local proliferation and extrahepatic recruitment of liver macrophages (Kupffer cells) in partial body-irradiated rats, *J. Leuk. Biol.*, 39, 687, 1986.
44. **Daemen, T., Huitema, S., Koudstaal, J., Scherphof, G. L., and Hardonk, M. J.,** Population kinetics of rat liver macrophages after intravenous administration of liposome-encapsulated MDP, in *Cells of the Hepatic Sinusoid*, Vol. 2, E. Wisse, D. L. Knook, and K. Decker, Eds., Kupffer Cell Foundation, Rijswijk, The Netherlands, 1989, 400.
45. **Schneeberger-Keeley, E. and Burger, E.,** Intravascular macrophages in cat lungs after open chest ventilation, *Lab. Invest.*, 22, 361, 1970.
46. **Hardonk, M. J., Dijkhuis, F. W. J., Grond, J., Koudstaal, J., and Poppema, S.,** Migration of Kupffer cells, in *Cells of the Hepatic Sinusoid*, Vol. 1, A. Kirn, D. L. Knook, and E. Wisse, Eds., Kupffer Cell Foundation, Rijswijk, The Netherlands, 1986, 1.

Chapter 2

MODES OF COMMUNICATION BETWEEN KUPFFER CELLS AND HEPATOCYTES UNDER NORMAL AND PATHOLOGICAL CONDITIONS

T.C. Peterson

TABLE OF CONTENTS

ISBN 0-8493-6109-5

I. INTRODUCTION

Kupffer cell-hepatocyte communication, i.e., factor-mediated communication, receptor-mediated endocytosis, gap junctional, and other forms of intercellular interaction in normal and diseased conditions falls within the scope of this chapter. Factor-mediated communication involves release of monokines and other cytokines from Kupffer cells as a result of Kupffer cell activation. These cytokines then interact with hepatocytes, altering hepatocyte function. One result of interaction between Kupffer cell and hepatocytes is the inhibition of the cytochrome P-450 enzyme system in the hepatocytes. The interaction of Kupffer cells and hepatocytes controls the decrease of cytochrome P-450 in the hepatocytes. The potentially important interactive role of Kupffer cells in fibrosis is being investigated by several groups and will be reviewed here. Inhibitors of communication and their role in understanding cellular interactions will also be reviewed.

II. RECEPTOR MEDIATED CELL COMMUNICATION

The most common form of intercellular communication between Kupffer cells and hepatocytes involves receptor-mediated communication. Kupffer cells and sinusoidal endothelial cells have Fc receptors[1] and ingest immune complexes.[2] Fc receptors are important for the metabolism of soluble immune complexes.[3] Recent studies suggest that another type of liver cell, the sinusoidal endothelial cell, may also participate in the metabolism of the immune complexes.[4] Fc receptors appear to be preserved in the liver in the presence of acute injury until the necrotic foci are infiltrated by inflammatory cells.[5] In chronic injury with D-galactosamine, Fc receptors are lost and Kupffer cells decrease in the periportal areas. The ingestion of circulating immune complexes by Kupffer cells and endothelial cells in normal and injured livers shows the same pattern of distribution as the Fc receptors.[5] These authors felt that the enhancement of Fc receptor activity in the endothelium and the increase in Kupffer cells may compensate for the loss of sinusoids in severe liver damage.

Recent evidence[6] suggests that human Thomsen-Friedenreich (T) and Tn antigens specifically bind to the galactosamine/*N*-acetyl galactosamine (Gal/GalNAC) receptors of rat Kupffer cells and undergo endocytosis via the coated pit/vesicle pathway. These authors have reported that isolated rat Kupffer cells in the absence of antibodies bind and are taken up by endocytosis T/Tn antigens adsorbed on gold particles. Binding and uptake results suggested that the Gal/GalNAC receptor on Kupffer cells is the relevant system for recognition of T/Tn.[6] The significance of these results may lie in the understanding that Kupffer cells are the first barrier to be encountered by tumor cells, and this may prove to be important because the liver is a frequent target for metastasis.[7]

Receptors for lactate dehydrogenase (LD) may be present on liver macrophages,[8] and endocytosis may mediate the clearance of LD-5. These authors have also shown that the rapid plasma clearance in rats of LD-5, alcohol dehydrogenase, and several other enzymes is mainly due to the receptor-mediated endocytosis by macrophages from liver, spleen, and bone marrow,[9,10] and recently Smit and co-workers[9] have reported that LD-1 is also cleared by endocytosis and subsequent breakdown of the enzyme in the liver. At this point, it is unknown what feature of the enzymes is recognized by the endocytosing cells, and the receptor on the macrophages is as of yet unidentified.

Carcinoembryonic antigen (CEA) is metabolized chiefly in the liver. It is a multistep process involving both the Kupffer cells and the hepatocytes.[11] CEA is removed from the circulation by the Kupffer cells via a specific receptor. The binding of CEA conjugate to Kupffer cells is very specific and can be inhibited by an excess of unlabeled CEA.[11] Results suggest that CEA uptake by human Kupffer cells is a saturable and concentration-dependent mechanism.

Rat Kupffer cells possess high-affinity binding sites for platelet activating factor (PAF).[12] These authors report that the receptor was functionally active and was shown to mediate arachidonic acid release and eicosanoid production in Kupffer cells, and that these events were inhibited by receptor antagonists. The rat Kupffer cells had specific receptors which were involved in signaling mechanisms which may be important in the production of other autacoid-type mediators. The receptors were lost from the plasma membrane in unstimulated cells and were reformed continuously.[13] The lost receptors were not recycled regardless of the presence or absence of the receptor agonist. Exposure to the agonist accelerated the loss of the receptor. Ligand-mediated loss of surface receptors is fairly common and has been shown for β-adrenergic agonists,[14] human calcitonin,[15] interleukin-2 (IL-2),[16] tumor necrosis factor (TNF),[17] and insulin.[18]

Several receptors are involved in the uptake and catabolism of circulating glycoproteins and glycosoaminoglycans by Kupffer cells. Following uptake, the endocytosed material is degraded in lysosomes.[19] Hyaluronan is taken up by the liver, chiefly by the sinusoidal endothelial cells, and is degraded to *N*-acetylglucosamine, which is then phosphorylated and deacetylated. The activity of this deacetylase is high in endothelial cells and also in Kupffer cells, but is absent in hepatocytes.[19] These authors suggest that the *N*-acetylglucosamine is not degraded in the hepatocyte, but rather reused in the synthesis of new molecules. The high activity of the deacetylase in the sinusoidal endothelial cells and Kupffer cells suggest that the *N*-acetylglucosamine is probably not reused in these cell types.

Immunoglobulin A (IgA) is actively transported from blood to bile, but the receptor for IgA is not detected on hepatocytes.[20] The transfer occurs via the bile duct lining cells. The receptor for polymeric IgA (pIgA) is present on hepatocytes in human liver[21] as detected using a monoclonal antibody. The presence of this receptor on the human hepatocyte suggests that the liver may be very important in the removal of pIgA from the blood to the bile. Recently, Kupffer cells have been reported to be involved in the clearance of IgA aggregates in rats.[22]

The clearance of insulin in the liver is mediated by insulin receptors.[23] In glucose-intolerant obese subjects, the number of insulin receptors in liver cells is reduced.[24] High free fatty acid levels correlate with a decreased clearance of insulin in the rat liver.[25] Free fatty acid inhibits insulin binding and degradation in hepatocytes.[26] These results suggest that the high free fatty acid levels in the portal blood of obese subjects reduces insulin binding and degradation. Insulin stimulates the activity of protamine kinase in isolated rat hepatocytes.[27] The physiological role of the regulation of the cytosolic protamine kinase is not yet known, but the high sensitivity and rapid response suggested to these authors that the insulin-stimulated increase in protamine kinase may have an important role in the insulin signal transduction chain in hepatocytes. The transport and uptake of lysosomal enzymes may be mediated by the insulinlike growth factor-II (IGF-II) and IGF-II/mannose-6-phosphate receptors.[28] The role of IGF-II in hepatic growth is unclear. IGF-II/mannose-6-phosphate receptors are increased in hepatocytes from regenerating rat liver, suggesting that IGF-II receptors are related to hepatocyte growth rates.[29] In parallel to an increase in receptor levels, the cells from regenerating rat livers showed an increased sensitivity to IGF-II, further suggesting a role for IGF-II in liver regeneration.

Internalization of glucagon by isolated hepatocytes is receptor mediated.[30] These authors show that glucagon internalization stops after 90 min in isolated hepatocytes, whereas asialofetuin internalization, which uses a different receptor, continues — probably due to receptor recycling or rapid *de novo* synthesis. Further, these authors suggest that internalization of glucagon as a glucagon-receptor complex is accompanied by a loss of unoccupied receptors. Whether or not these unoccupied receptors are internalized, as has been reported for epidermal growth factor (ECF) receptors, is presently unknown.[31] In the β-adrenergic system, loss of high-affinity receptors appears to be due to phosphorylation.[32] Autophosphorylation of the insulin receptor system leads to desensitization of the receptor.[33] This differs from glucagon, where the receptor number is decreased, but the affinity remains unchanged. Desensitization of β-adrenergic receptors in hepatocytes involves both homologous and heterologous desensitization.[34] Homologous desensitization involves a decreased response to the same agent, whereas heterologous desensitization involves decreased responsiveness to agents chemically unrelated to the initial stimulus. These authors show that stimulation of protein kinase C blocks the actions due to activation of adenylate cyclase-linked receptors (β-adrenoreceptors and glucagon receptors) and leads to heterologous desensitization. The authors also show that the homologous and heterologous types of β-adrenergic desensitization are additive and that the heterologous type is markedly reduced by pretreatment with pertussis toxin. Early studies showed pertussis would block some forms of desensitization and that pertussis toxin would block glucagon desensitization in hepatocytes.[35] The mechanism for the action of pertussis toxin is unknown. Angiotensin-II (ANG II) and vasopressin (AVP) may play a role in the drop in systemic blood pressure which occurs in pregnancy.[36] Receptors for ANG II and AVP are present on isolated rat liver cells, and during pregnancy in the rat there is a functional uncoupling of the ANG II and AVP receptors in the liver.[37] The mechanism involved in the functional uncoupling is presently unknown, but these effects may explain the systemic fall in blood pressure which occurs in pregnancy.

Hepatocytes have receptors for EGF. Receptors for EGF can be upregulated by compounds such as butyrate,[38] and glucocorticoids increase EGF binding in hepatocytes.[39] Glucocorticoids and butyrate act synergistically to enhance the expression of EGF receptors on hepatocytes.[40] Due to the common presence of dexamethasone in culture media, this synergistic activity of glucocorticoids and butyrate could have wide significance. Dexamethasone was shown to be required for butyrate to be effective, and butyrate was shown to increase the effect of dexamethasone and to counteract the inhibition by insulin on the dexamethasone-induced upregulation of the EGF receptors.[40] The mechanism for these actions is presently unknown.

Endotoxins are cleared through the liver. Bacterial lipopolysaccharides (LPS) are potent endotoxins.[41,42] Studies report that Kupffer cells are largely involved in the clearance of LPS,[43] yet conflicting results have been reported indicating that parenchymal cells are also equally involved.[44] Hepatocytes have high-affinity receptors for LPS, and the receptors recognize the inner core region of bacterial LPS and bind to the heptose 3-deoxy-D-manno-2-octulosonic acid (KDO) region of LPS, not lipid A.[45] The structural determinant important for IL-1 induction by polysaccharides is located in the heptose-KDO disaccharide.[46] At present, it is unknown whether or not macrophagelike cells and hepatocytes have similar receptors for the inner core region of LPS.

Rat hepatocytes have binding sites for low-density lipoprotein (LDL). One of these binding sites is similar to the classical LDL receptor which is upregulated by newborn calf serum and insulin[47] and is retarded by dexamethasone.[48] Recent results reported by these authors indicate the cholesterol esterification must be inhibited for downregulation of LDL receptor activity to occur.[49]

Laminin receptors are present on rat hepatocytes.[50] Laminin is a cell-adhesion protein which appears to be important in maintaining tissue architecture and has been identified on neurons and muscle cells.[51,52] Three binding sites of laminin have previously been identified on hepatocytes.[53] Hepatocyte receptors which recognize two binding sites in the laminin molecule have been characterized.[50] The function of the laminin receptor on rat hepatocyte remains to be elucidated. The role of adhesion molecules in human liver grafts during rejection is an area of intense interest. Adhesion receptor molecules have recently been studied in biopsies taken during complications following liver transplant and compared to biopsies of liver taken at the time of transplant.[54] These authors report a *de novo* induction of intercellular adhesion molecule (ICAM-1) on parenchymal liver cells during immune reactions after liver transplant. ICAM-1 was found to be inducible on hepatocyte membranes, and endothelial and Kupffer cells during acute rejection, viral infection, ore sepsis. The regulation of expression of ICAM-1 is presently unknown. The induction of ICAM-1 on hepatocyte membranes and endothelia was down-regulated with successful treatment of the clinical complications. The expression of ICAM-1 on hepatocytes in human liver grafts was found to be restricted to the sites of infiltration and cell lysis. The role of adhesion molecules in human liver rejection remains to be elucidated.

III. RECEPTOR MEDIATED COMMUNICATION IN DISEASE

In liver disease, there is a reduced uptake and clearance of compounds by the liver. One possible mechanism for this reduced clearance is a change in receptor function in hepatocytes. The induction of cirrhosis in an animal model using carbon tetrachloride leads to a reduction in receptor-ligand interaction in hepatocytes.[55] Elevated serum concentrations of EGF,[56] insulin,[57] and glucagon[58] are explained by reduced uptake and clearance by the diseased liver. These authors[55] have reported for the first time that this alteration in uptake and clearance is related to hepatocellular function, i.e., reduced receptor-ligand interaction. These authors report that the number of EGF and asialorosomucoid high-affinity binding sites are markedly reduced in cirrhotic livers, whereas those for insulin are unchanged in hepatocytes.

The role of the hepatic asialoglycoprotein receptor (ASGP) in chronic active hepatitis was investigated,[59] and these authors suggest that the ASGP receptor is preferentially detected on periportal liver cells when assessed by perfusion with antibody to the ASGP receptor. Though all hepatocytes synthesize the ASGP receptor, its preferential expression at high density in the periportal cells may provide an explanation for the periportal liver damage which occurs in chronic active hepatitis patients who have high titer of circulating anti-ASGP receptor, through the involvement of these antibodies in antibody dependent cellular cytotoxic reactions or in complement-mediated cytolysis.

Subcellular dissection of hepatic trafficking pathways has been used to understand receptors, channels, pumps, and enzymes distributed on the cell surface of hepatocytes.[60-62] Subcellular fractionation allows examination of the exocytic and endocytic pathways that operate in hepatocytes. Subcellular fractionation of liver allowed analysis of endosomes, golgi, lysosomes, sinusoidal membranes, bile canalicular plasma membranes, lateral plasma membranes, and gap junctions. Studies using the regenerating liver as a model indicate that within 24 h after hepatectomy the endocytic compartment involved in uptake of nutrients, hormones, and growth factors from the blood may undergo remodeling, whereas the compartment involved in storing intracellular receptors is maintained.[65] These authors suggest that the major endocytic compartment identified in the hepatocyte biliary axis indicates the metabolic importance of this area of the cell. Cells (including hepatocytes and macrophage cells) take up solutes by fluid phase or receptor-mediated endocytosis.[63] Ligands are taken up by cells via binding to specific receptors and entrance through a coated pit pathway.[64]

IV. CELL COMMUNICATION VIA CYTOKINES: INTERLEUKINS

Many cytokines elicit their effects on hepatocytes and Kupffer cells. Kupffer cells synthesize and release several cytokines. IL-6 was first described as a secretory product of fibroblasts[65] and has recently been shown to be synthesized and produced by monocytes and endothelial cells.[66-68] One of the major physiological functions of IL-6 is the induction of the hepatic acute-phase response.[68] Following recent reports suggesting that human recombinant IL-6 binds to hepatocyte membranes, the IL-6 receptor (IL-6-R) has been detected on human hepatocytes.[69] The expression of IL-6-R is different in human primary hepatocytes than in the monocytes. Expression of IL-6-R is stimulated by both IL-1 and IL-6 in human hepatocytes, whereas IL-6-R is downregulated by IL-1 and IL-6 in monocytes. Therefore, a tissue-specific downregulation of IL-6-R is evident from these studies and may have a role in noninflammatory conditions.

IL-1 is a cytokine released from monocytes and macrophages, including Kupffer cells, and regulates the production of other cytokines as well as the stimulating acute-phase protein synthesis in the liver.[70] IL-1 has been shown to inhibit cytochrome P-450 levels in hepatocytes.[71-73] Primary cultured rat hepatocytes express IL-1 receptor, and this IL-1 receptor mediates the dose-dependent decrease of cytochrome P-450 by IL-1.[74] IL-1 can self-control its own production and action.[75] These results suggest that prostaglandin E_2 and IL-1 inhibitor released by Kupffer cells may be involved in the negative self-control in regulating IL-1 production and action. These authors suggest that the inhibitory effect of these agents on LPS-induced IL-1 production by Kupffer cells is mediated via the elevated second messenger, cyclic adenosine monophosphate (cAMP). IL-1 inhibitor has been reported by several investigators.[76-79] An inhibitor to IL-1 with a high molecular weight, as well as a low-molecular weight inhibitor of IL-1, have been described.[80] The high-molecular weight inhibitor acts specifically toward the IL-1 action since the inhibition could be overcome with addition of high concentrations of IL-1 to the assay.[75] Both IL-1 and IL-6, and glucocorticoids cooperatively and separately, regulate the production of mouse acute-phase proteins.[81] The *in vivo* effect of IL-1 on fibronectin synthesis is mediated by IL-6. Though human blood monocytes normally express the IL-6-R, treatment of monocytes with endotoxin, IL-1β, or IL-6 will result in decrease of IL-6-R messenger RNA levels.[69] On the other hand, treatment of hepatocytes with IL-6 or IL-1 resulted in increased IL-6-R messenger RNA levels. In noninflammatory conditions, monocytes strongly express the IL-6-R, whereas in inflammatory conditions the IL-6-R appears to change from the monocyte population to that of the hepatocytes.[67] IL-6 is the major regulator of acute-phase protein synthesis in adult human hepatocytes,[84] whereas IL-1β and TNF have a moderate effect on the positive acute-phase proteins. The actions of IL-1β and TNF may in fact be mediated through the actions of IL-6, i.e., they stimulate other cell types such as fibroblasts to synthesize IL-6.[84,85] The functions of IL-6 as an exocrine hormone in inflammation have been reviewed.[86] Hepatocytes undergoing acute-phase response require exogenous IL-6. Though the hepatocyte is capable of producing IL-6, it does not do so under normal conditions nor while undergoing acute-phase response; only when it is placed in culture does it express IL-6. The physiological significance of IL-6 expression under culture conditions is unknown. Human hepatoma cells and primary hepatocytes produce IL-6 activity, and express IL-6 proteins and mRNA in a similar fashion to fibroblasts; the hepatoma-derived IL-6 can stimulate acute-phase protein production in hepatoma cells.[87] The hepatocyte-stimulating activities derived from monocytes are due to IL-6, IL-1, and TNF molecules.[88] The minor hepatocyte-stimulating factor activity is due to the IL-1 and TNF activity. Each of these cytokines has a receptor on the hepatocyte which regulates the production of acute-phase proteins. These acute-phase proteins are regulated differently by the different cytokines, i.e., IL-6 alone will stimulate a subset of acute-phase protein genes, and in combination with other cytokines will stimulate another subset of acute-phase protein genes. IL-6 is one of several cytokines that affect acute-phase gene expression in human hepatocytes and human macrophages.[89] IL-6 is produced by many cells, including fibroblasts, epithelial cells, and mononuclear phagocytes in response to several activators, including IL-1β, TNF, platelet-derived growth factor (PDGF), interferon-β (IFN-β), LPS, and phorbol esters.[90] IL-6 stimulates the acute-phase response in hepatocytes, but also in other cells such as blood monocytes and other macrophages. The acute-phase response includes the elevation of several proteins, such as C-reactive protein and serum amyloid A, with moderate increases in fibrinogen and alpha$_1$-antitrypsin as well as complement proteins factor B and C3, while other plasma proteins such as albumin and transferrin decrease.[89] This author concludes that several cytokines — IL-1β, TNF, and IFN-β_2 — may have the same net effect on a single acute-phase gene by distinct and separable cell-surface receptors and signal transduction pathways. The action of IL-6 has been reported to be potentiated by phorbol myristate acetate (PMA), a known activator of kinase C, in such a way that PMA and IL-6 appear to have additive effects. The results further suggest that IL-6, PMA, and glucagon stimulate liver cells during the acute-phase response by different pathways.[91]

V. CELL COMMUNICATION VIA GROWTH FACTORS

Another cytokine, transforming growth factor alpha (TGF-alpha), is an important growth-regulating polypeptide which appears to bind to cell-surface EGF receptors[92] and is synthesized and secreted by hepatoma cell lines.[93] TGF-alpha mRNA, which is present in normal rat liver,[94] has recently been shown to stimulate proto-oncogene c-jun expression in rat hepatocytes and is mitogenic in cultures of adult rat hepatocytes.[95] This suggests that TGF may have an important role in the control of hepatic growth, regeneration, and gene expression. EGF, when it first appears, is a polypeptide that stimulates DNA synthesis in hepatocytes in culture,[96] and thus may have a role in regenerative growth in liver. The constitutive high-affinity EGF receptors do not elicit the proliferate response to EGF.[97] These studies indicated a loss of the constitutive high-affinity EGF receptors as the proliferate response increases. This is different from most cell types in which the binding to high-affinity EGF receptors elicits the mitogenic response. Zinc increases EGF-stimulated DNA synthesis in primary mouse hepatocytes and suggests a need to substitute adequate amounts of zinc in serum-free culture cells.[98] TGF-β (β1 and β2), on the other hand, strongly inhibit hepatocyte proliferation,[99,100] and TGF-β_1 is produced by nonparenchymal liver cells in culture.[101] The inhibitory effects of TGF-β_2 on hepatocyte proliferation is modified by bovine serum and α-2 macroglobulin, possibly by binding of these substances to TGF-β_2.[102] A factor is present in the serum of partially hepatectomized rats which induces primary hepatocytes to enter S phase, and these authors suggest a possible role for prostaglandins. It is well established that adult hepatocytes in primary culture are unable to replicate.[104] Several factors have been identified which can stimulate DNA synthesis and proliferation of hepatocytes, including EGF,[105] platelet growth factor,[106] and various extracts of the serum[107] and extracts from hepatic cytosol taken after partial hepatectomy.[108] The growth-promoting activity of serum from partially hepatectomized animals was markedly decreased by addition of inhibitors of prostaglandin synthesis.[103] A stimulator substance obtained from weanling rat liver which stimulates DNA synthesis is a protein of molecular weight 15,000 to 20,000.[109] Another polypeptide growth factor for hepatocytes, human hepatopoietin A, whose mitogenic activity was totally prevented in the presence of TGF-β, has a high molecular weight in the order of two major bands, at 65,000 and 35,000 molecular weight.[110] A cofactor is provided by conditioned medium or by a cell extract that enables a hepatomitogen to act on hepatocytes in primary culture, i.e., similar to hepatocyte *in vivo,* hepatocytes in primary culture need to be initiated before they can respond to hepatomitogen.[111] The initiation *in vivo* was usually done by hepatectomy.[112] The hepatomitogen was ineffective in increasing DNA synthesis in control or sham-operated rats. The identity of the cofactor present in hepatoma cell-conditioned medium is presently unknown.

VI. KUPFFER CELL INITIATED COMMUNICATION

Kupffer cells have an integral role in the whole concept of intercellular communication in the liver because it is the Kupffer cells which are the first cells in contact with foreign substances in the blood. Kupffer cells phagocytose a wide variety of substances by binding to specific sites, either Fc, C3b, mannose, galactose, or apolipoprotein B.[113] Receptors for fibronectin-containing particles,[114] insulin, and glucagon[115] are also present. Binding sites for bacteria, yeast viruses and so-called foreign body receptors are also present.[116] The endocytosed substances are then transported to the lysosomes for metabolism. Many substances are endocytosed by Kupffer cells.[117] As a result of phagocytosis of bacteria, viruses, colloidal compounds, immune complexes, and other materials, Kupffer cells synthesize and secrete several factors.[116] These mediators include prostaglandins, IL, TNF, and other cytokines. These factors in turn can act on nearby hepatocytes or can act back on the Kupffer cells themselves to modulate various functions of the hepatic cells. A role for endotoxin-responsive macrophages in hepatic injury has been suggested, such that the oxidative free radicals of LPS-responsive macrophages contribute to the pathogenesis of galactosamine-induced hepatic injury.[118] Endotoxin elicits the secretion of many factors from hepatic macrophages, including oxygen free radicals which have been implicated in tissue injury.[119] The mechanism for galactosamine-induced liver injury is unknown, but recent studies suggest roles for endotoxin,[116,120] TNF,[121] and leukotrienes,[122] though their mechanisms are unknown. Several inhibitors of leukotriene synthesis were effective in reducing the histological signs of liver injury induced by galactosamine.[122]

The complete elimination of Kupffer cells from the liver occurs following an intravenous injection of liposome-encapsulated dichloromethylene diphosphonate (C12MDP), with a rapid reappearance of these Kupffer cells in the second week after treatment.[123] The absence of macrophages from the liver for a 1-week period following treatment with the C12MDP liposomes allowed study of the functional aspect of these cells. A selective depletion of Kupffer cells in mice occurs with intact ricin.[124] Ricin significantly reduced Kupffer cell numbers in mice by 33% with no evidence of hepatic parenchymal damage. The effect persisted for at least 3 d with recovery evident within 1 week. Higher concentrations of ricin were found to lead to systemic toxicity and nonspecific uptake. These authors reported that intraportal injections of ricin were more effective than intraperitoneal doses. Depletion of populations of liver macrophages may prove to be an important tool for understanding the function of these cells in intercellular communication. The A-chain of ricin, which selectively depleted Kupffer cells, was mannosylated and therefore presumably bound to the mannose receptor on the Kupffer cell. Neomannosylated vesicles have been used as tools for targeting bioactive molecules to the Kupffer cells.[125] Neomannosylated vesicles showed a ninefold increase in binding compared to control vesicles. These may prove to be important tools for targeting molecules to macrophages (including the Kupffer cells) in macrophage-related diseases, or as the authors suggest in targeting liposomes containing endotoxin which will interact with macrophages to improve regression of solid tumors. An interaction between metastatic colon carcinoma cells and syngeneic Kupffer cells has been reported.[126] Binding was found to be increased by pretreatment of the cells with phorbol 12-myristate 13-acetate and inhibited by colchicine and cytochalasin B, even to a larger extent by D-mannose and *N*-acetyl-D-galactosamine. These results may prove to be important in studies of metastasis of tumor cells to the liver when it was found that the Kupffer cells showed enhanced binding to the colon carcinoma cells, but did not show enhanced cytolysis of this cell. Liver phagocytes acquire augmented cytotoxic activity against tumor cells following treatment with pyran copolymer.[127] The rat liver nonparenchymal cells, unlike the mouse, did not require pretreatment with pyran copolymer to induce this cytotoxic activity. Liver macrophages from *Listeria monocytogenes*-infected mice exert enhanced prostaglandin E release and tumor cytostatic activity *in vitro*.[128] Recent studies suggest that exposure to liver *L. monocytogenes* results in depression of murine hepatic mixed-function oxydase.[129] These authors suggest that the depression during infection with *L. monocytogenes* could result from activation of the Kupffer cells since massive phagocytosis occurs during infection, and as a result of this phagocytosis, factors could be released from Kupffer cells which depress P-450, similar to that reported earlier for dextran sulfate.[71] These authors suggest that IL-1 or hemolysin may be the mediators involved.

VIII. KUPFFER CELL HEPATOCYTE COMMUNICATION AND CYTOCHROME P-450

Activation of Kupffer cells results in release of soluble mediators which can affect hepatocyte function. Activation of Kupffer cells by dextran sulfate resulted in the depression of cytochrome P-450-dependent drug biotransformation in hepatocytes.[131] Using isolated parenchymal and nonparenchymal cells, these authors reported that a soluble factor was released from Kupffer cells which depressed hepatocyte cytochrome P-450 and related enzymes *in vitro,* and suggested that the depression observed *in vivo* after administration of dextran sulfate could be mediated by release of factors from Kupffer cells which depressed cytochrome P-450 and related enzyme activity in nearby hepatocytes. The involvement of other cells, including lymphocytes and other mediators such as IFN, was eliminated in earlier studies.[72] These early studies suggested an intercellular communication between Kupffer cells and hepatocytes whereby activation of Kupffer cells resulted in changes in functional activities of the hepatocytes.[131] This mechanism applies to the phagocytosis of other compounds, including phagocytosis of latex particles by Kupffer cells. Latex particles administered to animals depressed cytochrome P-450 activity, but histology of the livers of these animals indicated uptake of latex particles by Kupffer cells, while hepatocytes appeared free of latex particles though the cytochrome P-450-dependent activity was localized largely in the hepatocyte.[131] These authors reported that activation of Kupffer cells by latex particles led to release of soluble factors which could be isolated, and when incubated with hepatocytes depressed cytochrome P-450 and related enzyme activities in the hepatocytes.

The cytochrome P-450-dependent monooxygenase system is responsible for metabolism of most drugs, carcinogens, xenobiotics, and endogenous compounds.[133] The activity of the cytochrome P-450 system in the liver is altered by many factors which affect host defense mechanisms, including viral infections and agents that activate the reticuloendothelial system.[131] Early studies indicated a possible involvement of the reticuloendothelial system in the depression of cytochrome P-450 by agents such as Bacillus Calmette-Guerin[134] and *Corynebacterium parvum*.[135] The phagocytosis of particulate matter by Kupffer cells and by other macrophages was found to decrease hepatocyte cytochrome P-450 and arylhydrocarbon hydroxylase (AHH) activity by a factor secreted from the Kupffer cells or other macrophages.[130] These factors could be isolated and would depress hepatocyte enzyme activity *in vitro*.[71] The mechanism for the depression of the cytochrome P-450 activity was suggested to be due to production of IL-1.[71] Several reports using recombinant IL-1 or monocyte condition medium (MCM) corroborate the effect of IL-1 on hepatocyte cytochrome P-450 and related enzyme activities.[72,73,136] Preincubation of MCM with anti-IL-1 prevents the depressive action of the MCM on cytochrome P-450.[137] A role for reactive oxygen intermediates in the depression of cytochrome P-450 by MCM has also been suggested, because pretreatment of MCM with catalase significantly reduced the inhibitory action of MCM on cytochrome P-450-mediated AHH activity.[136] The effect of oxygen free radicals and reactive oxygen intermediates in depressing cytochrome P-450 has been previously reported.[130] Factors released from Kupffer cells and other macrophages decrease not only the hepatocyte cytochrome P-450 and related enzymes, but also act back on the Kupffer cells or macrophages themselves and inhibit enzyme activity in the Kupffer cells and other macrophages. Drug-metabolizing enzymes such as AHH have been detected in Kupffer cells[133,141] and peripheral blood monocytes.[140-143] The contribution of the macrophage drug-metabolizing enzyme to the total enzyme pool is small, but has been shown to be a good index of the status of the drug-metabolizing enzymes in the liver.[143] AHH activity was detectable in macrophages from rat, human, pig, and mouse macrophages and was depressed in response to phagocytic activation.[143] Monocyte AHH activity is decreased in patients with liver disease, and the authors suggested a possible role for macrophage factors (including IL-1) in the depression of AHH activity.[142] Activated macrophages may play a role in the effect of phenobarbital on the induction of hepatic cytochrome P-450.[144] Phenobarbital induced an accumulation of activated macrophages in the liver, which produced elevated quantities of hydrogen peroxide. The authors suggested that the direct action of phenobarbital on hepatocyte masks the suppressive effects of factors released from the activated macrophages on cytochrome P-450. It is also known that the cytochrome P-450 isozyme induced by phenobarbital is distinct from the isozyme that mediates AHH activity, and the induction of the phenobarbital-sensitive P-450 isozyme could indeed mask a depression in the cytochrome P-450-mediated AHH activity. The interaction (i.e., cooperation) between Kupffer cells and hepatocyte via soluble mediators (either cytokines or activated oxygen intermediates) has a net effect of altering drug-metabolizing enzymes in the hepatocytes. In liver disease, the Kupffer cells are activated as a result of phagocytosis of necrotic hepatocellular debris.[145] As a result of activation, Kupffer cells release cytokines which could inhibit hepatocyte drug-metabolizing enzyme activity. Macrophages from patients with liver disease secrete factors which depress hepatocyte drug-metabolizing enzymes.[142] This factor may be IL-1, because its activity is attenuated by anti-IL-1.[137] The AHH inhibitory activity was also reduced by catalase, suggesting that activated oxygen intermediates (e.g., hydrogen peroxide) may also be involved.

VIII. KUPFFER CELL HEPATOCYTE INTERACTIONS IN DISEASE AND HOMEOSTASIS

Communication between Kupffer cells and hepatocytes in liver disease may also be responsible for the initiation of fibrosis. It is well known that Kupffer cells and other macrophages release cytokines, and MCM obtained from patients with liver disease is fibrogenic.[146,147] The fibrogenic activity of the MCM could not be attenuated by preincubation with anti-IL-1 or with catalase, suggesting that IL-1 and activated oxygen intermediates were not involved in the fibrogenic activity of the cytokines released from these cells.[147] Thus, in the cellular communication between Kupffer cells (or other macrophages) with hepatocytes in liver disease, several cytokines appear to mediate different responses, i.e., IL-1 and reactive oxygen intermediates appear to be involved in the depression of cytochrome P-450-mediated drug-metabolizing enzyme activity, whereas another cytokine appears to be involved in the fibrogenic activity. The fibrogenic properties of PDGF are well established.[148] Recent results suggest that PDGF is involved in the fibrogenic activity in MCM of patients with liver disease.[149] Exogenous PDGF is fibrogenic, and preincubation of MCM obtained from patients with liver disease with anti-human PDGF reduced the fibrogenic activity of the MCM.[150] PDGF does not affect hepatocyte AHH activity, again suggesting that the interaction between Kupffer cells (or other macrophages) and hepatocytes in liver disease involves several cytokines which have distinct actions.[150]

Interactions and cooperation between Kupffer cells and hepatocytes in other hepatic functions have also been reported. PAF affects liver metabolism, i.e., it induces glycogenolysis,[151] but PAF will not affect glycogenolysis in isolated parenchymal cells,[152] suggesting that something is missing in the *in vitro* preparation. The induction of glycogenolysis in the liver by PAF is mediated by prostaglandin D_2 from Kupffer cells, i.e., the primary interaction of PAF is with the Kupffer cells with subsequent release of prostaglandin and its effect on glucose production in parenchymal cells.[153] The interaction of PAF with Kupffer cells has been corroborated.[154] These authors demonstrate that PAF increased inositol phosphate production and cytosolic-free calcium concentration in the cultured rat Kupffer cell, indicating that PAF exerts effects on the Kupffer cell. In response to stimulation with the calcium ionophore A23186, Kupffer cells can synthesize and release PAF,[155] further suggesting that the reticuloendothelial cells play an important role in the hepatic metabolism and activity of PAF.

Another interaction between Kupffer cells and hepatocytes[155] mediated by prostaglandins released from activated Kupffer cells has been reported.[156] Administration *in vivo* of PMA induced ornithine decarboxylase, but if added *in vitro* to parenchymal cells, this enzyme was not affected. These results indicate that PMA activates Kupffer cells to release prostaglandins which induce ornithine decarboxylase in parenchymal cells.

Kupffer cells play an intermediatory role in the stimulation of DNA synthesis in hepatocytes following partial hepatectomy.[157] Kupffer cells release factors as a result of partial hepatectomy, and these factors are involved in stimulating DNA synthesis in the hepatocytes at an early stage of liver regeneration. The identity of these factors is presently unknown. Activation of Kupffer cells also results in production of L-arginine metabolites that release cell-associated iron and inhibit hepatocyte protein synthesis.[158] Kupffer cell cytotoxicity to hepatocytes in cocultures requires L-arginine,[159] again supporting the hypothesis that toxic L-arginine metabolites contribute to liver cell damage and may be an important mechanism in liver cell damage seen in patients with sepsis. In response to inflammatory products of Kupffer cells following exposure to LPS, hepatocytes produced nitrogen oxides from L-arginine.[160] Thus the Kupffer cells and the hepatocytes appear to interact in such a fashion that the Kupffer cells have an important role in the control of hepatocyte protein synthesis, particularly in conditions where inflammatory stimuli exist. The mediators of the interaction between Kupffer cells and hepatocytes have not been identified, but are unlikely to be IL-1 or TNF.[161] A Kupffer cell factor produced by LPS which inhibits hepatocyte protein synthesis can be transferred with supernatant obtained from Kupffer cells triggered by LPS, and this activity can be blocked by dexamethasone.[162] Further studies are necessary to identify the nature of the factor involved in this cellular communication between Kupffer cells and hepatocytes.

IX. HEPATOCYTE INITIATED COMMUNICATION

Recent studies indicate that interactions between Kupffer cells and hepatocytes can also be mediated by factors released from hepatocytes which act on Kupffer cells.[163] Hepatocytes produce a soluble mediator that enhances Kupffer cell tumoricidal activity. Earlier studies[144] indicated that acetaminophen hepatotoxicity involved an interaction between Kupffer cells and hepatocytes, whereby the hepatocytes released factors which resulted in recruitment of macrophages into the liver. *N*-acetylcysteine prevents acetaminophen hepatotoxicity and inhibits PDGF,[164] proposing a role for PDGF and cell communication in this toxicity. These reports suggest that cooperation and communication between Kupffer cells and hepatocytes in the liver occurs in both directions between the cells and plays an important role in normal physiology as well as the pathological conditions of the liver.

X. COMMUNICATION VIA GAP JUNCTIONS

Another form of intercellular communication in the liver is via special plasma membrane structures called gap junctions. The major gap junction polypeptide in the liver is 32 kDa.[165] Gap junctions allow direct communication between cells in the liver.[167] Gap junctions allow communication between adjacent cells and allow passage of ions and molecules up to 1000 Da.[167] Gap junctions are thought to have an important function in the regulation of cell proliferation and differentiation, and the inhibition of gap junction intercellular communication has been studied in tumor development.[168] Inhibition of gap junctional intercellular communication by cigarette smoke condensate and phorbol ester has been reported.[169] Growth factors, including EGF and PDGF, inhibit gap junctional communication in mammalian cells.[170] Since Kupffer cells can release these factors, they may have a role in this form of communication. Gap junctions represent cell to cell communication that requires cell contact,[171] in contrast to receptor-mediated endocytosis which does not require cellular contact. Gap junctional communication mediates the activation of cAMP-dependent protein kinase by hormone-induced signals passed from receptor-bearing cells to receptorless partners.[172,173] Regulation of gap junctions by protein kinases is of major significance.[174] The function and regulatory features of gap junction between hepatocytes has been reviewed.[173] More studies are needed to understand the functioning of gap junctions. Gap junctions can be solubilized by detergent and reconstituted to restore gap junctionlike structures which will allow study of elements involved in their functioning.[175] A recent report indicates that cAMP delays the disappearance of gap junctions between pairs of rat hepatocytes in primary culture.[176] The authors suggest that this is due to an extension of the lifetime of the mRNA encoding the main gap junction protein, and cAMP may act by decreasing removal of junctions from appositional membranes. Albumin secretion and junctional communication in hepatocytes are modulated by coculturing hepatocytes with sinusoidal liver cells.[177] Though the mechanism for this interaction is unknown, the authors suggest that secretion of soluble factors may be important.

XI. INVESTIGATING CELL COMMUNICATION *IN VITRO*

Most studies of hepatocytes in cell interactions are done *in vitro*. Hepatocytes in suspension or in culture are different from hepatocytes *in vivo*. Hepatocytes in the intact liver build up cell rods that are in contact with the blood plasma and the bile.[179] These cells *in vivo* exhibit a definite polarity which is lost when cells are isolated. Therefore, the transport processes in single cells may differ greatly from transport in cells *in vivo*. Hepatocytes maintained in primary culture reestablish their polarity, and the cell to cell contacts are rebuilt and the bile canaliculi develop.[178] Isolated hepatocytes maintain several transport systems, including those for bile acids,[180] amino acids,[181] sugars,[182] halloidin,[183] ouabain,[184] and bumetanide,[185] whereas transport of methotrexate[186] and asialoorosomucoid[187] decline rapidly while in culture. The mechanism for loss of transport systems in cultured hepatocytes is unknown, but could relate to the loss of cytochrome P-450 activity which has been observed in cultured hepatocytes.[188] Recent studies[189] suggest that the extracellular matrix is important in determining tissue-specific expression in cells in culture;[189] the transcription of the albumin gene was induced in hepatocytes in culture due to extracellular matrix proteins. In earlier studies, albumin synthesis and secretion was maintained in hepatocytes for several weeks when the cells were plated on a basement membranelike matrix derived from a mouse sarcoma tumor,[190] while hepatocytes plated on type I collagen show a rapid loss of albumin production.[193] The mechanism for the interaction between extracellular matrix and cell surface and its role in increased transcription of the albumin gene is presently unknown, although laminin may have an important role.[190] A recent report suggests that dipeptidylpeptidase IV, a cell-surface glycoprotein, is important in fibronectin-mediated interactions of hepatocytes with extracellular matrix.[192]

XII. CELLULAR COMMUNICATION INVOLVING OTHER LIVER CELLS

Other intercellular forms of communication (in addition to the interaction between Kupffer cells and hepatocytes) are present in the liver. The role of rat liver epithelial cells in the induction of the acute-phase response in hepatocytes has been investigated using cocultures of hepatocytes and epithelial cells.[193] These authors showed that hepatocytes in coculture respond to cytokines, suggesting that the coculture allows hepatocytes to maintain the differentiated functional stage and allows the study of the action of cytokines for longer periods, because coculture allows hepatocytes to remain in culture longer. In coculture, rat hepatocytes were able to respond to repeated inflammatory stimuli, further suggesting that in a typical pure hepatocyte culture something is missing such that they are not typical of the rat cell *in vivo* and hence do not reproduce exactly the response that is seen *in vivo*.[194] An interaction between Kupffer cells and hepatic lipocytes has been reported.[195] Lipocytes can be transformed into more fibroblastic cells in animal models of liver disease.[196,197] Monocytes from patients with liver disease secrete fibrogenic factors, including PDGF.[147,150] Kupffer cell-conditioned medium obtained from a rat model of alcoholic liver injury stimulated collagen formation of lipocytes, and results suggested that a combination of a high-fat diet and ethanol resulted in lipocytes which were biochemically activated to produce more collagen and sensitized to be stimulated more effectively by Kupffer cell-derived factors.[195] These authors suggest that a high-fat diet was responsible at least in part for the sensitization of the lipocytes. The nature of the interaction between lipocyte, hepatocytes, and Kupffer cells is not totally understood, but the role of cytokines in this interaction is an area of intense research.

XIII. INHIBITORS OF CELL COMMUNICATION

In the study of cellular interactions within the liver between Kupffer cells, hepatocytes, and lipocytes, this interaction can be investigated with the aid of inhibitors such as the use of monoclonal antibodies to specific cytokines *in vitro*. The depression of cytochrome P-450-mediated drug metabolism by MCM of patients with liver disease via IL-1 was determined with the aid of an antibody to human IL-1.[139] Reduction of the fibrogenic activity of MCM using an antibody to human PDGF suggested that PDGF was probably involved in the fibrogenic activity of MCM obtained from patients with liver disease.[150] Vanadate interferes with at least two steps in the endocytosis of asialoorosomucoid in rat hepatocytes: (1) it reduces the surface receptors 70% and (2) it prevents transport of the ligand from endosomes to lyosomes.[198] Such an inhibitor could provide important information on the role of receptor-mediated endocytosis in hepatocytes. Tumor promoters inhibit gap junctional intercellular communication. Cholesterol epoxides inhibit intercellular communication.[199] A recent study suggests that inhibitors of adenosine diphosphate-ribosyl transferase will suppress the mitogenic actions exerted by tumor promoters on primary neonatal rat hepatocytes, providing a further compound to test the role of gap junction in intercellular communication in the liver.[200] Diazepam and the structurally similar drug clonazepam were investigated for their ability to effect cell to cell communication in the liver and to act as tumor promoters.[201] Though structurally similar, diazepam was a strong tumor promoter in mouse liver and significantly inhibited mouse hepatocyte gap junctional intercellular communication, while clonazepam had no effect on gap junctional intercellular communication and was inactive as a liver tumor promoter. This is in contrast to the effect of clonazepam on cytochrome P-450IIB1. The induction of P-450IIB1 was considered to be an important predictor of susceptibility to liver tumor promotion,[201,202] and clonazepam was reported to be a potent inducer of hepatic P-450IIB1-mediated enzyme activities.[201] The ability of a compound to inhibit gap junction-mediated intercellular communication may be more important than its ability to induce P-450IIB1 in determining whether or not that compound is a tumor promoter.[200] Thus, studies of intercellular communication are aided with the use of compounds which can inhibit specific forms of intercellular communication in the liver. Compounds which block phagocytic activation of Kupffer cells or destroy Kupffer cells, e.g., ricin, would be important in investigating the role of Kupffer cells as an intermediatory cell in the effects of compounds on hepatocytes. As better and more selective inhibitors of various mediators of intercellular communication become available, the intricate patterns of intercellular communication in normal and in diseased states will become better understood.

REFERENCES

1. **Crofton, R. W., Dieselhoff-den Dulk, M. M., and van Furth, R.,** The origin, kinetics and characteristics of the Kupffer in the normal steady state, *J. Exp. Med.,* 148, 1, 1978.
2. **Skogh, T., Blomhoff, R., Eskild, W., and Berg, T.,** Hepatic uptake of circulating IgG immune complexes, *Immunology,* 55, 585, 1985.
3. **Kurlander, R. J., Ellison, D. M., and Hall, J.,** The blockade of Fc receptor-mediated clearance of immune complexes in vivo by a monoclonal antibody (2,4G2) directed against Fc receptor on murine leukocytes, *J. Immunol.,* 133, 855, 1984.
4. **Muro, H., Shirasawa, H., Takahashi, Y., Maeda, M., and Nakamura, S.,** Localization of FC receptors on liver sinusoidal endothelium: a histological study by electron microscopy, *Acta Pathol. Jpn.,* 38, 291, 1988.

5. **Ito, I., Hiroyuki, M., Kosugi, I., and Shirasawa, H.,** Alterations in FC receptor activity in sinusoidal endothelial cells and Kupffer cells during D-galactosamine (GalN)-induced liver injury in rats, *Virchows Arch. B Cell Pathol.*, 58, 417, 1990.
6. **Schlepper-Schafer, J. and Springer, G. F.,** Carcinoma autoantigens T and Tn and their cleavage products interact with Gal/GalNAc-specific receptors on rat Kupffer cells and hepatocytes, *Biochim. Biophys. Acta,* 1013, 266, 1989.
7. **Schlepper-Schafer, J., Holl, N., Kolb-Bachofen, V., Friedrich, E., and Kolb, H.,** Role of carbohydrates in rat leukemia cell-liver macrophage cell contacts, *Biol. Cell,* 52, 253, 1984.
8. **Smit, M. J., Beekhuis, H., Duursma, A. M., Bouma, J. M. W., and Gruber, M.,** Catabolism of circulating enzymes: plasma clearance, endocytosis, and breakdown of lactate dehydrogenase-1 in rabbits, *Clin. Chem.*, 34, 2475, 1988.
9. **Smit, M. J., Duursma, A. M., Bouma, J. M. W., and Gruber, M.,** Receptor-mediated endocytosis of lactate dehydrogenase M4 by liver macrophages: a mechanism for elimination of enzymes from plasma. Evidence for competition by creatine kinase MM, adenylate kinase, malate, and alcohol dehydrogenase, *J. Biol. Chem.*, 262, 13020, 1987.
10. **Bouma, J. M. W. and Smith, M. J.,** Elimination of enzymes from plasma in the rat, *Adv. Clin. Enzymol.*, 6, 111, 1988.
11. **Toth, C. A., Thomas, P., Proitman, S., and Zamcheck, N.,** A new Kupffer cell receptor mediating plasma clearance of carcinoembryonic antigen by the rat, *Biochem. J.*, 204, 377, 1982.
12. **Chao, W., Liu, H., Debuysere, M., Hanahan, D. J., and Olson, M. S.,** Identification of receptors for platelet-activating factor in rat Kupffer cells, *J. Biol. Chem.*, 264, 13591, 1989.
13. **Chao, W., Heling, L., Hanahan, D. J., and Olson, M. S.,** Regulation of platelet-activating factor receptors in rat Kupffer cells, *J. Biol. Chem.*, 264, 20448, 1989.
14. **Benovic, J. L., Bouvier, M., Caron, M. G., and Lefkowitz, R. J.,** *Annu. Rev. Cell Biol.*, 4, 405, 1988.
15. **Wright, D. R., Ivey, J. L., and Tashijian, A. H.,** Self induced loss of calcitonin receptors in bone: a possible explanation for "escape", *Clin. Res.*, 24, 461, 1976.
16. **Smith, K. A. and Cantrell, D. A.,** Interleukin-2 regulates its own receptors, *Proc. Natl. Acad. Sci. U.S.A.*, 82, 864, 1985.
17. **Watanabe, N., Kuriyama, H., Sone, H., Neda, H., Yamauchi, N., Maeda, M., and Nitsu, Y.,** Continuous internalization of tumor necrosis factor receptors in a human myosarcoma cell line, *J. Biol. Chem.*, 263, 10262, 1988.
18. **Lipson, K. E., Kolhatkar, A. A., and Donner, D. B.,** Cell surface proteolysis of the hepatic insulin receptor, *J. Biol. Chem.*, 263, 10495, 1988.
19. **Campbell, P., Thompson, J. N., Fraser, J. R. E., Laurent, T. C., Pertoft, H., and Roden, L.,** N-Acetylglucosamine-6-phosphate deacetylase in hepatocytes, Kupffer cells and sinusoidal endothelial cells from rat liver, *Hepatology,* 11, 199, 1990.

20. **Mullock, B. M., Shaw, L. J., Fitzharris, B., Peppard, J., Hamilton, M. J. R., Simpson, M. T., Hunt, T. M., and Hinton, R. H.,** Sources of proteins in human bile, *Gut,* 26, 500, 1985.
21. **Perez, J. H., Wight, D. G. D., Wyatt, J. I., Schaik, M. V., Mullock, B. M., and Luzio, J. P.,** The polymeric immunoglobulin A receptor is present on hepatocytes in human liver, *Immunology,* 68, 474, 1989.
22. **Bogers, W. M., Gorter, A., Janssen, D., et al.,** The involvement of Kupffer cells in the clearance of IgA aggregation in rats, *Scand. J. Immunol.,* 31, 679, 1990.
23. **Duckworth, W. C.,** Insulin degradation: mechanisms, products are significant, *Endocrinol. Rev.,* 9, 319, 1988.
24. **Arner, P., Einarsson, K., and Livingstone, J.,** Studies of the human liver insulin receptor in non-insulin-dependent diabetes mellitus, *J. Clin. Invest.,* 77, 1717, 1986.
25. **Stromblad, G. and Bjorntorp, P.,** Reduced hepatic insulin clearance in rats dietary-induced obesity, *Metabolism,* 35, 323, 1978.
26. **Svedberg, J., Bjorntorp, P., Smith, U., and Lonnroth, P.,** Free-fatty acid inhibition of insulin binding, degradation, and action in isolated rat hepatocytes, *Diabetes,* 39, 570, 1990.
27. **Reddy, S. A. G., Amick, G. D., Cooper, R. H., and Damuni, Z.,** Insulin stimulates the activity of a protamine kinase in isolated rat hepatocytes, *J. Biol. Chem.,* 265, 7748, 1990.
28. **Dahms, N. M., Lobel, P., and Kornfeld, S.,** Mannose 6-phosphate receptors and lysosomal enzyme targeting, *J. Biol. Chem.,* p. 264, 1989.
29. **Scott, C. D. and Baxter, R. C.,** Insulin-like growth factor-II/mannose-6-phosphate receptors are increased in hepatocytes from regenerating rat liver, *Endocrinology,* 126, 2543, 1990.
30. **Horwitz, E. M. and Gurd, R. S.,** Quantitative analysis of internalization of glucagon by isolated hepatocytes, *Arch. Biochem. Biophys.,* 267, 758, 1988.
31. **Beguinot, L., Hanover, J. A., Ito, S., Richert, N. D., Willingham, M. D., and Pastan, I.,** Phorbol esters induce transient internalization without degradation of unoccupied epidermal growth factor receptors, *Proc. Natl. Acad. Sci. U.S.A.,* p. 2274, 1985.
32. **Sibley, D. R., Strasser, R. H., Caron, M. G., and Lefkowitz, R. J.,** Homologous desensitization of adenylate cyclase is associated with phosphorylation of the β-adrenergic receptor, *J. Biol. Chem.,* 260, 3883, 1985.
33. **Crettaz, M., Jialal, I., Kasuga, M., and Kahn, C. R.,** Homologous desensitization of adenylate cyclase is associating with phosphorylation of the β adrenergic receptor, *J. Biol. Chem.,* 259, 11543, 1984.
34. **Hernandez-Sotomayor, S. M., Maciase-Silva, M., Plebanski, M., and Garcia-Sainz, A.,** Homologous and heterologous B-adrenergic desensitization in hepatocytes. Additivity and effect of pertussis toxin, *Biochim. Biophys. Acta,* 972, 311, 1988.
35. **Heyworth, C. M., Hanski, E. M., and Houslay, M. D.,** Islet activating protein blocks glucagon desensitization in intact hepatocytes, *Biochem. J.,* 222, 189, 1984.
36. **Aoi, W. D., Gable, R. E., Cleary, P. C., Young, M., and Weinberger, M. H.,** The antihypertensive effect of pregnancy in spontaneously hypertensive rats, *Proc. Soc. Exp. Biol. Med.,* 153, 13, 1976.

37. **Massicotte, G., Codere, L., Chiasson, J. L., Thibault, G., Schiffrin, E. L., and St-Louis, J.,** Regulation of ANG II and AVP receptors in isolated hepatocytes of pregnant rats, *Amer. J. Physiol,* 258, 597, 1990.
38. **Gladhaug, I. P., Refsnes, M., Sand, T. E., and Christoffersen, T.,** Effects of butyrate on epidermal growth factor receptor binding, morphology, and DNA synthesis in cultured rat hepatocytes, *J. Cancer Res.,* 48, 6560, 1988.
39. **Lin, Q., Blaisdell, J., O'Keefe, E., and Earp, H. S.,** Insulin inhibits the glucocorticoid mediated increase in hepatocyte EGF binding, *J. Cell Physiol.,* 119, 267, 1984.
40. **Gladhaug, I. P. and Christoffersen, T.,** n-Butyrate and dexamethasone synergistically modulate the surface expression of epidermal growth factor receptors in cultured rat hepatocytes, *FEBS Lett.,* 243, 21, 1989.
41. **Braude, A. I., Carey, F. J., and Zalesky, M.,** Studies with radioactive endotoxin II correlation of physiological effects with distribution of radioactivity in rabbits, *J. Clin. Invest.,* 34, 858, 1955.
42. **Yamaguchi, Y., Yamaguchi, K., Babb, J. L., and Gans, H. J.,** In vivo quantitation of the rat liver's ability to eliminate endotoxin from portal vein blood, *Reticuloendothel. Soc.,* 32, 409, 1982.
43. **Ruiter, D. J., van der Meulen, J., Brouwer, A., Hummel, M. J. R., Mauw, B. J., van der Ploeg, J. C. M., and Wisse, E.,** Uptake by liver cells of endotoxin following its intravenous injection, *Lab. Invest.,* 45, 38, 1981.
44. **Van Bossuyt, H., Zanger, R. B., and Wisse, E.,** *J. Hepatol.,* 7, 325, 1988.
45. **Parent, J. B.,** Membrane receptors on rat hepatocytes for the inner core region of bacterial lipopolysaccharides, *J. Biol. Chem.,* 26, 3455, 1990.
46. **Lebbar, S., Cavaillon, J. M., Canoff, M., Leden, A., Brady, H., Sarfati, R., and Haiffner-Cavaillon, N.,** Molecular requirement for interleukin-1 induction by LPS stimulated human monocytes; involvement of KDO region, *Eur. J. Immunol.,* 16, 87, 1986.
47. **Salter, A. M., Fisher, S. C., and Brindley, D. N.,** Binding of low-density lipoprotein to monolayer cultures of rat hepatocytes is increased by insulin and decreased by dexamethasone, *FEBS Lett.,* 220, 159, 1987.
48. **Salter, A. M., Fisher, S. C., and Brindley, D. N.,** Interactions of triiodothyronine, insulin and dexamethasone on the binding of human LDL to rat hepatocytes in monolayer culture, *Atherosclerosis,* 71, 77, 1988.
49. **Salter, A. M., Al-seeni, M., Brindley, D. N., and Middleton, B.,** Inhibition of cholesterol esterification in rat hepatocytes is necessary for down-regulation of low-density-lipoprotein receptor activity, *Biochem. Soc. Trans.,* 17, 112, 1989.
50. **Forsberg, E., Paulsson, M., Timpl, R., and Johansson, S.,** Characterization of a laminin receptor on rat hepatocytes, *J. Biol. Chem.,* 265, 6376, 1990.
51. **Martin, G. R. and Timpl, R.,** Laminin and other basement membrane components, *Annu. Rev. Cell Biol.,* 3, 57, 1987.
52. **Menko, A. S. and Boetiger, D.,** Occupation of the extracellular matrix receptor, integrin, is a control point for myogenic differentiation, *Cell,* 51, 51, 1987.
53. **Timpl, R., Johansson, S., Van Delden, V., Oberaumer, I., and Hook, M.,** Characterization of protease resistant fragments of laminin mediating attachment and spreading of rat hepatocytes, *J. Biol. Chem.,* 25, 8922, 1983.

54. **Steinhoff, G., Behrend, M., and Wonigeit, K.,** Expression of adhesion molecule on lymphocytes/monocytes and hepatocytes in human liver grafts, *Hum. Immunol.,* 28, 123, 1990.
55. **d'Arville, C. N., Le, M., Kloppel, T. M., and Simon, F. R.,** Alterations in the functional expression of receptors on cirrhotic rat hepatocytes, *Hepatology,* 9, 6, 1989.
56. **d'Arville, C. N., Kloppel, T. M., and Le, M.,** Carbon tetrachloride treatment alters asialoorosomucoid (ASOR) and epidermal growth factor (EGF) binding to isolated rat hepatocytes while not affecting insulin binding, *Hepatology,* 6, 1204, 1986.
57. **Johnston, D. G., Alberti, K. G. M. M., Binder, C., et al.,** Hormonal and metabolic changes in hepatic cirrhosis, *Horm. Metab. Res.,* 14, 34, 1982.
58. **Marco, J., Diego, J., Villanueva, M. L., et al.,** Elevated plasma glucagon levels in cirrhosis of the liver, *N. Engl. J. Med.,* 289, 1107, 1973.
59. **McFarlane, B. M., Sipos, J., Gove, C. D., McFarlane, I. G., and Williams, R.,** Antibodies against the hepatic asialoglycoprotein receptor perfused in situ preferentially attach to periportal liver cells in the rat, *Hepatology,* 11, 408, 1990.
60. **Evans, W. H.,** A biochemical dissection of the functional polarity of the plasma membrane of the hepatocyte, *Biochim. Biophys. Acta,* 604, 27, 1980.
61. **Evans, W. H. and Enrich, C.,** Liver plasma membrane domains and endocytic trafficking, *Biochem. Soc. Trans.,* 17, 619, 1989.
62. **Evans, W. H., Nawab, A., and Enrich, C.,** Membrane compartmentation and trafficking in hepatocytes, *Biochem. Soc. Trans.,* 18, 137, 1989.
63. **Goldstein, J. L., Brown, M. S., Anderson, R. G. W., Russell, D. W., and Schneider, W. J.,** *Annu. Rev. Cell Biol.,* 1, 1, 1985.
64. **Oka, J. A. and Wigel, P. H.,** The pathways for fluid phase and receptor mediated endocytosis in rat hepatocytes are different but thermodynamically equivalent, *Biochem. Biophys. Res. Commun.,* 159, 488, 1989.
65. **Sehgal, P. B. and Sagar, A. D.,** Heterogeneity of poly(I)-poly(C)-induced human fibroblast interferon mRNA species, *Nature (London),* 288, 95, 1980.
66. **Horii, Y., Maraguchi, A., Suematsu, S., Matsuda, T., Yoshizaki, K., Hirano, T., and Kishimoto, T.,** Regulation of BSF-2/IL-6 production by human mononuclear cells, *J. Immunol.,* 141, 1529, 1988.
67. **Jirik, F. R., Podor, T. J., Hirano, T., Kishimoto, T., Loskutoff, D. J., Carson, D. A., and Lotz, M.,** Bacterial lipopolysaccharide and inflammatory mediators augment IL-6 secretion by human endothelial cells, *J. Immunol.,* 142, 144, 1989.
68. **Gauldie, J., Richards, C., Harnish, D., Lansdorp, P., and Baumann, H.,** Interferon B2/B-cell stimulatory factor type 2 shares identity with monocyte-derived hepatocyte-stimulating factor and regulates the major acute phase protein response in liver cells, *Proc. Natl. Acad. Sci. U.S.A.,* 84, 7251, 1987.
69. **Bauer, J., Bauer, T. M., Thomas, K., Taga, T., Lengyel, G., Hirano, T., Kishimoto, T., Acs, G., Mayer, L., and Wolfgang, G.,** Regulation of interleukin 6 receptor expression in human monocytes and monocyte-derived macrophages, *J. Exp. Med.,* 170, 1537, 1989.
70. **Bauer, J., Weber, W., and Tran-Thi, T. A.,** Murine IL-1 stimulates a_2-macroglobulin synthesis in rat hepatocyte primary culture, *FEBS Lett.,* 190, 271, 1985.

71. **Peterson, T. C. and Renton, K. W.,** The role of lymphocytes, macrophages and interferon in the depression of drug metabolism by dextrane sulfate, *Immunopharmacology,* 11, 21, 1986.
72. **Ghezzi, P., Saccardo, B., Villa, P., Bianchi, M., and Dinarello, C. A.,** Role of interleukin in the depression of liver drug metabolism by endotoxin, *Infect. Immun.,* 54, 837, 1986.
73. **Okuno, S., Sugita, K., Arai, M., and Eto, S.,** Effective of recombinant interleukin on hepatic drug metabolism in rat, *Hepatology,* 8-5, 1424, 1988.
74. **Sujita, K., Okuno, F., Tanaka, Y., Hirano, Y., Inamoto, Y., Eto, S., and Arai, M.,** Effect of interleukin 1 (IL-1) on the levels of cytochrome P-450 involving IL-1 receptor on the isolated hepatocytes of rat, *Biochem. Biophys. Res. Commun.,* 168, 1217, 1990.
75. **Shirahama, M., Ishibashi, H., Tsuchiya, Y., Kurokawa, S., Hayashida, K., Okumura, Y., and Niho, Y.,** Kupffer cells may autoregulate interleukin 1 production by producing interleukin 1 inhibitor and prostaglandin E_2, *Scand. J. Immunol.,* 28, 719, 1988.
76. **Liao, Z., Grimshaw, R. S., and Rosenstreich, D. L.,** Identification of a specific IL-1 inhibitor in the urine of febrile patients, *J. Exp. Med.,* 159, 126, 1984.
77. **Arend, W. P., Joslin, F. G., and Massoni, R. J.,** Effects of immune complexes on production by human monocytes of interleukin 1 or an interleukin 1 inhibitor, *J. Immunol.,* 134, 3868, 1985.
78. **Scala, G., Kuang, Y. D., Hall, R. E., Muchmore, A. V., and Oppenheim, J. J.,** Accessory cell function of human B cells. I. Production of both interleukin 1-like activity and an interleukin 1 inhibitory factor by an EBV-transformed human B cell line, *J. Exp. Med.,* 159, 1637, 1984.
79. **Tiku, K., Tiku, M. L., Liu, S., and Skosey, J. L.,** Normal human neutrophils are a source of a specific interleukin 1 inhibitor, *J. Immunol.,* 136, 3686, 1986.
80. **Roberts, N. J., Prill, A. H., and Mann, T. N.,** Interleukin 1 and interleukin 1 inhibitor production by human macrophages exposed to influenza virus or respiratory syncytial virus. Respiratory syncytial virus is a potent inducer of inhibitor activity, *J. Exp. Med.,* 163, 511, 1986.
81. **Prowse, K. R. and Baumann, H.,** Interleukin-1 and interleukin-6 stimulate acute-phase protein production in primary mouse hepatocytes, *J. Leuk. Biol.,* 45, 55, 1989.
82. **Hagiwara, T., Suzuki, H., Kono, I., Kashiwagi, H., Akiyama, Y., and Onozaki, K.,** Regulation of fibronectin synthesis by interleukin-1 and interleukin-6 in rat hepatocytes, *Am. J. Pathol.,* 136, 39, 1990.
83. **Castell, J. V., Gomex-Lechon, M. J., David, M., Andus, T., Geiger, T., Trullenque, R., Fabra, R., and Heinrich, P. C.,** Interleukin-6 is the major regulator of acute phase protein synthesis in adult human hepatocytes, *FEBS Lett.,* 242, 237, 1989.
84. **Kohase, M., May, L. T., Tamm, I., Vilcek, J., and Seghal, P. B.,** A cytokine network in human diploid fibroblasts interactions of β-interferons, tumor necrosis factor, platelet-derived growth factor IL-1, *Mol. Cell Biol.,* 7, 273, 1987.

85. **Moshage, H. S., Roelofs, H. M. J., Van Pelt, J. F., Hazenberg, B. P. C., Van Leeuwen, M. A., Limburg, P. C., Aarden, L. A., and Yap, S. H.,** The effect of interleukin-1, interleukin-6 and its interrelationship on the synthesis of serum amyloid A- and C-reactive protein in primary cultures of adult human hepatocytes, *Biochem. Biophys. Res. Commun.*, 155, 112, 1988.
86. **Gauldie, J., Northemann, W., and Fey, G. H.,** IL-6 functions as an exocrine hormone in inflammation, *J. Immunol.*, 144, 3804, 1990.
87. **Lotz, M., Suraw, B. L., Carson, D. A., and Jirik, F. R.,** Hepatocytes produce interleukin-6[a], *Ann. N.Y. Acad. Sci.*, 595, 509, 1990.
88. **Gauldie, J., Richards, C., Northemann, W., Fey, G., and Baumann, H.,** IFNB2/BSF2/IL-6 is the monocyte-derived HSF that regulates receptor-specific acute phase gene regulation in hepatocytes, *Ann. N.Y. Acad. Sci.*, 595, 46, 1990.
89. **Perlmutter, D. H.,** IFNB2/IL6 is one of several cytokines that modulate acute phase gene expression in human hepatocytes and human macrophages, *Ann. N.Y. Acad. Sci.*, 595, 332, 1990.
90. **Sehgal, P. B. and May, L. T.,** Human interferon-beta 2, *J. Interferon Res.*, 7, 521, 1987.
91. **Kurdowska, A., Bereta, J., and Koj, A.,** Comparison of the action of interleukin-6, phorbol myristate acetate, and glucagon on the acute phase protein production and amino acid uptake by cultured rat hepatocytes, *Ann. N.Y. Acad. Sci.*, 595, 506, 1990.
92. **Delarco, J. E. and Todaro, G. J.,** Growth factors from murine sarcoma virus-transformed cells, *Proc. Natl. Acad. Sci. U.S.A.*, 75, 4001, 1978.
93. **Luetteke, N. C., Michalopoulos, G., Teixido, J., Gilmore, R., Massague, J., and Lee, D. C.,** Characterization of high molecular weight transforming growth factor-a produced by rat hepatocellular carcinoma cells, *Biochemistry*, 27, 6487, 1988.
94. **Lee, D. C., Rose, T. M., Webb, N. R., and Todaro, G. J.,** Cloning and sequence analysis of a cDNA for rat transforming growth factor-a, *Nature*, 313, 489, 1985.
95. **Brenner, D. A., Koch, K. S., and Leffert, H. L.,** Transforming growth factor-a simulates proto-oncogene c-jun expression and a mitogenic program in primary cultures of adult rat hepatocytes, *DNA*, 8-4, 279, 1989.
96. **Baribault, H., Leroux-Nicollet, I., and Marceau, N.,** Differential responsiveness of cultured suckling and adult rat hepatocytes to growth promoting factors: entry into S phase and mitosis, *J. Cell Physiol.*, 122, 105, 1985.
97. **Wollenberg, G. K., Harris, L., Farber, E., and Hayes, M. A.,** Inverse relationship between epidermal growth factor induced proliferation and expression of high affinity surface epidermal growth factor receptors in rat hepatocytes, *Lab. Invest.*, 60, 254, 1989.
98. **Kobusch, A. B. and Bock, K. W.,** Zinc increases EGF-stimulated DNA synthesis in primary mouse hepatocytes, *Biochem. Pharmacol.*, 39, 555, 1990.
99. **Carr, B. I., Hayashi, I., Branum, E. L., and Moses, H. L.,** Inhibition of DNA synthesis in rat hepatocytes by platelet derived type beta transforming growth factor, *Cancer Res.*, 46, 2330, 1986.
100. **Strain, A. J., Frazer, A., Hill, D. J., and Milner, R. D.,** Transforming growth factor B inhibits DNA synthesis in hepatocytes isolated from normal and regenerating rat liver, *Biochem. Biophys. Res. Commun.*, 145, 436, 1987.

101. **Braun, L., Mead, J. E., Panzica, M., Mikumo, R., Bell, G. I., and Fausto, N.,** Transforming growth factor B mRNA increases during liver regeneration: a possible paracrine mechanism of growth regulation, *Proc. Natl. Acad. Sci. U.S.A.*, 85, 1539, 1988.
102. **LaMarre, J., Wollenberg, G. K., Gauldie, J., and Hayes, M. A.,** a2-Macroglobulin and serum preferentially counteract the mitoinhibitory effect of transforming growth factor-B2 in rat hepatocytes, *Lab. Invest.*, 62, 545, 1990.
103. **Little, P., Skouteris, G. C., Ord, M. G., and Stocken, L. A.,** Serum from partially hepatectomized rats induces primary hepatocytes to enter S phase: a role for prostaglandins?, *J. Cell Sci.*, 91, 549, 1988.
104. **Bonney, R. J., Becker, J. E., Walker, P. R., and Potter, V. R.,** Primary monolayer cultures of adult rat liver parenchymal cells suitable for study of regulation of enzyme synthesis, *In Vitro,* 9, 399, 1974.
105. **McGowan, J. A., Strain, A. J., and Bucher, N. L. R.,** DNA synthesis in primary cultures of adult rat hepatocytes in a defined medium: effects of epidermal growth factor, insulin, glucagon and cyclic-AMP, *J. Cell Physiol.*, 108, 353, 1981.
106. **Strain, A. J., McGowan, J. A., and Bucher, N. L. R.,** Stimulation of DNA synthesis in primary cultures of adult rat hepatocytes by rat platelet associated substance(s), *In Vitro,* 18, 108, 1982.
107. **Michalopoulos, G., Cianciulli, H. D., Novotny, A. R., et al.,** Liver regeneration studies with rat hepatocytes in primary culture, *Cancer Res.*, 42, 4673, 1982.
108. **LaBrecque, D. R. and Resch, L. A.,** Preparation and partial characterization of hepatic regenerative stimulator substance (SS) from rat liver, *J. Physiol.*, 248, 273, 1975.
109. **Fleig, W. E. and Hoss, G.,** Partial purification of rat hepatic stimulator substance and characterization of its action on hepatoma cells and normal hepatocytes, *Hepatology,* 9, 240, 1988.
110. **Zarnegar, R. and Michalopoulos, G.,** Purification and biological characterization of human hepatopoietin A, a polypeptide growth factor for hepatocytes, *Cancer Res.*, 49, 3314, 1989.
111. **Ove, P., Francavilla, A., Coetzee, M. L., Makowka, L., and Starzl, T. E.,** Response of cultured hepatocytes to a hepatomitogen after initiation by conditioned medium or other factors, *Cancer Res.*, 49, 98, 1989.
112. **LaBrecque, D. R. and Pesch, L. A.,** Preparation and partial characterization of hepatic regenerative stimulator substance (Ss) from rat liver, *J. Physiol.*, 248, 273, 1975.
113. **Knook, D. L. and Wisse, E., Eds.,** *Sinusoidal Liver Cells,* Elsevier, Amsterdam, 1982, 519.
114. **Rieder, H., Birmelin, M., and Decker, K.,** in *Sinusoidal Liver Cells,* D. L. Knook and E. Wisse, Eds., Elsevier, Amsterdam, 1982, 193.
115. **Kreusch, J.,** Thesis, Faculty of Medicine, University of Freiburg (FRG), 1980.
116. **Wake, K., Decker, K., Kirn, A., Knook, D. L., McCuskey, R. S., Bouwens, L., and Wisse, E.,** Cell biology and kinetics of Kupffer cells in the liver, *Intl. Rev. Cytol.*, 118, 173, 1989.
117. **Jones, E. A. and Summerfield, J. A.,** *Kupffer Cells, The Liver: Biology and Pathobiology,* 2nd ed., Raven Press, New York, 1988, 683.

118. **Shiratori, Y., Mitsugu, T., Keni, H., Kawase, T., Shina, S., and Sugimoto, T.,** Role of endotoxin-responsive macrophages in hepatic injury, *Hepatology,* 11, 183, 1990.
119. **Klebanoff, S. J. and Slivka, A.,** Monocyte and granulocyte-mediated tumor cell destruction, *J. Clin. Invest.,* 93, 489, 1980.
120. **Camara, D. S., Caruana, J. A., Schwartz, K. A., Montes, M., and Nola, J. P.,** D-galactosamine liver injuy: absorption of endotoxin and protective effect of small bowel and colon resection in rabbits, *Proc. Soc. Exp. Biol. Med.,* 23, 64, 1976.
121. **Lehmann, V., Freudenberg, M. S., and Galanos, C.,** Lethal toxicity of lipopolysaccharide and tumor necrosis factor in normal and D-galactosamine-treated mice, *J. Exp. Med.,* 165, 657, 1987.
122. **Keppler, D., Hagmann, W., Rapp, S., Denzlinger, C., and Koch, H. K.,** The relation of leukotrienes to liver injury, *Hepatology,* 5, 883, 1985.
123. **Van Rooijen, N., Kors, N., Ende, M., and Dijkstra, C. D.,** Depletion and repopulation of macrophages in spleen and liver of rat after intravenous treatment with liposome-encapsulated dichloromethylene diphosphonate, *Cell Tissue Res.,* 260, 215, 1990.
124. **Zenilman, M. E., Fiani, M., Stahl, P. D., Brunt, E. M., and Flye, M. W.,** Selective depletion of Kupffer cells in mice by intact ricin, *Transplantation,* 47, 200, 1988.
125. **Muller, C. D. and Schuber, F.,** Neo-mannosylated liposomes: synthesis and interaction with mouse Kupffer cells and resident peritoneal macrophages, *Biochim. Biophys. Acta,* 986, 97, 1989.
126. **Gjoen, T., Selijelid, R., and Kolset, S. O.,** Binding of metastatic colon carcinoma cells to liver macrophages, *J. Leuk. Biol.,* 45, 362, 1989.
127. **Decker, T., Lohmann-Matthes, M. L., Karch, U., Peters, T., and Decker, K.,** Comparative study of cytotoxicity, tumor necrosis factor, and prostaglandin release after stimulation of rat Kupffer cells, murine Kupffer cells, and murine inflammatory liver macrophages, *J. Leuk. Biol.,* 54, 139, 1989.
128. **Leser, H. G., Gerok, W., Gemsa, D., and Kaufmann, S. H. E.,** Increased prostaglandin E release and tumor cytostasis by resident Kupffer cells during listeria monocytogenes infection, *Immunobiology,* 178, 224, 1988.
129. **Azri, S. and Renton, K. W.,** Depression of murine hepatic mixed function oxidase during infection with listeria monocytogenes, *J. Pharmacol. Exp. Ther.,* 243, 1089, 1987.
130. **Peterson, T. C. and Renton, K. W.,** Depression of cytochrome P-450 dependent drug biotransformation in hepatocytes after the activation of the reticuloendothelial system by dextran sulfate, *J. Pharmacol. Exp. Ther.,* 229, 299, 1984.
131. **Renton, K. W. and Peterson, T. C.,** The influence of infection on the metabolism of foreign compounds, in *Foreign Compound Metabolism,* J. Caldwell and G. Paulson, Eds., Taylor Francis Ltd., Philadelphia, 1984, 289.
132. **Peterson, T. C. and Renton, K. W.,** Kupffer cell factor mediated depression of hepatic parenchymal cell cytochrome P-450, *Biochem. Pharmacol.,* 35, 1491, 1986.
133. **Nebert, D. W., Eisen, H. J., Negishi, M., Lang, M. A., and Hjelmelend, L. M.,** Genetic mechanisms controlling the induction of polysubstrate monooxygenase (P-450) activities, *Annu. Rev. Pharmacol. Toxicol.,* 21, 431, 1981.

134. **Farquhar, D., Loo, T. L., Gulterman, J. U., Hersh, E. M., and Luna, M. A.,** Inhibition of drug metabolizing enzymes in the rat after Bacillus Calmette-Guerin (BCG) treatment, *Biochem. Pharmacol.,* 25, 1529, 1976.
135. **Soyka, L. F., Stephens, C., MacPherson, B. R., and Foster, R. S.,** Role of mononuclear phagocytes in decreased hepatic drug metabolism following administration of Corynebacterium parvum, *Int. J. Immunopharmacol.,* 1, 101, 1979.
136. **Shedlofsky, S. I., Cohen, D. A., McClean, C. T., Kaplan, A. M., Swim, A. T., and Freidman, D. W.,** Effect of interleukin 1 on cytochrome P-450 levels in cultured hepatocytes, *J. Leuk. Biol.,* 37, 742, 1986.
137. **Peterson, T. C.,** Interleukin-1 and free radicals in liver disease, *Cytokine,* 1, 140, 1989.
138. **Cantrell, E. and Bresnick, E.,** Benzpyrene hydroxylase activity in isolated parenchymal and non-parenchymal cells of rat liver, *J. Cell Biol.,* 52, 316, 1972.
139. **Bast, R. C., Okuda, T., Plotkin, E., Tarone, R., Rapp, H. J., and Gellboin, H. V.,** Development of an assay for arylhydrocarbon hydroxylase in human peripheral blood monocytes, *Cancer Res.,* 36, 1967, 1976.
140. **Peterson, T. C. and Williams, C. N.,** Influence of liver disease on monooxygenase activity in cultured monocytes, *J. Clin. Invest. Med.,* 8, 3, 1985.
141. **Peterson, T. C. and Williams, C. N.,** Depression of peripheral blood monocyte arylhydrocarbon hydroxylase activity in patients with liver disease: possible involvement of macrophage factors, *Gastroenterology,* 90, 1756, 1986.
142. **Peterson, T. C. and Williams, C. N.,** Depression of peripheral blood monocyte arylhydrocarbon hydroxylase activity in patients with liver disease: possible involvement of macrophage factors, *Hepatology,* 7, 333, 1987.
143. **Peterson, T. C.,** Drug metabolizing enzymes in rat, mouse, pig and human macrophages, *Biochem. Pharmacol.,* 36, 3911, 1987.
144. **Laskin, D. L., Robertson, M., Pilaro, A. M., and Laskin, J. D.,** Activation of liver macrophages following phenobarbital treatment of rats, *Hepatology,* 8, 1051, 1988.
145. **Scheuer, P. J.,** *Liver Biopsy Interpretation,* 3rd ed., Williams & Wilkins, Baltimore, 1980, 36.
146. **Peterson, T. C. and Isbrucker, R.,** Anti-human PDGF inhibits fibrogenic activity of MCM from patients with liver disease, *Hepatology,* 12, 909, 1990.
147. **Peterson, T. C. and Isbrucker, R.,** Fibrogenesis, IL-1 activated oxygen and liver disease, *Cytokine,* 1, 408, 1989.
148. **Ross, R., Raines, E. W., and Bowen-Pope, D. F.,** The biology of platelet-derived growth factor, *Cell,* 46, 155, 1986.
149. **Peterson, T. C. and Isbrucker, R.,** Role of IL-1β and PDGF in the AHH depressive and fibrogenic activity of MCM from patients with liver disease, *J. Leuk. Biol.,* Suppl. 1, 52, 1990.
150. **Peterson, T. C. and Isbrucker, R.,** Fibroproliferation in liver disease: role of monocyte factors, *Hepatology,* 15, 191, 1992.
151. **Buxton, D. B., Fisher, R. A., Hanahan, D. J., and Olson, M. S.,** Platelet-activating factor-mediated vasoconstruction and glycogenolysis in the perfused rat, *J. Biol. Chem.,* 261, 644, 1986.

152. **Fisher, R. A., Shukla, S. D., Debuysere, M. S., Hanahan, D. S., and Olson, M. S.,** The effect of acetylglyceryl ether phosphorylcholine on glycogenolysis and phosphatidyl-inositol of 4.5-bisphosphate metabolism in rat hepatocytes, *J. Biol. Chem.,* 259, 8685, 1984.
153. **Kuiper, J., De Rije, Y. B., Zijlstra, F. J., Van Waas, M. P., and Van Berkel, T. J. C.,** The induction of glycogenolysis in the perfused liver by platelet activating factor is mediated by prostaglandin D_2 from Kupffer cells, *Biochem. Biophys. Res. Commun.,* 157, 1288, 1988.
154. **Fisher, R. A., Sharma, R. V., and Bhalla, C. R.,** Platelet-activating factor increases inositol phosphate production and cytosolic free Ca^2 concentrations in cultured rat Kupffer cells, *FEBS Lett.,* 251, 22, 1989.
155. **Chao, W., Siafaka-Kapadai, A., Olson, M. S., and Hanahan, D. J.,** Biosynthesis of platelet-activating factor by cultured rat Kupffer cells stimulated with calcium ionophore A23187, *Biochem. J.,* 257, 829, 1989.
156. **Kuiper, J., Kamps, J. A. A. M., and Van Berkel, T. J. C.,** Induction of ornithine decarboxylase in rat liver by phorbol ester is mediated by prostanoids from Kupffer cells, *J. Biol. Chem.,* 264, 6874, 1989.
157. **Katsumoto, F., Miyazaki, K., and Nakayama, F.,** Stimulation of DNA synthesis in hepatocytes by Kupffer cells after partial hepatectomy, *Hepatology,* 9, 405, 1989.
158. **Billiar, T. R., Curran, R. D., Stuehr, D. J., Ferrari, F. K., and Simmons, R. L.,** Evidence that activation of Kupffer cells results in production of L-arginine metabolites that release cell-associated iron and inhibit hepatocyte protein synthesis, *Surgery,* 106, 364, 1989.
159. **Billiar, T. R., Curran, R. D., West, M. A., Hofmann, K., and Simmons, R. L.,** Kupffer cell cytotoxicity to hepatocytes in coculture requires L-arginine, *Arch. Surg.,* 124, 1416, 1989.
160. **Curran, R. D., Billiar, T. R., Stuehr, D. J., Hofmann, K., and Simmons, R. L.,** Hepatocytes produce nitrogen oxides from L-arginine in response to inflammatory products of Kupffer cells, *J. Exp. Med.,* 170, 1769, 1989.
161. **West, M. A., Billiar, T. R., Mazuski, J. E., Curran, R. D., Cerra, F. B., and Simmons, R. L.,** Endotoxin modulation of hepatocyte secreted and cellular protein synthesis is mediated by Kupffer cells, *Arch. Surg.,* 123, 1400, 1986.
162. **West, M. A., Billiar, T. R., Curran, R. D., Hyland, B. J., and Simmons, R. L.,** Evidence that rat Kupffer cells stimulate and inhibit hepatocyte protein synthesis *in vitro* by different mechanisms, *Gastroenterology,* 96, 1572, 1989.
163. **Curran, R. D., Billiar, T. R., Kispert, P. H., Bentz, B. G., May, M. T., and Simmons, R. L.,** Hepatocytes enhance Kupffer cell-mediated tumor cell cytostasis *in vitro, Surgery,* 106, 126, 1989.
164. **Peterson, T. C. and Isbrucker, R.,** Inhibition of PDGF-driven fibroproliferation, *Cytokine,* 3, 502, 1991.
165. **Goodenough, D. A.,** Bulk isolation of mouse hepatocyte gap junctions, *J. Cell Biol.,* 61, 557, 1974.
166. **Evans, W. H. and Rahman, S.,** Gap junctions and intercellular communication: topology of the major junctional protein of rat liver, *Membr. Struct. Dyn.,* 17, 983, 1989.
167. **Simpson, I., Rose, B., and Loewenstein, W. R.,** Size limit of molecules permeating the junctional membrane channels, *Science,* 195, 294, 1977.

168. **Trosko, J. E., Chang, C. C., and Metcalf, A.,** Mechanisms of tumor promotion: potential role of intercellular communication, *Cancer Invest.*, 1, 511, 1983.
169. **Van der Zandt, P. T. J., de Feijter, A. W., Homan, E. C., Spaaij, C., de Haan, L. H. J., van Aelst, A. C., and Jongen, W. M. F.,** Effects of cigarette smoke condensate and 12-O-tetradecanoylphorbol-13-acetate on gap junction structure and function in cultured cells, *Carcinogenesis,* 11, 883, 1990.
170. **Maldonado, P. E., Rose, B., and Lowenstein, W. R.,** Growth factors modulate junctional cell-to-cell communication, *J. Membr. Biol.*, 106, 203, 1988.
171. **Spray, D. C., Saez, J. C., and Hertzberg, E. L.,** Gap junctions between hepatocytes: structural and regulatory features, in *The Liver: Biology and Pathobiology,* 2nd ed., I. M. Arias, W. B. Jackoby, H. Popper, D. Schachter, and D. A. Shafritz, Eds., Raven Press, New York, 1988, 851.
172. **Lawrence, T. S., Beers, W. H., and Gilula, N. B.,** Transmission of hormonal stimulation by cell-to-cell communication, *Nature,* 272, 50, 1978.
173. **Fletcher, W. H. and Greenan, J. R. T.,** Receptor mediated action without receptor occupancy, *Endocrinology,* 116, 1660, 1985.
174. **Stagg, R. B. and Fletcher, W. H.,** The hormone-induced regulation of contact-dependent cell-cell communication by phosphorylation, *Endocr. Rev.*, 11, 302, 1990.
175. **Mazet, F. and Mazet, J. L.,** Restoration of gap junction-like structure after detergent solubilization of the proteins from liver gap junctions, *Exp. Cell Res.*, 188, 312, 1990.
176. **Saez, J. C., Gregory, W. A., Watanabe, T., Dermietzel, R., Hertzber, E. L., Reid, L., Bennett, M. V. L., and Spray, D. C.,** cAMP delays disappearance of gap junctions between pairs of rat hepatocytes in primary culture, *J. Physiol.*, 257, C1, 1989.
177. **Goulet, F., Normand, C., and Morin, O.,** Cellular interactions promote tissue-specific function, biomatrix deposition and junctional communication of primary cultured hepatocytes, *Hepatology,* 8, 1010, 1988.
178. **Follmann, W., Ernst, P., and Kinne, R. K. H.,** Alterations of bile acid and bumetanide uptake during culturing of rat hepatocytes, *Am. Physiol. Soc.*, 700, 1990.
179. **Maurice, M., Rogier, E., Cassio, D., and Feldmann, G.,** Formation of plasma membrane domains in rat hepatocytes and hepatoma cell lines in culture, *J. Cell Sci.*, 90, 79, 1988.
180. **Schwarz, L. R., Burr, R., Schwenk, M., Pfaff, E., and Greim, H.,** Uptake of taurocholic acid into isolated rat liver cells, *Eur. J. Biochem.*, 55, 617, 1975.
181. **Kilberg, M. S.,** Amino acid transport in isolated rat hepatocytes, *J. Membr. Biol.*, 69, 1, 1982.
182. **Baur, H. and Heldt, H. W.,** Transport of hexoses across the liver cell membrane, *Eur. J. Biochem,* 74, 397, 1977.
183. **Petzinger, E., Ziegler, K., and Frimmer, M.,** Occurrence of a multispecific transporter for the hepatocellular accumulation of bile acids and various cyclopeptides, in *Bile Acids and the Liver,* G. Paumgartner, A. Stiehl, and W. Gerok, Eds., MTP, Lancaster, UK, 1987, 111.
184. **Eaton, D. L. and Klaassen, C. D.,** Carrier-mediated transport of ouabain in isolated hepatocytes, *J. Pharmacol. Exp. Ther.*, 205, 480, 1978.

185. **Petzinger, E., Muller, N., Follmann, W., Deutscher, J., and Kinne, R. K. H.,** Uptake of bumetanide into isolated rat hepatocytes and primary liver cell cultures, *Am. J. Physiol.,* 256, G78, 1989.
186. **Galivan, J.,** Transport of methotrexate by primary cultures of rat hepatocytes: stimulation of uptake in vitro by the presence of hormones in the medium, *Arch. Biochem. Biophys.,* 206, 113, 1981.
187. **Tarentino, A. L. and Galivan, J.,** Membrane characteristics of adult rat liver parenchymal cells in primary monolayer culture, in *In Vitro Cell Dev. Biol.,* 16, 833, 1980.
188. **Schwarz, L. R., Gotz, R., Wolff, T. H., and Wiebel, F. J.,** Monooxygenase and glucuronyltransferase activities in short term cultures of isolated rat hepatocytes, *FEBS Lett.,* 98, 203, 1979.
189. **Caron, J. M.,** Induction of albumin gene transcription in hepatocytes by extracellular matrix proteins, *Mol. Cell. Biol.,* 10, 1239, 1990.
190. **Bissell, D. M., Arenson, M., Maher, J. J., and Roll, F. J.,** Support of cultured hepatocytes by a laminin-rich gel, *J. Clin. Invest.,* 79, 801, 1987.
191. **Ben-Ze'ev, A. and Amsterdam, A.,** Regulation of cytoskeletal proteins involved in cell contact formation during differentiation of granulosa cells on extracellular matrix, *Proc. Natl. Acad. Sci. U.S.A.,* 83, 2894, 1986.
192. **Piazza, G. A., Callahan, H. M., Mowery, J., and Hixson, D. C.,** Evidence for a role of dipeptidyl peptidase IV in fibronectin mediated interactions of hepatocytes with extracellular matrix, *Biochem. J.,* 26, 327, 1989.
193. **Conner, J., Vallet-Collom, I., Daveau, M., Delers, F., Hiron, M., Lebreton, J. P., and Guillouzo, A.,** Acute-phase-response induction in rat hepatocytes co-culture with rat liver epithelial cells, *Biochem. J.,* 266, 683, 1990.
194. **Koj, A.,** in *The Acute Phase Response to Injury and Infection,* A. H. Gordon and A. Koj, Eds., Elsevier Science Publishers, Amsterdam, 73, 1985.
195. **Matsuoka, M., Zhang, M. Y., and Tsukamoto, H.,** Sensitization of hepatic lipocytes by high-fat diet to stipulatory effects of Kupffer cell-derived factors: implication in alcoholic liver fibrogenesis, *Hepatology,* 11, 173, 1990.
196. **McGee, J. O. and Patrick, R. S.,** The role of perisinusoidal cells in hepatic fibrogenesis; electron microscopic study of acute carbon tetrachloride liver injury, *Lab. Invest.,* 26, 429, 1972.
197. **Minato, Y., Hasumura, Y., and Takeuchi, J.,** The role of fat-storing cells in Disse space fibrogenesis in alcoholic liver disease, *Hepatology,* 87, 188, 1984.
198. **Berg, T., Gudmundsen, O., and Kindberg, G. M.,** Effects of vanadate on receptor-mediated endocytosis of asialoglycoproteins in isolated rat hepatocytes, *Biochem. Soc. Trans.,* 17, 1083, 1989.
199. **Chang, C. C., Jones, C., Trosko, J. E., Peterson, A. R., and Sevania, A.,** Effect of cholesterol epoxides on the inhibition of intercellular communication and on mutation induction in Chinese hamster V79 cells, *Mutat. Res.,* 206, 471, 1988.
200. **Romano, F., Menapace, L., and Armato, U.,** Inhibitors of ADP-ribosyl transferase suppress the mitogenic actions exerted by tumor promoters, but not these evoked by peptide mitogens, in primary neonatal rat hepatocytes, *Carcinogenesis,* 9, 2147, 1988.

201. **Diwan, B. A., Lubet, R. A., Nims, R. W., Klaunig, J. E., Weghorst, C. M., Henneman, J. R., Ward, J. M., and Rice, J. M.,** Lack of promoting effect of clonazepam on the development of N-nitrosodiethylamine-initiated hepatocellular tumors in mice is correlated with its inability to inhibit cell-to-cell communication in mouse hepatocytes, *Carcinogenesis,* 10, 1719, 1989.
202. **Lubet, R. A., Nims, R. W., Ward, J. M., Hagiwara, A., and Rice, J. M.,** Lack of effect of phenobarbital on hepatocellular carcinogenesis initiated by N-nitrosodiethylamine or methylazoxymethanol acetate in male Syrian golden hamsters, *Toxicol. Appl. Pharmacol.,* 86, 298, 1986.

Chapter 3

EXPERIMENTAL MODELS FOR STUDYING THE INTERACTION OF KUPFFER CELLS AND HEPATOCYTES

B.G. Harbrecht, T.R. Billiar, and R.D. Curran

TABLE OF CONTENTS

ISBN 0-8493-6109-5

I. INTRODUCTION

Hepatic parenchymal cells are estimated to account for 80 to 90% of the total liver cell mass and play a central role in both normal and abnormal physiology and metabolism.[1] The nonparenchymal cell (NPC) population that accounts for the remainder of liver cells consists of Kupffer cells (KC) as well as endothelial cells, fat-storing cells, pit cells, and cells of the biliary ductal system. KC make up approximately one third of the NPC population,[2] and endothelial cells make up the majority of the remaining NPC. KC are the predominant cell type in the fixed macrophage system, and through phagocytosis function prominently in clearing the portal circulation of blood-borne particulate matter, including lipopolysaccharide (LPS).[3-5] Functional differences exist between KC and other mononuclear phagocytes which have been hypothesized to be due, in part, to their unique position in the hepatic sinusoid.[5,6] Investigations into the specific factors involved in this unique environment as well as the interaction between parenchymal cells and NPC have been performed for many years.

While *in vivo* experiments and the use of isolated perfused liver preparations have proven quite useful in the study of hepatic metabolism, these techniques do have limitations.[7] The development within the past 20 years of techniques for the isolation, purification, and culture of the separate cellular populations of the liver in relatively high yield and purity has significantly expanded the number of methods available for studying not only KC and hepatocyte (HC) physiology separately, but KC:HC interactions as well. In this section, we will review some of the concepts in the isolation and culture of both HC and KC as well as describe some *in vitro* techniques for studying KC:HC interaction. We will also discuss methods for the *in vivo* alteration of KC number and function as a method of studying KC:HC interaction.

II. HEPATOCYTE ISOLATION

Early efforts at obtaining HC from experimental animals for use in isolated culture involved the mechanical dispersion of the cells from the reticular stroma, resulting in a significant amount of damage to the parenchymal cells.[7] This problem was alleviated by the introduction of enzymatic methods for cell dispersion using collagenase or a combination of collagenase and hyaluronidase.[8] Another important technical advance was the development of the continuous recirculating perfusion system for the application of collagenase.[9] Detailed studies by Seglan[7,10] on the optimal perfusion buffers as well as other technical aspects greatly increased the cell yield, purity, and viability obtained with this method. Modifications of this technique have been developed,[11-13] but the basic concepts have remained standard. While the principle source of HC for culture continues to be the rat, these techniques have been applied to other experimental animals including mice, guinea pigs, and frogs, and have found use in both adult and fetal animals.[11-13] A diagram of the Seglan method[7] for use in the rat as previously modified in our laboratory[14] is shown in Figure 1.

The essential feature of this method for HC isolation involves a two-step liver perfusion. After anesthetizing the rat and exposing the abdominal viscera, the portal vein is cannulated with a 16-gauge catheter to provide inflow and the inferior vena cava is divided to establish drainage. The liver is initially perfused with a calcium-free buffer containing EDTA or EGTA. Absence of calcium in this initial perfusion allows for optimal cell dispersion.[7]. While perfusing, the attachments to the liver are severed and the liver is placed in an *ex vivo* recirculating perfusion apparatus. The second step consists of perfusion with 0.05% collagenase for 30 min. Table 1 lists the components of the perfusion solutions.[15] After collagenase perfusion, the liver is placed in a petri dish containing the desired culture media. The capsule of the liver can be removed by peeling it back from the organ and the cells separated from the fibrous skeleton by gently combing with forceps or scissors.

Both parenchymal cells and NPC will be present in the cell suspension produced by the above steps. The remaining steps of the procedure involve purification of the parenchymal cell population. Differential centrifugation is the simplest method for separating HC from NPC and takes advantage of the greater density of HC compared to NPC. A series of short, low-speed (50 g) centrifugations is performed to sediment most if not all of the heavier HC, thus establishing the fact that the purity of the HC isolation will be inversely proportional to HC yield.[7] The use of metrizamide, percoll, or dextran-polyethylene glycol gradients have also been applied to purifying suspensions of HC.[7,16,17] Coating of the culture plates with gelatin or collagen prior to plating the purified HC suspension promotes attachment and adherence.[7] The standard culture media and its added supplements are listed in Table 2, as previously described.[15] The cells will flatten and form aggregates during the first 24 h of culture. Sufficient attachment will occur within the first 2 to 4 h of culture to allow a washing step at this point to remove nonviable or nonadherent cells if desired.[18] HC may be maintained in culture for a week or more, but will have a progressive cell loss with time.[11] Addition of dexamethasone or dimethyl sulfoxide to the culture media has been used to improve HC viability.[19,20] It has been our practice to culture the HC for 24 h in isolation prior to further manipulations.[14]

Confirmation of the purity of the cell population may be obtained by a variety of techniques. Simple morphologic criteria may be used to identify HC. These include their relatively large size (20 to 25 μm), typical histologic appearance, as well as the abundance of smooth endoplasmic reticulum, glycogen, and the presence of numerous microvilli as seen on electron microscopy.[7,21] Metabolic studies that are based on the ability of HC to synthesize glycogen and secrete specific proteins may also be used,[21] but are generally less specific in determining the number of contaminating cells.

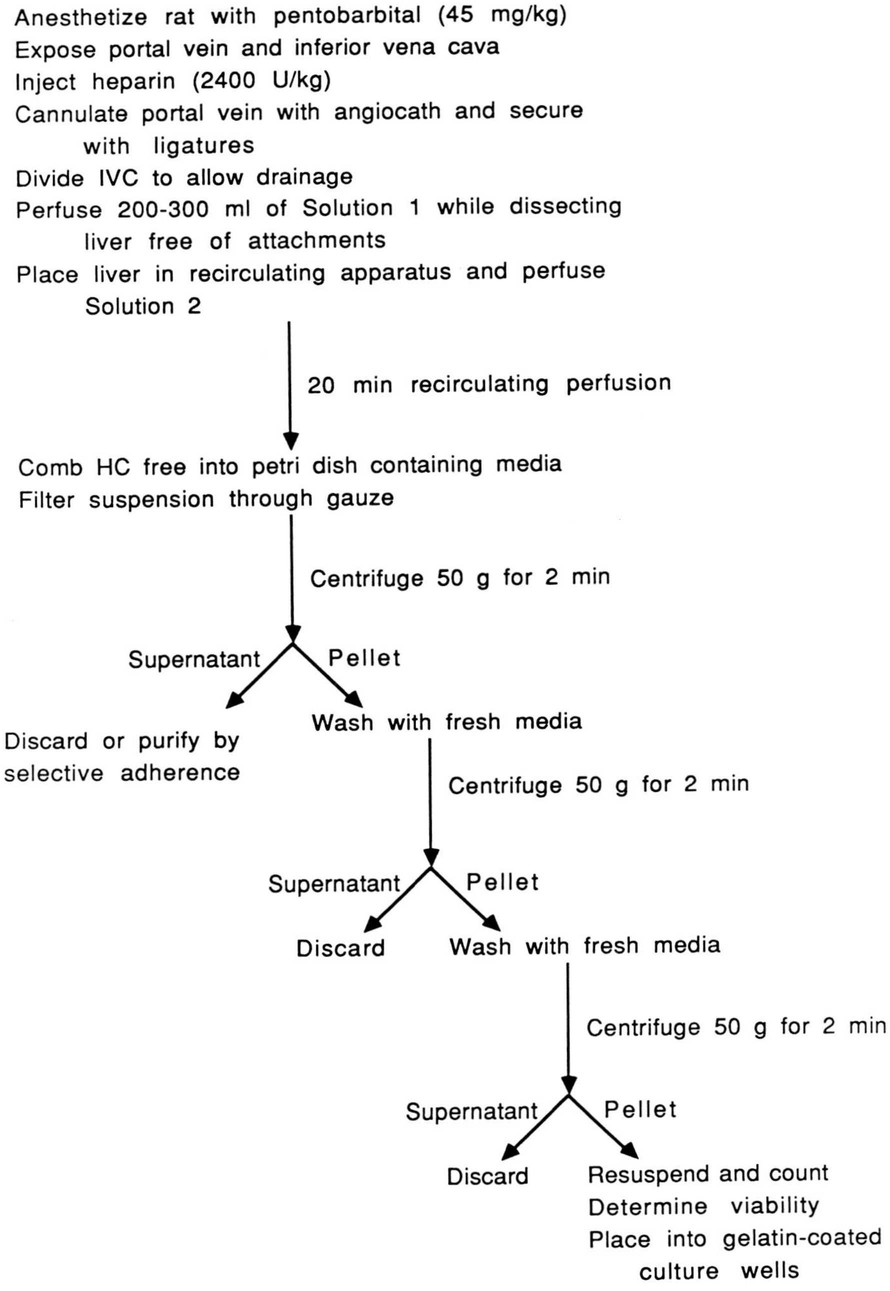

Anesthetize rat with pentobarbital (45 mg/kg)
Expose portal vein and inferior vena cava
Inject heparin (2400 U/kg)
Cannulate portal vein with angiocath and secure with ligatures
Divide IVC to allow drainage
Perfuse 200-300 ml of Solution 1 while dissecting liver free of attachments
Place liver in recirculating apparatus and perfuse Solution 2

↓ 20 min recirculating perfusion

Comb HC free into petri dish containing media
Filter suspension through gauze

↓ Centrifuge 50 g for 2 min

Supernatant → Discard or purify by selective adherence
Pellet → Wash with fresh media

↓ Centrifuge 50 g for 2 min

Supernatant → Discard
Pellet → Wash with fresh media

↓ Centrifuge 50 g for 2 min

Supernatant → Discard
Pellet → Resuspend and count
Determine viability
Place into gelatin-coated culture wells

FIGURE 1

TABLE 1
Hepatocyte Isolation Solutions

Perfusion solution 1		Perfusion solution 2	
NaCl	0.145 *M*	Collagenase	0.05 g/100 U
KCl	6.7 m*M*	Albumin	1 g/100 U
HEPES buffer	10 m*M*	NaCl	67 m*M*
EGTA	2.4 m*M*	KCl	6.7 m*M*
		HEPES buffer	100 m*M*
		CaCl	4.8 m*M*

TABLE 2
Hepatocyte Culture Media

William's medium E	(Gibco® Laboratories)
Insulin	10^{-6} *M*
L-Glutamine	2 m*M*
HEPES buffer	15 m*M*
Penicillin	100 U/ml
Streptomycin	100 μg/ml
Calf serum	2.5 to 10%

III. KUPFFER CELL ISOLATION

Advances in techniques to isolate and purify KC developed in parallel with the advances made in HC isolation. The use of pronase to selectively destroy parenchymal cells results in a purified population of NPC.[22,23] Similarities in density between KC and the more prevalent endothelial cells, however, make efforts to separate these cell types using density gradient centrifugation difficult unless changes are made in the density of one cell population, as in the use of iron or colloids.[24] The application of centrifugal elutriation to NPC suspensions has allowed the separation of KC and endothelial cells with satisfactory purity.[24-26] A modification of the method of Emeis and Planque[23] for use with rats, as performed in our laboratory, is presented in Figure 2.

KC isolation also employs a two-step liver perfusion. The liver is cannulated, as in the HC isolation, and perfused with Gey's balanced salt solution (GBSS) to clear blood elements. A brief (50 ml) perfusion with 0.2% pronase follows, after which the liver is excised and minced into 2-mm fragments. The fragments are then continuously stirred in pronase at 37°C for 1 h. Some investigators maintain the pH of this solution at 7.2 with 1N NaOH. We have found, though, that this step may be omitted. DNAse is added (1 ml of 0.5 mg/ml solution) at intervals to prevent cell clumping. After 1 h, the NPC are pelleted and subjected to hypotonic sodium chloride to lyse contaminating erythrocytes. After washing with fresh media, the KC fraction is separated

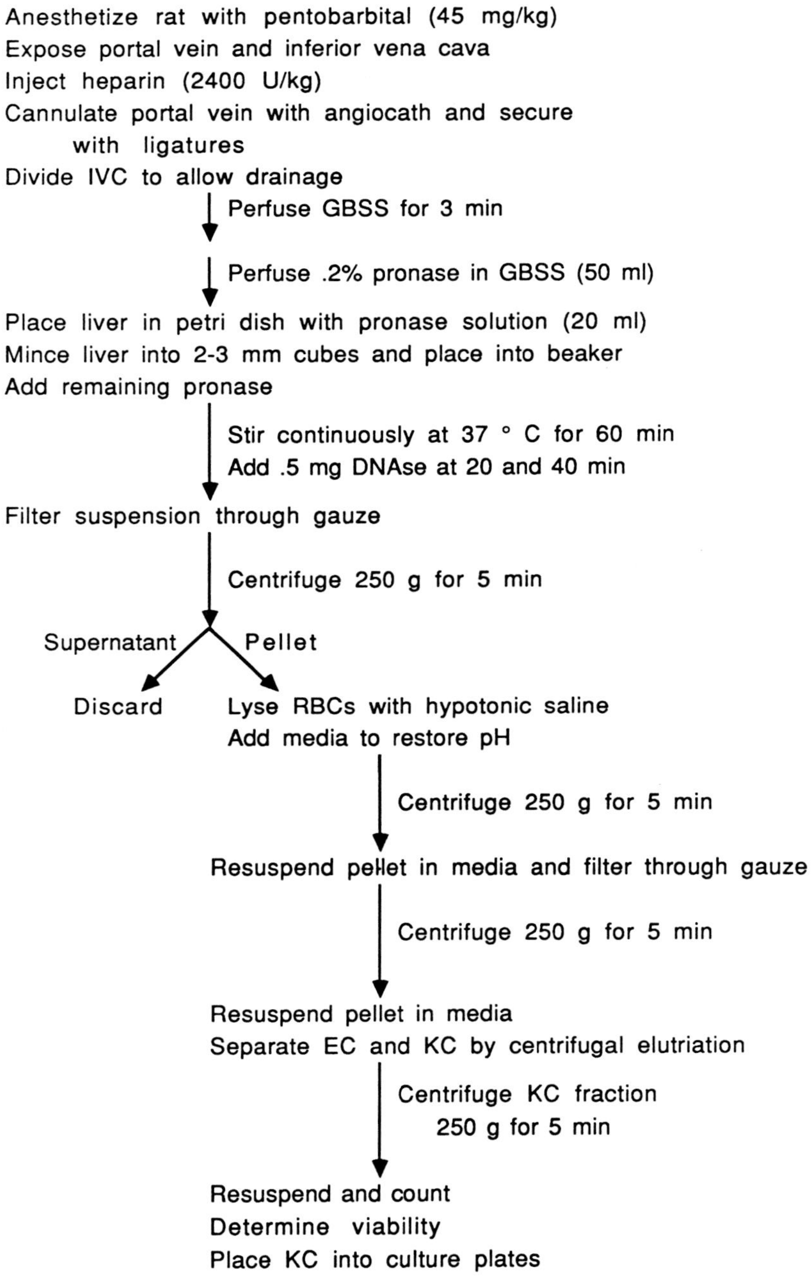

Anesthetize rat with pentobarbital (45 mg/kg)
Expose portal vein and inferior vena cava
Inject heparin (2400 U/kg)
Cannulate portal vein with angiocath and secure
with ligatures
Divide IVC to allow drainage

↓ Perfuse GBSS for 3 min

↓ Perfuse .2% pronase in GBSS (50 ml)

Place liver in petri dish with pronase solution (20 ml)
Mince liver into 2-3 mm cubes and place into beaker
Add remaining pronase

↓ Stir continuously at 37 ° C for 60 min
Add .5 mg DNAse at 20 and 40 min

Filter suspension through gauze

↓ Centrifuge 250 g for 5 min

Supernatant → Discard

Pellet → Lyse RBCs with hypotonic saline
Add media to restore pH

↓ Centrifuge 250 g for 5 min

Resuspend pellet in media and filter through gauze

↓ Centrifuge 250 g for 5 min

Resuspend pellet in media
Separate EC and KC by centrifugal elutriation

↓ Centrifuge KC fraction
250 g for 5 min

Resuspend and count
Determine viability
Place KC into culture plates

FIGURE 2

by centrifugal elutriation following the method of Garvey and Heil[26] using a commercially available device from Beckman® Instruments (Palo Alto, CA, JE-6 rotor). Centrifugal elutriation, also called counterflow centrifugation, is a method of cell separation that utilizes a rotating chamber into which cell suspensions are pumped and withdrawn.[27] Separation into different cell populations is based on the balance between outwardly directed centrifugal forces on the cells and inwardly directed buoyancy and fluid dynamic forces.[27,28] Protein is included in the elutriation solution to decrease damaging shear forces on the cell.[27,28] If the separation is performed at a constant rotor speed at room temperature, adequate separation of endothelial cells and KC can be obtained by varying the speed at which cells are pumped into the separation chamber.[28]

The predominant contaminating cell in KC suspensions will be endothelial cells.[26] Estimation of the purity of the KC population can be made utilizing a variety of techniques that emphasize the different features between KC and endothelial cells.

A. CYTOCHEMISTRY

While both KC and endothelial cells stain positively for nonspecific esterase, only KC stain for peroxidase activity.[23] It has been reported that human endothelial cells are peroxidase positive,[29] so this technique would not be applicable if human tissue has been used as a source of KC.

B. PHAGOCYTOSIS

KC demonstrate uptake of a number of phagocytosable particles, including sheep red blood cells, latex particles, and bacteria.[25] Carbon particles may also be used, although endothelial cells have been shown to endocytose small amounts of this substance.[30]

C. MORPHOLOGY

Both endothelial cells and KC demonstrate Fc receptors, but only KC possess complement (C) receptors.[16] These cell types may also be differentiated by the use of radiolabeled ligands such as albumin/ovalbumin that bind to endothelial cells, but not to KC.[16] Electron microscopy may be the best method for differentiating KC from endothelial cells. KC are approximately 10 μm in diameter, possess numerous heterogeneous lysosomal structures, and contain extensive microvilli.[31] Endothelial cells are 6 to 9 μm in diameter, contain a vacuolated, spongelike cytoplasm, and have few microvilli.[28] Recently, monoclonal antibodies have been developed that identify rat KC *in vivo,* and these techniques may eventually be applied to *in vitro* use as well.[32]

While the use of pronase is relatively simple and results in a high yield of KC free of contaminating parenchymal cells, it may also alter and destroy receptor proteins on the surface of the KC, requiring a period in culture for receptor renewal prior to use.[16] Alternate methods that avoid the use of pronase utilize collagenase for cell dispersion and either treatment with enterotoxin to destroy parenchymal cells followed by elutriation[33] or the use of two-step percoll gradients and selective adherence[2,16] to separate KC from parenchymal cells and endothelial cells, respectively. While the purity of the cell population is generally satisfactory, the yield with these methods is lower than with methods that utilize pronase/elutriation.[16] We prefer to isolate KC by pronase/elutriation followed by 24 to 48 h of culture to allow recovery from the separation, at which time greater than 90% of Fc and C receptors will be expressed.[3]

IV. *IN VIVO* TECHNIQUES FOR STUDYING KC:HC INTERACTION

While models of inflammation or organ failure may yield much information on the effects of specific changes on the liver as a whole organ, they generally yield little information about the more specific area of KC:HC interaction. These interactions may be altered or masked by effects on other cell populations active in inflammation, including endothelial cells, circulating monocytes, neutrophils, and fibroblasts.

Specific techniques have been developed, however, to specifically study KC:HC interactions *in vivo*. The use of frog virus 3 (FV3) has proven especially useful. This virus selectively infects KC and endothelial cells *in vivo*, leading to altered HC function and eventually to HC necrosis.[34] FV3 may also be taken up by KC and endothelial cells *in vitro*[35,36] and studied in single cell cultures.

Specific substances may be injected *in vivo* that either have effects specifically on KC or affect the KC population significantly, and thus can be used to study KC function. Ricin, a toxin purified from castor beans, is one such substance. When given in proper doses, it inhibits ribosomal function.[37] When given by intraportal injection, this toxin depletes the number of KC found histologically by >80% at 24 h, and this depletion lasts for 72 h.[38,39] Hepatic damage may be evident, however, at high doses,[38] so care must be taken when using this substance.

Gadolinium chloride is a raw earth metal that may also be useful to study KC:HC interaction. This substance interferes with KC phagocytosis.[40] It has been used to study the role of KC in both transplantation immunology and inflammation.[41,42]

In contrast to the mechanisms of inhibiting KC function mentioned above, a number of stimuli have been demonstrated to result in an increase in mononuclear phagocytes in the liver, including endotoxin,[43] zymosan,[44,45] Freund's adjuvant,[46] or *Corynebacterium parvum*.[47] While disagreements may exist concerning whether this mononuclear cell increase represents local proliferation of resident KC or the recruitment of extrahepatic phagocytes,[43,44-49] these liver macrophages provide useful information on the mechanisms involved in hepatic pathophysiology during inflammation.[47]

TABLE 3
***Corynebacterium parvum* Model of Acute Hepatic Necrosis**

C. parvum	Mouse	50 mg/kg
	Rat	28 mg/kg
LPS	Mouse	50 μg/kg
	Rat	100 μg/kg

Our laboratory has found the use of *C. parvum* (also referred to as *Propionibacterium acnes*) to be useful for the study of mechanisms involved in hepatic injury during inflammation.[50] This preparation has also proven useful for *ex vivo* studies of both parenchymal cells and NPC and their function after inflammatory stimuli.[47,51,52] *C. parvum* is one of a number of organisms that displays potent reticuloendothelial system (RES) stimulating properties.[53] Mice injected with *C. parvum* develop hepatic macrophage granulomas and a marked sensitivity to subsequent endotoxin injection.[51,54] This preparation has also been used as an immunostimulant in the study of wound infection[55] and cytokine production.[56]

For the study of inflammatory hepatic damage with *C. parvum,* the most common experimental animals are either the mouse or the rat. While the general phenomenon of marked and rapid hepatic necrosis is similar in both animals, subtle differences may exist.[50,51,54] The most notable differences occur in the rate of mortality, which begins 6 to 8 h following endotoxin in rats,[54] while in mice it may be delayed up to 24 h.[51] In addition, both the dose of *C. parvum* used to prime the RES and the amount of LPS used as a secondary stimulus vary (Table 3).

The protocol for *C. parvum*-induced hepatic necrosis is given in Figure 3. The *C. parvum* injection will result in RES stimulation if given either i.v. or i.p.,[53] and we have found the i.v. administration results in reliable stimulation in mice.[50] Acute hepatic necrosis may be precipitated 7 d after the initial injection by injection of a small amount of LPS, which will eventually lead to hypoglycemia and death in nearly all animals.[51,54]

The *C. parvum* vaccine itself may be obtained commercially (Burroughs Wellcome® Company, Research Triangle Park, NC). Live strains may be

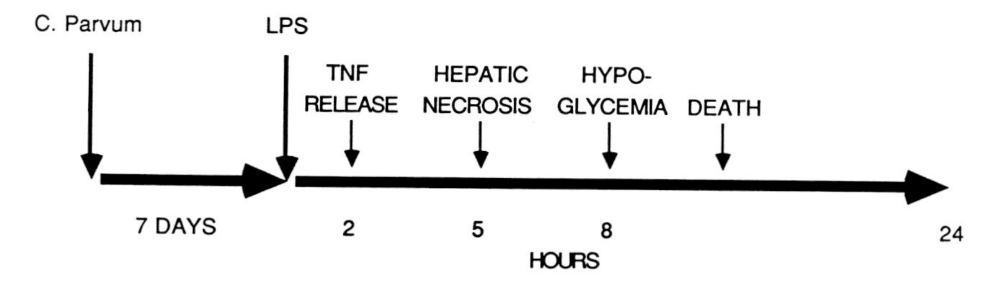

FIGURE 3. *C. parvum*-induced acute hepatic necrosis.

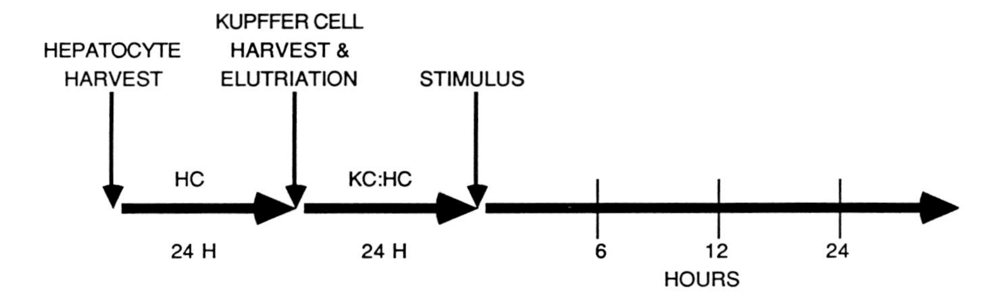

FIGURE 4. Kupffer cell:hepatocyte coculture.

purchased from a number of sources and used to prepare individual preparations following standard laboratory techniques for the culture of anaerobic bacteria.[57] We have followed the method of Cummins and Linn[53] and found it useful in the preparation of large amounts of the vaccine. However, differences may be found in RES-stimulating ability among different *C. parvum* strains,[53] and this may require modification of the doses of *C. parvum* used in order to establish reproducible levels of hepatic necrosis.[75]

Between the extremes on the spectrum of inhibition or exaggeration of KC function mentioned above are studies that reveal how *in vivo* alterations alter KC function *ex vivo*.[58-61] These studies have been reviewed in detail in other portions of this monograph, and will not be elaborated upon further here.

V. *IN VITRO* TECHNIQUES FOR STUDYING KC:HC INTERACTION

In vitro culture of KC and HC together has been performed for many years.[62] In addition to extending the viability of HC cultures,[62] these techniques provide a useful model for the study of complex cellular interactions that may occur during inflammation.[14] These studies have confirmed the hypothesis that activated KC may exert profound influences on the metabolism of neighboring HC.[63]

The KC:HC coculture model used in this laboratory is shown in Figure 4.[64] HC are harvested on day 1, as described earlier in this chapter, and cultured in isolation for 24 h. On day 2, KC are collected and added to the established HC culture in fresh media. The KC:HC coculture is incubated for an additional 24 h, and on day 3 the desired experimental conditions are established. In our laboratory, KC- or macrophage-mediated changes in total HC protein synthesis in response to LPS has been frequently studied as a potential factor involved in hepatic failure associated with sepsis.[14,15,65] However, numerous other aspects of KC:HC interaction may also be studied with this model, including KC-mediated HC toxicity,[63] the effects of or changes in cytokine or eicosanoid production,[6,63,64] nitrogen oxide production,[66-68] and secretion of specific HC proteins.[69] Specific details regarding these findings will be reviewed in subsequent chapters in this book.

While KC:HC cocultures provide information regarding the bidirectional cellular communications involved in KC:HC interactions, supernatant transfer studies allow one to investigate unidirectional cellular influences. These techniques have allowed investigators to study the effects that KC products exert on HC function.

Conditioned media from KC cultures that have been stimulated with LPS, or from resting KC cultures, have been used to study HC metabolism.[66-71] Examples of HC functions that have been shown to be effected by transferable KC products include nitrogen oxide synthesis,[66,67] iron metabolism,[70] protein phosphorylation,[32] and the secretion of fibronectin[72] and other acute-phase proteins.[73]

That KC:HC interactions are bidirectional has been demonstrated by studies which reveal that the presence of HC alters KC function.[74] Conditioned media from HC cultures has also been shown to alter KC function, specifically KC eicosanoid synthesis.[6]

VI. CONCLUSION

Technical developments within the last 20 years have permitted the isolation and culture of both HC and KC with high yield and purity. These advances have allowed detailed study of the factors involved in KC-induced changes in HC function as well as HC-induced changes in KC function. The use of KC:HC cocultures as well as transfer of conditioned media from one cell type to the other have elucidated the degree to which these cell types may influence each other. A number of *in vivo* techniques have also been developed which either accentuate or depress KC function and thus contribute to the understanding of *in vivo* KC:HC interactions. As these and other techniques continue to be refined, further advances in the study of KC:HC interactions can be anticipated.

REFERENCES

1. **Zahlten, R. N., Rogoff, T. M., and Steer, C. S.,** Isolated Kupffer cells, endothelial cells, and hepatocytes as investigative tools for liver research, *Fed. Proc., Fed. Am. Soc. Exp. Biol.,* 40, 2460, 1981.
2. **Smedsrod, B., Pertoft, H., Eggertsen, G., and Sundstrom, C.,** Functional and morphological characterization of cultures of Kupffer cells and liver endothelial cells prepared by means of density separation in percoll and selective substrate adhesion, *Cell Tissue Res.,* 241, 639, 1985.

3. **Crofton, R. W., Diesselhoff-den Dulk, M. M. C., and van Furth, R.,** The origin, kinetics, and characterization of the Kupffer cells in the normal steady state, *J. Exp. Med.*, 148, 1, 1978.
4. **Mathison, J. C. and Ulevitch, R. J.,** The clearance, tissue distribution, and cellular localization of intravenously injected lipopolysaccharide in rabbits, *J. Immunol.*, 123, 2133, 1979.
5. **Laskin, D. L., Sirak, A. A., Pilaro, A. M., and Laskin, J. D.,** Functional and biochemical properties of rat Kupffer cells and peritoneal macrophages, *J. Leuk. Biol.*, 44, 71, 1988.
6. **Billiar, T. R., Lysz, T. W., Curran, R. D., Bentz, B. G., Machiedo, G. W., and Simmons, R. L.,** Hepatocyte modulation of Kupffer cell prostaglandin E_2 production *in vitro*, *J. Leuk. Biol.*, 47, 304, 1990.
7. **Seglen, P. O.,** Preparation of isolated rat liver cells, *Methods Cell Biol.*, XIII, 29, 1976.
8. **Howard, R. B., Christensen, A. K., Gibbs, F. A., and Pesch, L. A.,** The enzymatic preparation of isolated intact parenchymal cells from rat liver, *J. Cell Biol.*, 35, 675, 1967.
9. **Berry, M. N. and Friend, D. S.,** High-yield preparation of isolated rat liver parenchymal cells: a biochemical and fine structure study, *J. Cell Biol.*, 43, 506, 1969.
10. **Seglen, P. O.,** Preparation of rat liver cells. III. Enzymatic requirements for tissue dispersion, *Exp. Cell Res.*, 82, 391, 1973.
11. **Pitot, H. C. and Sirica, A. E.,** Methodology and utility of primary cultures of hepatocytes from experimental animals, *Methods Cell Biol.*, XXIB, 441, 1980.
12. **Fry, J. R.,** Preparation of mammalian hepatocytes, *Methods Enzymol.*, 77, 130, 1981.
13. **Klaunig, J. E., Goldblatt, P. J., Hinton, D. E., Lipsky, M. M., Chacko, J., and Trump, B. F.,** Mouse liver culture. I. Hepatocyte isolation, *In Vitro*, 17, 913, 1981.
14. **West, M. A., Billiar, T. R., Curran, R. D., Hyland, B. J., and Simmons, R. L.,** Kupffer cell-mediated alterations of hepatocyte protein synthesis. V. Evidence for separate mechanisms in increases and decreases of protein synthesis, *Gastroenterology*, 96, 1572, 1989.
15. **West, M. A., Keller, G. A., Cerra, F. B., and Simmons, R. L.,** Killed Escherichia coli stimulates macrophage-mediated alterations in hepatocellular function during *in vitro* coculture: a mechanism of altered liver function in sepsis, *Infect. Immun.*, 49, 563, 1985.
16. **Smedsrod, B. and Pertoft, H.,** Preparation of pure hepatocytes and reticuloendothelial cells in high yield from a single rat liver by means of percoll centrifugation and selective adherence, *J. Leuk. Biol.*, 38, 213, 1985.
17. **Walter, H., Krob, E. J., Ascher, G. S., and Seaman, G. V. F.,** Partition of rat liver cells in aqueous dextran-polyethylene glycol phase systems, *Exp. Cell Res.*, 82, 15, 1973.
18. **Starke, P. E. and Farber, J. L.,** Ferric iron and superoxide iron are required for the killing of cultured hepatocytes by hydrogen peroxide, *J. Biol. Chem.*, 260, 10099, 1985.

19. **Isom, H. C., Secote, T., Georgoff, I., Woodworth, C., and Mummaw, J.,** Maintenance of differentiated rat hepatocytes in primary culture, *Proc. Natl. Acad. Sci. U.S.A.*, 82, 3252, 1985.
20. **Guillouzo, A.,** Plasma protein production by cultured adult hepatocytes, in *Isolated and Cultured Hepatocytes,* G. Guillouzo and C. Guguen-Guillouzo, Eds., John Libby Eurotext, Ltd., Paris, 1986, 155.
21. **Drochmans, P., Wanson, J. C., May, C., and Bernaert, D.,** Ultrastructure and metabolic studies of isolated and culture hepatocytes, in *Hepatotrophic Factors,* R. Porter and J. Whelan, Eds., Elsevier, New York, 1978, 7.
22. **Munthe-Kass, A. C., Berg, T., Seglen, P. O., and Seljelid, R.,** Mass isolation and culture of rat Kupffer cells, *J. Exp. Med.*, 141, 1, 1975.
23. **Emeis, J. J. and Planque, B.,** Heterogeneity of cells isolated from rat liver by pronase digestion: ultrastructure, cytochemistry, and cell culture, *J. Reticuloend. Soc.*, 20, 11, 1976.
24. **Knook, D. L., Blansjaar, N., and Sleyster, E. C.,** Isolation and characterization of Kupffer and endothelial cells from the rat liver, *Exp. Cell Res.*, 109, 317, 1977.
25. **Knook, D. L. and Sleyster, E. C.,** Preparation and characterization of Kupffer cells from rat and mouse liver, in *Kupffer Cells and Other Liver Sinusoidal Cells,* E. Wisse and D. L. Knook, Eds., Elsevier, Amsterdam, 1977, 273.
26. **Garvey, J. S. and Heil, M. F.,** Separation and collection of rat liver macrophages, *Methods Enzymol.*, 108, 285, 1984.
27. **Sanderson, R. J. and Bird, K. E.,** Cell separation by counterflow centrifugation, *Methods Cell Biol.*, XV, 1, 1977.
28. **Zahlten, R. N., Hagler, H. K., Nejtek, M. E., and Day, C. J.,** Morphological characterization of Kupffer and endothelial cells of rat liver isolated by counterflow elutriation, *Gastroenterology,* 75, 80, 1978.
29. **Zafrani, E. S., Vasconcelos, A. W., Gourndin, M. F., Pinaudeau, Y., and Reyes, F.,** Is detection of endogenous peroxidatic activity helpful in distinguishing Kupffer cells from endothelial cells in normal adult human liver?, in *Sinusoidal Liver Cells,* D. L. Knook and E. Wisse, Eds., Elsevier, New York, 1982, 505.
30. **Widmann, J. J., Cotran, R. S., and Fahimi, H. D.,** Mononuclear phagocytes (Kupffer cells) and endothelial cells: identification of two functional cell types in rat liver sinusoids by endogenous peroxidase activity, *J. Cell Biol.*, 52, 159, 1972.
31. **Drochmans, P., Sleyster, E. C., Penasse, W., Wanson, J. C., and Knook, D. L.,** Morphology of isolated and cultured sinus-lining cells of rat liver, in *Kupffer Cells and Other Liver Sinusoidal Cells,* E. Wisse and D. L. Knook, Eds., Elsevier, Amsterdam, 1977, 131.
32. **Sugihara, S., Martin, S. R., Hsuing, C. K., Maruiwa, M., Block, K. J., Moscicki, R. A., and Bhan, A. K.,** Monoclonal antibodies to rat Kupffer cells: anti-KCA-1 distinguishes Kupffer cells from other macrophages, *Am. J. Pathol.*, 136, 345, 1990.
33. **Blomhoff, R., Smedsrod, B., Eskild, W., Granum, P. E., and Berg, T.,** Preparation of isolated liver endothelial cells and Kupffer cells in high yield by means of an enterotoxin, *Exp. Cell Res.*, 150, 194, 1984.

34. **Kirn, A., Gut, J. P., Bingen, A., and Steffan, A. M.,** Murine hepatitis induced by frog virus 3: a model for studying the effect of sinusoidal cell damage on the liver, *Hepatology,* 3, 105, 1983.
35. **Gendrault, J. L., Steffan, A. M., and Bingen, A.,** Penetration and uncoating of frog virus 3 (FV3) in cultured rat Kupffer cells, *Virology,* 112, 375, 1981.
36. **Steffan, A. M., Lecerf, F., and Kelb, F.,** Isolement et culture de cellules endotheliales de fois humain et murin, *CR Acad. Sci. (D) (Paris),* 292, 809, 1981.
37. **Sirpe, F. and Barbieri, L.,** Ribosome inactivating proteins up to date, *FEBS Lett.,* 195, 1, 1986.
38. **Zenilman, M. E., Fiani, M., Stahl, P. D., Brunt, E. M., and Flye, M. W.,** Selective depletion of Kupffer cells in mice by intact ricin, *Transplantation,* 47, 200, 1989.
39. **Zenilman, M. E., Fiani, M., Stahl, P., Brunt, E., and Flye, M. W.,** Use of ricin A chain to selectively deplete Kupffer cells, *J. Surg. Res.,* 45, 82, 1988.
40. **Husztik, E., Lazar, G., and Parducz, A.,** Electron microscopic study of Kupffer cell phagocytosis blockade induced by gadolinium chloride, *Br. J. Exp. Pathol.,* 61, 624, 1980.
41. **Callery, M. P., Kamei, T., and Flye, M. W.,** Kupffer cell blockade increases mortality during intra-abdominal sepsis despite improving systemic immunity, *Arch. Surg.,* 125, 36, 1990.
42. **Callery, M. P., Kamei, T., and Flye, M. W.,** Kupffer cell blockade inhibits induction of tolerance by the portal venous route, *Transplantation,* 47, 1092, 1989.
43. **Pilaro, A. M. and Laskin, D. L.,** Accumulation of activated mononuclear phagocytes in the liver following lipopolysaccharide treatment of rats, *J. Leuk. Biol.,* 40, 29, 1986.
44. **Bouwens, L. and Wisse, E.,** Proliferation, kinetics, and fate of monocytes in rat liver during a zymosan-induced inflammation, *J. Leuk. Biol.,* 37, 531, 1985.
45. **Bouwens, L., Kook, K. L., and Wisse, E.,** Local proliferation and extrahepatic recruitment of liver macrophages (Kupffer cells) in partial-body irradiated rats, *J. Leuk. Biol.,* 39, 687, 1986.
46. **Bugelski, P. J.,** Sequential histochemical staining for resident and recruited macrophages, *J. Leuk. Biol.,* 38, 687, 1985.
47. **Tanner, A., Keyhani, A., Reiner, R., Holdstock, G., and Wright, R.,** Proteolytic enzymes released by liver macrophages may promote hepatic injury in a rat model of hepatic damage, *Gastroenterology,* 80, 647, 1981.
48. **Bouwens, L. and Wisse, E.,** On the dual origin of the Kupffer cell, in *Sinusoidal Liver Cells,* D. L. Knook and E. Wisse, Eds., Elsevier, New York, 1982, 165.
49. **Gali, R. P., Sparkes, R. S., and Golde, D. W.,** Bone marrow origin of hepatic macrophages (Kupffer cells) in humans, *Science,* 201, 937, 1978.
50. **Billiar, T. R., Curran, R. D., Harbrecht, B. G., Stuehr, D. J., Demetris, A. J., and Simmons, R. L.,** Modulation of nitrogen oxide synthesis *in vivo*: N^G monomethyl-L-Arginine inhibits endotoxin-induced nitrite/nitrate biosynthesis while promoting hepatic damage, *J. Leuk. Biol.,* 48, 565, 1990.

51. **Mizoguchi, Y., Tsutsui, H., Miyajima, K., Sakagami, Y., Seki, S., Kobayashi, K., Yamamoto, S., and Morisawa, S.,** The protective effects of prostaglandin E_1 in an experimental massive hepatic cell necrosis model, *Hepatology,* 7, 1184, 1987.
52. **Billiar, T. R., Curran, R. D., Stuehr, D. J., Stadler, J., Simmons, R. L., and Murray, S. A.,** Inducible cytosolic enzyme activity for the production of nitrogen oxides from L-arginine in hepatocytes, *Biochem. Biophys. Res. Commun.,* 168, 1034, 1990.
53. **Cummins, C. S. and Linn, D. M.,** Reticulostimulating properties of killed vaccines of anaerobic coryneforms and other organisms, *J. Natl. Cancer Inst.,* 59, 1697, 1977.
54. **Arthur, M. J. P., Bentley, I. S., Tanner, A. R., Kowalski Saunders, P., Millward-Sadler, G. H., and Wright, R.,** Oxygen-derived free radicals promote hepatic injury in the rat, *Gastroenterology,* 89, 1114, 1985.
55. **Calhoun, K., Trachtenberg, L., Hart, K., and Polk, H. C.,** Corynebacterium parvum: immunomodulation in local bacterial infections, *Surgery,* 87, 52, 1980.
56. **Green, S., Dobrjansky, A., Chiasson, M. A., Carswell, E., Schwartz, M. K., and Old, L. J.,** Corynebacterium parvum as the priming agent in the production of tumor necrosis factor in the mouse, *J. Natl. Cancer Inst.,* 59, 1519, 1977.
57. *Anaerobe Laboratory Manual,* 3rd ed., L. V. Holdemann and W. E. Moore, Eds., Virginia Polytechnic Institute and State University, Blacksburg, 1975.
58. **Billiar, T. R., West, M. A., Hyland, B. J., and Simmons, R. L.,** Splenectomy alters Kupffer cell response to endotoxin, *Arch. Surg.,* 123, 306, 1988.
59. **Billiar, T. R., Maddaus, M. A., West, M. A., Curran, R. D., Wells, C. A., and Simmons, R. L.,** Intestinal gram-negative bacterial overgrowth *in vivo* augments the *in vitro* response of Kupffer cells to endotoxin, *Ann. Surg.,* 208, 532, 1988.
60. **Billiar, T. R., Maddaus, M. A., West, M. A., Dunn, D. L., and Simmons, R. L.,** The role of intestinal flora on the interactions between nonparenchymal cells and hepatocytes in coculture, *J. Surg. Res.,* 44, 397, 1988.
61. **Billiar, T. R., Bankey, P. E., Svingen, B. A., Curran, R. D., West, M. A., Holman, R. T., Simmons, R. L., and Cerra, F. B.,** Fatty acid intake and Kupffer cell function: fish oil alters eicosanoid and monokine production to endotoxin stimulation, *Surgery,* 104, 343, 1988.
62. **Wanson, J. C., Drochmans, P., Mosselmans, R., and Knook, D. L.,** Symbiotic culture of adult hepatocytes and sinus-lining cells, in *Kupffer Cells and Other Liver Sinusoidal Cells,* E. Wisse and D. L. Knook, Eds., Elsevier, Amsterdam, 1977, 141.
63. **Billiar, T. R., Curran, R. D., West, M. A., Hofmann, K., and Simmons, R. L.,** Kupffer cell cytotoxicity to hepatocytes in coculture requires L-arginine, *Arch. Surg.,* 124, 1416, 1989.
64. **Curran, R. D., Billiar, T. R., West, M. A., Bentz, B. G., and Simmons, R. L.,** Effect of interleukin-2 on Kupffer cell activation, *Arch. Surg.,* 123, 1373, 1988.
65. **Keller, G. A., West, M. A., Cerra, F. B., and Simmons, R. L.,** Macrophage-mediated modulation of hepatocyte protein synthesis: effects of dexamethasone, *Arch. Surg.,* 121, 1199, 1986.

66. **Curran, R. D., Billiar, T. R., Stuehr, D. J., Hofmann, K., and Simmons, R. L.,** Hepatocytes produce nitrogen oxides from L-arginine in response to inflammatory products of Kupffer cells, *J. Exp. Med.*, 170, 1769, 1989.
67. **Curran, R. D., Billiar, T. R., Stuehr, D. J., Ochoa, J. B., Harbrecht, B. G., Flint, S. G., and Simmons, R. L.,** Multiple cytokines are required to induce hepatocyte nitrogen oxide production and inhibit total protein synthesis, *Ann. Surg.*, 212, 462, 1990.
68. **Billiar, T. R., Curran, R. D., Stuehr, D. J., West, M. A., Bentz, B. G., and Simmons, R. L.,** An L-arginine-dependent mechanism mediates Kupffer cell inhibition of hepatocyte protein synthesis *in vitro, J. Exp. Med.*, 169, 1467, 1989.
69. **West, M. A., Billiar, T. R., Mazuski, J. E., Curran, R. D., Cerra, F. B., and Simmons, R. L.,** Endotoxin modulation of hepatocyte secretory and cellular protein synthesis is mediated by Kupffer cells, *Arch. Surg.*, 123, 1400, 1988.
70. **Sibille, J. C., Kondo, H., and Aisen, P.,** Interactions between isolated hepatocytes and Kupffer cells in iron metabolism: a possible role for ferritin as an iron carrier protein, *Hepatology*, 8, 296, 1988.
71. **Casteleijn, E., Kuiper, J., van Rooij, H. C. J., Koster, J. F., and van Berkel, T. J. C.,** Conditioned media of Kupffer cells and endothelial cells influence protein phosphorylation in parenchymal liver cells, *Biochim. J.*, 252, 601, 1988.
72. **Lanser, M. E. and Brown, G. E.,** Stimulation of rat hepatocyte fibronectin production by monocyte-conditioned medium is due to interleukin-6, *J. Exp. Med.*, 170, 1781, 1989.
73. **Baumann, H.,** Hepatic acute phase reaction in vivo and in vitro, *In Vitro Cell. Devel. Biol.*, 25, 115, 1989.
74. **Curran, R. D., Billiar, T. R., Kispert, P. H., Bentz, B. G., May, M. T., and Simmons, R. L.,** Hepatocytes enhance Kupffer cell-mediated tumor cell cytostasis *in vitro, Surgery*, 106, 126, 1989.
75. **Harbrecht, B.,** unpublished observation.

Section II

Nonseptic Pathologic States

Chapter 4

ANTITUMOR DEFENSE SYSTEM OF THE LIVER

S.-P. Tzung and S.A. Cohen

TABLE OF CONTENTS

ISBN 0-8493-6109-5

I. LIVER AS A TUMOR CELL-KILLING ORGAN

The liver, a major component of the reticuloendothelial system, has long been recognized for its role as a scavenger. When traversing the liver, effete cells, foreign or altered antigens, and microbial products such as endotoxins derived from portal blood are cleared and detoxified.[1,2] During the past decade, however, increasing attention has been drawn to the liver as an organ involved in the defense against tumor cell metastasis.

The liver is a common metastatic site for a variety of tumor cells, particularly cancers of gastrointestinal origin. Colon cancer metastasizes to the liver in 70% of patients with advanced disease (Dukes' stage D).[3] Although patients with solitary metastasis can be occasionally cured, the overwhelming majority succumb to their disease.[4] An understanding of the interaction between metastatic tumor cells and the liver is therefore essential if one hopes to design new therapeutic strategies for this common clinical problem. The liver possesses a dual blood supply, the portal vein and the hepatic artery. The portal vein is the route by which splanchnic tumors metastasize, while the hepatic artery provides access to tumors originating from outside the splanchnic circulation. Both vasculature systems branch further in the liver and ultimately drain into liver sinusoids. Since liver sinusoids, especially the periportal ones, are very narrow (4.15 μm in diameter),[5] it is reasonable to presume that any invading tumor cell must come into close contact with cells lining the sinusoid prior to subsequent extravasation and proliferation in the liver parenchyma.

An earlier experiment by Riccardi et al.,[6] where radiolabeled YAC-1 and RBL-5 lymphoma cells were inoculated intravenously into the mice, demonstrated a rapid *in vivo* clearance of tumor cells from the liver, indicating that the liver can be an efficient filter system for certain tumor cells. Due to the relative difficulty in harvesting lymphoid cells from the liver, it was not until the technique of *in situ* enzyme perfusion was adopted that our laboratory[7-9] and others[10,11] were able to show that nonparenchymal liver cells (NPC), or sinusoidal liver cells, from nonimmunized rodents had a potent ability to kill tumor cells *in vitro*. Moreover, when quantitatively compared with the spleen, the liver contained a much higher organ capacity (= lytic units × cell numbers) for tumoricidal activity.[11] At least three distinct NPC populations, including Kupffer cells (KC), natural killer (NK) cells, and promonocytes[12,13] have been demonstrated to lyse or suppress the growth of a variety of tumor cells *in vitro* under various experimental conditions. These NPC adhere to the sinusoidal endothelial linings, and experimental evidence from rodent models strongly suggests that they constitute the first-line defense against invading tumor cells.

II. POTENTIAL ANTITUMOR MECHANISMS OF THE LIVER

A. KUPFFER CELLS

KC comprise the largest population of fixed tissue macrophages in the body, accounting for more than half of the mononuclear phagocyte system. The mitotic activity of KC in the normal steady state is very low, with a labeling index of less than 1% after ^{3}H-thymidine pulse.[14] A long-held hypothesis has been that KC are terminally differentiated cells incapable of division, and KC hyperplasia observed in the liver following administration of inflammatory stimuli is derived from recruitment of bone marrow precursor cells. Further support for the monocytic origin of KC comes from studies in long-term surviving male to female liver homografts, showing that 3.5 months after hepatic transplantation, KC are of host origin.[15] More recent experimental evidence, however, indicates that KC are capable of self-replicating and migrating to other tissue compartments.[16,17] There exists an F4/80$^+$ macrophage precursor cell population in the normal liver that proliferates in response to colony-stimulating factor 1 (CSF-1) *in vitro*.[18] Using whole-body irradiation vs. partial-body irradiation (liver shielded) plus splenectomy, Bouwens et al.[17] have estimated that following zymosan injection, local proliferation accounted for at least 60% of the observed KC hyperplasia, whereas extrahepatic recruitment played a lesser role. Although hyperplastic response in the liver is very common after administration of various biological response modifiers or under conditions of tumor growth or hepatitis, the relative contribution of local replication vs. extrahepatic recruitment in many cases remains unclear.

The phagocytic ability of KC has been known for more than a century. Even before the name KC was coined in the early 1900s, several investigators had reported the uptake of dye particles by one type of "connective tissue liver cells", described as "spindle-shape with several processes — between the liver cells", which was clearly one of the earliest documentations of the function and morphology of KC.[19] The antitumor activity of KC is a relatively recent finding. After injection of TA3 (mammary adenocarcinoma) or B16 (melanoma) cells into the portal vein of mice, within 30 min many tumor cells were seen to be completely or partly surrounded by thin electron-lucent processes of KC.[20] Other tumors such as lymphosarcoma MB V1 A and GRSL leukemia cells were not appreciably affected. These results seemed to be consistent with the *in vitro* observations reported by Gardner et al.[21] Killing of P815 plasmacytoma cells by *in vivo* lipopolysaccharide (LPS)-activated rat KC initially involved phagocytosis of intact tumor cells, as evidenced by light and electron microscopy and by uptake of ^{51}Cr-labeled cells. In contrast, killing of N1S1 hepatoma and RBL-1 leukemia cells was mediated exclusively through extracellular cytolysis. In general, however, most investigators found that the *in vitro* destruction of tumor targets by KC, like monocyte and peritoneal macrophages, occurs by a nonphagocytic process and requires cell contact and prolonged incubation (18 to 72 h). These results suggest that the cytotoxicity displayed by KC depend upon both the activation state of KC and the nature of the tumor target cells.

Variable observations have been reported regarding the cytotoxic ability of resident KC, i.e., those not artificially activated by biological response modifiers. We,[8] as well as Decker et al.,[22] have shown that resident KC isolated from C57BL/6 mice displayed no spontaneous killing of the allogeneic YAC-1 and P815 tumor cells in an 18-h assay. Xu et al.[23] reported similar findings in a syngeneic system. In contrast, the groups of Bodenheimer et al.[24] and Sherwood et al.,[25] working with rats and mice, respectively, found significant KC cytotoxicity toward two syngeneic hepatoma and one adenocarcinoma cell lines. This apparent discrepancy might be explained by: (1) species differences. Decker et al.[26] have conducted a comparative study on murine and rat KC. Resident rat KC, but not murine KC, exhibited natural cytotoxicity against all tumor targets tested. They concluded that resident rat KC were already in a primed state and functionally resembled the activated mouse liver macrophages induced by inflammatory stimuli. Recently, Roh et al.[27] reported that human KC required stimulation by LPS and interferon-gamma (IFN-γ) to become cytotoxic toward several solid tumor cell lines. Thus, human KC appear functionally more like mouse KC than rat KC; (2) The nature of tumor target cells. Different tumors vary in their susceptibility to the effector mechanisms of KC; (3) The method applied for KC isolation. Enzyme treatment with either pronase or collagenase is required to dissociate KC from endothelial linings. Pronase has been shown to induce membrane changes and strip off surface receptors.[28] We have observed that both the cytotoxic activity and cytokine production of KC were severely compromised after pronase treatment.[95] Likewise, Decker et al.[22] noted that KC freshly isolated by the pronase method were refractory to stimulation by any activating agent. A period of at least 72 h was required for KC to regain their normal function.

KC can be activated *in vitro* to become tumoricidal by a variety of agents, including LPS, IFN-γ, and free or liposome-encapsulated muramyl dipeptide (MDP).[24,25,29] Resident KC, despite manifesting little cytotoxicity in several studies, have been shown to exert a potent cytostatic effect against a large panel of tumor cells.[30] Therefore, in addition to a cytotoxic interaction with tumor cells in the intravascular phase of metastasis, these cells may also be able to modulate the growth of established metastatic foci. At present, there is little information regarding the exact role of resident KC in the defense of the host against liver metastases. However, based on the *in vitro* results of KC activation, various strategies designed to activate tissue macrophages *in vivo* were devised to determine whether or not these cells can be manipulated to increase their effectiveness against invading tumor cells. The *in vivo* activation of KC was observed following systemic administration of LPS,[21] pyran copolymer,[26] glucan,[25] maleic anhydride divinyl ether (MVE-2),[31] muramyl tripeptide (MTP),[32] and bacterial products such as *Propionibacterium acnes*,[8] OK-432,[33] and BCG.[32] Moreover, in animal models the degree of KC activation clearly correlated with the inhibition of liver metastases.[33] The use of liposomes as drug carriers is worth a brief comment here since these lipid vesicles are primarily taken up by liver and spleen macrophages when given parenterally.[34] Compared with free MDP, encapsulation of MDP into liposomes not only resulted in a 500-fold reduction in the amount of MDP needed to achieve the same level of cytotoxicity, but also increased the highest obtainable level of killing by more than twofold.[29] It appears that liposome-MDP after uptake by KC serves as a sustained drug-release vehicle and maintains a high intracellular MDP concentration for a prolonged period of time. Chemotherapy or immunotherapy employing liposomes as carriers would therefore appear most appealing for the treatment of primary or metastatic liver tumors.[33,35]

Besides functioning as cytotoxic effector cells against tumor targets, KC are able to secrete a multitude of biologically active mediators which then directly act upon tumor cells or activate other effector cell populations. In response to various stimuli, KC have been reported to produce interleukin 1 (IL-1),[36] tumor necrosis factor (TNF),[1,37] IL-6,[38] interferon-α/β (IFN-α/β),[39-41] a series of eicosanoids,[26,37] and active oxygen radicals.[42] While most of these observations were made in rodents, Leser et al.[43] have demonstrated recently that freshly isolated human KC, like their rodent counterparts, produced IL-1 in response to LPS, but not IFN-γ. IL-1, IL-6, TNF, and IFN-α/β not only are cytotoxic and/or cytostatic against certain tumor cells, but also modify the activity of NK cells, lymphokine-activated killer (LAK) cells, cytotoxic T lymphocytes, and macrophages themselves. Furthermore, the anatomical proximity may allow KC to modulate the responses of other resident liver cells through these secretory factors. For instance, IL-1, IL-6, and TNF together can induce hepatocytes to produce the full spectrum of acute-phase reactants.[43] TNF also binds to the surface receptor on endothelial cells, rendering them adhesive to granulocytes.[37] Conceivably, the interplay among different cell populations involving a complex cytokine circuit regulates normal physiological as well as pathological processes taking place in the liver. Further discussion on the aspect of cellular interaction will be elaborated upon later in this chapter.

B. NATURAL KILLER CELLS

NK activity is functionally defined as the rapid (3 to 6 h), nonmajor-histocompatibility (MHC)-restricted killing of certain tumor cells in the absence of prior sensitization.[45] In the peripheral blood and spleen, this activity is associated predominantly with a morphologically distinct cell type, the large granular lymphocytes (LGL). In 1982, our laboratory[7] first showed that NPC from untreated rats exerted strong natural cytotoxicity against YAC-1 and P815 tumor cells in a 4-h ^{51}Cr-release assay. This natural cytotoxicity was apparently due to a type of activated NK activity. Later, Kaneda et al.[45] found that pit cells, first described by Wisse et al.[46] as one component of the liver sinusoids, corresponded morphologically to LGL and carried out NK cytotoxicity. Therefore, pit cells, NK cells, and LGL refer largely to the same type of cells and will be used interchangeably here. Liver-associated NK cells adhere to or are embedded in the endothelial linings and are characterized by such prominent features as cell polarity, dense granules, and rod-core vesicles.[47] Similar to the spleen, murine hepatic NK activity differs among different strains (C3H/He > C57BL/6 > DBA/2) and also exhibits an age-related variation, reaching a maximum between 6 and 8 weeks of age and then declining.[48] As mentioned above, liver-associated NK cells rapidly lyse YAC-1 lymphoma cells *in vitro*. The killing of DHD-K12 colon adenocarcinoma cells, however, requires coincubation for 16 to 20 h.[49] Therefore, these NK cells also possess natural cytotoxic (NC) activity, defined as the cytotoxicity to solid tumor cells requiring at least 16 h incubation time. In contrast, KC are unable to kill either YAC-1 or DHD-K12 cells.

Experimental evidence suggests that NK cells are important effector cells in controlling hematopoiesis, infections (bacteria, parasites, and viruses), and surveillance against tumor cells.[50] The primary role of these cells in neoplasia appears to be directed against blood-borne tumor cells during the intravascular phase of tumor metastasis. The number of metastatic foci in the lung and liver increased drastically after intravenous injection of tumor cells if the NK activity of the animals was ablated by anti-asialo GM_1 antiserum (anti-$AsGM_1$).[51,52] On the other hand, reconstitution of NK-depleted rats with purified LGL restored both the tumor cell clearance and antimetastatic efficiency.[52] More recently, it was proposed that "organ-associated NK cells" in nonlymphoid organs such as liver and lung might play a role in preventing metastasis formation during the extravasation and early postextravasation phase of metastatic cascades.[53] Microscopic examination of the liver at various times following portal vein injection of tumor cells supports this *in vivo* defensive role of pit cells.[54] Pit cells adhere to YAC-1 tumor cells inside the sinusoids and their cytoplasmic projections protrude into the adjacent tumor cells 1 h after injection. Degenerative changes are also observed in some tumor cells.

In addition to harboring the largest macrophage population, liver also contains the largest proportion of NK cells in the body. We have examined NK activity in various organs.[9] Rat NPC contained an average of 13% LGL, whereas LGL accounted for only 2 to 6% of spleen cells and peripheral blood lymphocytes. The highest level of NK cytotoxicity was also observed in the liver LGL. These findings were later confirmed by Ito et al.[48] in mice. *In vivo,* liver NK cells can be further activated by various immunomodulators. A single injection of polyinosic-polycytidylic acid (polyI:C), IFN-α A/D, IFN-γ, IL-2, MVE-2, or bacterial products, such as *Propionibacterium acnes* and OK-432, drastically boosted NK activity in the liver as well as in the spleen, lung, and blood.[53,55-57] Again, the magnitude of NK augmentation in the liver was greater than in all other tissue sites examined. More importantly, the activation of hepatic NK activity was paralleled by the inhibition of liver metastasis. Interestingly, macrophage-activating agents such as liposome-encapsulated MTP-PE, which had no effect on NK activity *in vitro,* when administered *in vivo* markedly activated NK cells as well as tumoricidal macrophages in the liver.[57] One explanation is that hepatic NK cells are under the influence of neighboring KC and are stimulated by KC-derived cytokines, most likely IFN-α/β, as suggested by us previously.[40] Organ-associated NK activity can be augmented by two mechanisms, by further activation of the functional level of NK cells and by an increase in the number of LGL. After receiving various biological response modifiers, the total number of hepatic LGL recovered from treated mice reflected a 10- to 50-fold increase.[58] The origin of these LGL was recently investigated by Wiltrout et al.[59] MVE-2 and *P. acnes* enhanced liver NK activity largely through recruitment and accumulation in the liver of LGL newly derived from bone marrow. Conversely, in the case of IL-2, the relative contribution of local proliferation seemed to be comparable to bone marrow recruitment, perhaps due to the ability of IL-2 to directly induce LGL to proliferate.

NK activity is a functional definition. Although most of this activity is associated with the $CD3^-$ LGL, cellular heterogeneity clearly exists.[45] For example, the $CD3^+$, $Leu19^+$ T cells represent another small subset of the NK cell population. Recently, Lohmann-Matthes and co-workers[13] reported that a pool of immature macrophage precursor cells, or promonocytes, were present in normal mouse liver. These cells bear the macrophage-specific F4/80 antigen and respond to macrophage growth factor CSF-1 by differentiating into mature macrophages. Unlike mature macrophages, promonocytes are nonphagocytic, nonadherent, nonspecific esterase negative, and are characterized functionally by their NK activity. During isolation, they are copurified with LGL because of similar density, and a portion (about 30%) of them express the NK cell marker, NK1.1. Whether or not these promonocytes have an antimetastatic role *in vivo* has not yet been established.

C. HEPATOCYTES

Parenchymal liver cells, or hepatocytes, form the bulk of the liver mass, constituting an estimated 92.5% of total hepatocellular volume.[60] The physiological functions of these cells are probably the most diversified in the body, including metabolism, storage, detoxification of drugs and xenobiotics, and production of many plasma proteins. Traditionally, hepatocytes are not regarded as a component of the host immune defense system. In fact, they are targets for attack in some immune responses and are lysed by neutrophils[61] and LAK cells.[62] Besides vascular leak syndrome, hepatotoxicity is the other major side effect of adoptive immunotherapy with IL-2 and LAK cells. It has been suggested by Gately et al.[63] that IL-2-activated NK cells mediate the cytotoxicity either directly or indirectly against hepatocytes, since anti-$AsGM_1$

can alleviate the IL-2-induced hepatotoxicity. The fact that hepatocytes are frequent targets for cytotoxic effector cells does not preclude these cells from actively participating in immune responses. A number of hepatocyte-derived protein factors have been shown to possess immunomodulatory properties. These factors belong to one of the following three categories: (1) Normal secretory products. Various serum lipoproteins, in particular very low-density lipoproteins (VLDL), suppress mitogen-induced lymphocyte proliferation.[64] Since the inhibitory activity is apparent at physiological and subphysiological concentrations, Chisari[64] suggested that these immunoregulatory lipoproteins may function in the continuing maintenance of immunological homeostasis; (2) Inducible factors. Acute-phase proteins are produced under circumstances where physiological homeostasis is disturbed due to infection, tissue injury, immunological disorders, and tumor growth. Several acute-phase reactants such as α_1-acid glycoprotein,[65] α-fetoprotein,[66] and C-reactive protein (CRP)[67] suppress a number of immune functions *in vitro,* including: (a) blastogenic response to T and B cell mitogens; (b) antibody response to sheep red blood cells; and (c) induction of cytotoxic T cells against allogeneic tumor cells. Paradoxically, CRP delivered in multilamellar liposomes inhibits liver metastasis in C57BL/6 mice bearing MCA-38 colon adenocarcinoma cells, presumably through activation of NK-like cells.[68] Hepatocytes have been recently shown to produce IL-8 in response to the inflammatory mediators IL-1α, IL-1β, and TNF.[69] IL-8 is a potent neutrophil chemotactic factor and may be involved in neutrophil-mediated inflammatory process in the liver; (3) Cytosolic proteins. Arginase, a key enzyme in the urea cycle, inhibits mitogen-induced lymphocyte transformation.[70,71] It is normally not released by hepatocytes. However, the serum level of arginase increases significantly following liver injury.[70] It appears that arginase is released only after hepatocyte damage[71] and exerts its inhibitory effect through depletion of arginine, an essential nutrient for cell proliferation.

Obviously, hepatocytes not only play a passive role, but can potentially modify intrahepatic and systemic immune responses through release of mediators. Most of the hepatocyte-derived immunoregulatory factors appear to downregulate immune functions, perhaps to prevent the host immune responses from proceeding unchecked and thus to limit the extent of liver damage. However, host antitumor response mounted against growing or metastatic tumor cells could also be compromised by these factors. On the other hand, in addition to the immunosuppressive effect, many of these liver-derived factors potently inhibit the replication of tumor cells *in vitro*. Arginase depletes arginine and is a relatively nonspecific inhibitor. Another factor, the "invasion-inhibitory factor" (IIF), has been partially purified from normal rat liver.[72] IIF is unique in that it does not inhibit the growth of AH-130 hepatoma cells, but suppresses their migration and invasion through the mesothelial monolayer. Furthermore, a concomitant intraperitoneal injection of IIF with AH-130 tumor cells prevented the seeding of the peritoneum by these tumor cells.

In summary, hepatocytes can potentially produce a variety of mediators and alter the host-tumor relationships through modification of host immune responses and by a direct inhibition of the growth or invasion of tumor cells. Our current knowledge does not allow a prediction as to whether the balance will become tipped in favor of the host or the tumors *in vivo*. Further characterization of various hepatocyte-derived immunoregulatory factors is needed to address this important question.

III. *IN VIVO* MODELS FOR LIVER METASTASIS

As discussed above, isolated NPC under various experimental conditions exert cytostatic and/or cytotoxic activity against certain tumor cells. Although it is highly suggestive that these cells may assume an antitumor role *in vivo*, a simple extrapolation from *in vitro* findings to *in vivo* situations could be misleading. For example, Gresser et al.[73] had isolated an IFN-α/β-sensitive clone, FLC-745, and an IFN-α/β-resistant clone, FLC-3CL-8, from Friend erythroleukemia cells. Administration of IFN-α/β to mice bearing either FLC-745 or FLC-3CL-8 tumor resulted in a significant inhibition of tumor formation in the liver and a marked increase in survival. Thus, prediction of the therapeutic efficacy of IFN-α/β simply based on *in vitro* sensitivity to IFN led to an erroneous conclusion under their experimental conditions. Hence, more relevant animal models are needed to further explore the hepatic antitumor defense system *in vivo*.

In collaboration with Goldrosen,[74,75] we have developed a murine model of hepatic metastasis in which murine colon adenocarcinoma cells, MCA-38, are injected into the ileocolic vein (ICV) and subsequently observed to colonize the liver of C57BL/6 mice in a reproducible fashion. The hepatic tumor foci formed after this experimental procedure are evident microscopically 3 to 11 d after injection and are visible to the eye 14 to 21 d after injection. In this unique model, the number of metastatic foci can be quantitated easily and correlates inversely with the survival of the mice. Since the ICV route more closely follows the normal pathway of metastatic colon cancer cells and requires no splenectomy after tumor inoculation, it has an advantage over another murine model where liver foci were induced by intrasplenic injection of tumor cells.[76]

In an attempt to dissect the components of the hepatic antitumor defense system, we have applied this model to study the colonization of MCA-38 cells in immunocompetent, immunostimulated, and congenitally immunodeficient mice.[77] NPC from conventional C57BL/6 mice had minimal cytotoxicity against MCA-38 cells, but did exert a moderate (30 to 40%) cytostatic activity toward these cells (Table 1). Treatment with anti-AsGM$_1$ abrogated the tumoristatic effect and reduced the survival time from 35 to 17 d. The livers from T cell-deficient C57BL/6.nu/nu mice cleared tumor cells much faster than the liver of conventional mice (60 vs 27%, 24 h after tumor cell

TABLE 1
Experimental Liver Metastasis of MCA-38 Colon Adenocarcinoma Cells in Conventional Mice or Nude Mice

Mice	Treatment	Clearance[a]	Number of tumor foci[b] (mean)	Mean survival[b] (day)	Cytotoxic activity of NPC[c]	Cytostatic activity of NPC[c]
C57BL/6						
	None	27%	225	35	—	+
	Anti-AsGM$_1$		400	17	—	—
	Poly I:C		2	>64	+ +	+ + +
	Anti-AsGM$_1$ + Poly I:C		400	24	—	—
C57BL/6-nu/nu						
	None	60%	0	>60	+ +	+ + +
	Anti-AsGM$_1$		400	19	—	+

[a] ^{111}In-MCA-38 tumor cells (2.5 × 10^5) were injected into ileocolic vein (ICV) of mice and the amount of radioactivity in the liver was assessed 1 and 24 h later. The livers of C57BL/6 or C57BL/6.nu/nu mice incorporated a comparable amount of radioactivity at 1 h. Clearance = 1 radioactivity retained at 24 h/radioactivity retained at 1 h.

[b] Mice were divided into four groups. Each group received treatment of: (1) PBS; (2) 25 μl anti-AsGM$_1$; (3) 200 μg Poly I:C; or (4) anti-AsGM$_1$ + Poly I:C, on day −1, respectively. On day 0, liver-derived MCA-38 tumor cells (2.5 × 10^5) were injected into ICV of C57BL/6 or C57BL/6-nu/nu mice. Half of the mice were laparotomized on day 21 and the number of tumor foci on the liver was counted. Days of survival were recorded for the rest of the mice.

[c] Anti-AsGM$_1$ (25 μl) or Poly I:C (200 μg) was injected intraperitoneally to mice 1 d prior to sacrifice. NPC were collected and their cytotoxicity against MCA-38 was tested in a 5-h ^{51}Cr release assay. For cytostasis assay, NPC were cocultured with MCA-38 for 72 h and pulsed with [^{3}H]-TdR for 6 h.

injection). Moreover, NPC from nude mice exhibited cytotoxic as well as strong cytostatic effects against MCA-38, and the inoculated tumor cells failed to form metastatic foci in the liver. Both the antitumor potency of NPC and the prolonged survival were abrogated if the nude mice were treated with anti-$AsGM_1$. Results similar to those in the nude mice were observed if the conventional mice were treated with the IFN inducer, polyI:C, on either day 1 or day 8 after tumor injection. Taken together, these data indicate that $AsGM_1^+$ cells in the liver from conventional mice do not prevent the implantation of colon cancer cells, but have the capacity to lower the rate of focal tumor growth. These cells are nonadherent and carry out NK activity. In contrast, $AsGM_1^+$ liver cells from nude or polyI:C-treated mice appear to be "activated" and can suppress tumor implantation and early tumor growth. We found that these cells are also functioning in NK-deficient beige mice,[78] suggesting that they are different from the classical NK cells.

Experimental liver metastasis models are also important in developing effective treatment for liver tumors. Rosenberg et al.[79-81] have extensively experimented with a variety of immunotherapeutic regimens. Cytokines, including IL-1, IL-2, IL-4, IL-6, TNF, and IFN, alone or in combination with adoptive transfer of LAK cells or tumor-infiltrating lymphocytes, have been tested in murine models and have demonstrated a striking efficacy in reducing the number of both microscopic and macroscopic liver metastatic foci. Unfortunately, the great hope generated from these successful animal experiments has not been realized so far in cancer patients. The outcome of most clinical trials is less than satisfactory. Tumors responsive to IL-2/LAK therapy include renal cancer, melanoma, and some types of leukemia and lymphoma. Primary or metastatic liver tumors are usually refractory. In many clinical trials, the treatment was monitored only by assessing the cytotoxic activity or cytokine production of lymphoid cells that were easily accessible, such as peripheral blood lymphocytes, which might not reflect the actual immune responses in the organ where the tumor was located. Since tumor cells, at any stage of their progression, are constantly interacting with the surrounding environment, an understanding of the complex interaction among various host cellular components and the tumor cells is essential in elucidating the antitumor responses at the organ level. Such knowledge can then be utilized to optimally activate the organ-associated immune defense system required to effectively combat the tumor cells.

IV. INTERACTION OF NONPARENCHYMAL LIVER CELLS WITH HEPATOCYTES

During the past 2 decades, increasing attention has been paid to the cells residing in hepatic sinusoids. Morphologic studies first established that the liver sinusoids are normally inhabited by four types of cells: Kupffer, endothelial, fat-storing, and pit cells.[2,10] Subsequently, the development of various isolation techniques provided investigators with purified preparations of cells for functional characterization. More recently, the importance of the cooperation between different NPC and hepatocytes, under both physiological and pathological states, including metabolism, sepsis, and tumor metastasis, has been recognized. This cooperation is a complex bidirectional process involving a multitude of mediator. Sinusoidal liver cells and their neighboring hepatocytes therefore form not only a histological unit, but a functional one as well.

Previously, this laboratory[8] had shown that murine NPC lysed the standard NK target, YAC-1 cells, in 4 h and the NK-resistant target P815 cells in 18 h. A series of experiments were conducted and led to the conclusion that endogenous IFN-α/β production by KC augmented the lytic activity of liver-associated NK cells.[40] KC appeared to be in a heightened state of activation when compared to blood monocytes, splenic, and peritoneal macrophages. This was attributed to the effect of gut-derived endotoxin reaching the liver through the portal circulation. In response to minute levels of endotoxin normally present in portal serum or fetal bovine serum, KC produced greater than 20-fold more IFN-α/β and prostaglandin E_2 (PGE_2) in 24-h culture than did spleen cells.[82] IFN-α/β secreted by KC exerts pleiotropic effects in the liver. Gresser et al.[83] have demonstrated that antibody to IFN-α/β abrogates the resistance to the multiplication of leukemia cells in the liver of allogeneic mice. They suggested that endogenous IFN-α/β induced by tumor cells is a crucial component in hepatic defense against tumor cell proliferation. IFN-α/β also plays an autoregulatory role in the liver.[41] As shown in Table 2, addition of 200 U/ml anti-IFN-α/β antiserum significantly increases the amount of IL-1, TNF, and PGE_2, produced by NPC, indicating that the production of these KC-derived cytokines is downregulated by IFN-α/β. Since IL-1 and TNF are known inducers of acute-phase proteins, IFN-α/β may also dampen acute-phase responses.

In vitro culture of spleen cells or peripheral blood lymphocytes with IL-2 gives rise to LAK cells that are able to lyse both NK-sensitive and NK-resistant tumor cells.[84] LAK activity can also be recovered from various organs, including the liver, after repeated IL-2 administration to mice.[85] We have demonstrated, however, that KC produced IFN-α/β and PGE_2 in culture and profoundly inhibited LAK cell induction from NPC.[82] This *in vitro* observation was further corroborated as occurring *in vivo*. Although NPC from IL-2-treated mice displayed moderate cytotoxicity against NK-resistant tumor targets, a combination of indomethacin and anti-IFN-α/β antibody with IL-2 was required to generate maximal LAK activity from the liver. Despite the fact that our findings suggest an antagonistic role of IFN-α/β on the therapeutic effects of IL-2/LAK cells, the results from animal experiments have been variable. Compared with IL-2 treatment alone, combination with IFN-α inhibited the IL-2-induced lymphoid cell proliferation in the liver by nearly 70% on day 2, while a longer treatment schedule (7 d) resulted in an augmented

TABLE 2
Effect of IFN-α/β on the Production of IL-1, TNF, and PGE_2

In vitro treatment	Assay
	IL-1 (cpm × 10^{-3})[a]
—	29
Anti-IFN-α/β (200 U/ml)	147
	TNF (% specific lysis)[b]
—	8
LPS (1 μg/ml)	14
LPS (1 μg/ml) + Anti-IFN-α/β (200 U/ml)	31
	PGE_2 (pg/ml)[c]
—	2,520
IL-2 (10 U/ml) + Anti-IFN-α/β (200 U/ml)	4,260
	10,400

Note: NPC were isolated from C57BL/6 mice and cultured in 10% fetal bovine serum. LPS or IL-2 was added as stimulant in the presence or absence of anti-IFN-α/β antibody. Supernatant was removed 4, 18, and 24 h later for TNF, IL-1, and PGE_2 assay, respectively. A fourfold dilution of supernatant was used in these assays.

[a] IL-1 activity was assayed by a thymocyte coproliferation assay.
[b] TNF activity was determined by a 6-h ^{51}Cr release assay on WEHI-164 tumor cells which were pretreated with actinomycin-D (1 μg/ml) for 3 h.
[c] Measured by a radioimmunoassay kit (Du Pont®, Wilmington, DE).

proliferative response.[86] Synergistic antitumor effects against murine hepatic metastases were reported by Cameron et al.[79] when IFN-α was combined with IL-2; however, this synergy was observed only with the weakly immunogenic MCA-106 sarcoma and not with the nonimmunogenic MCA-102, suggesting an indirect role of IFN-α/β in this model. Since IFN-α/β has both direct antitumor effects and indirect immunomodulatory effects, it is conceivable that the actual role of IFN-α/β *in vivo* will depend on the treatment schedule as well as the site and the type of tumor. To summarize, IFN-α/β appears to play a central role in regulating various aspects of immune responses in the liver and serves as a good example to illustrate the complex cellular interaction mediated through different cytokines. Potentially, IFN-α/β could mediate host antitumor response through one or more of the following mechanisms: (1) effects on immune effector cells. NK and LAK cell activity are up- and downregulated by IFN-α/β, respectively; (2) regulation of expression of other cytokines. The production of many cytokines in the liver, including IL-1, TNF, PGE_2 and CSF, is suppressed in the presence of IFN-α/β;

(3) effects on tumor cells. The growth of many tumor cells is directly inhibited by IFN-α/β. IFN-α/β could also modify the expression of major histocompatibility antigens on tumor cell surface, thereby altering their sensitivity to lysis by immune effector cells;[87] and (4) effects on nonimmune cells. In this regard, Yasui et al.[88] recently demonstrated that IFN-α/β induced hepatocytes to produce a soluble factor that was tumoristatic to IFN-resistant Friend erythroleukemia cells. Even though the identity of this factor has not been reported, it nevertheless suggests another potential mechanism whereby IFN-α/β mediates antitumor effect *in vivo*.

Curran et al.[89,91] and Billiar et al.[90] have established a coculture model to study the interaction between KC and hepatocytes. They have shown that hepatocytes produced a soluble mediator that enhanced the tumoristatic and tumoricidal activity of KC. Furthermore, the secretion of other cytokines such as IL-1, TNF, and PGE_2 by KC was upregulated in the presence of hepatocyte culture supernatant. Our laboratory has also been interested in investigating the effects of various resident liver cells on the tumoricidal activity of NPC. Nonadherent NPC, devoid of KC, cultured with IL-2, gave rise to LAK cells.[82] Using IL-2-induced LAK cell induction from nonadherent NPC as a basis of comparison, we have examined the effects of KC and hepatocytes. Instead of incubating two different types of cells in the same well, coculture experiments were carried out in a two-chambered culture plate, where two types of cells under study were separated by a microporous membrane. A dynamic cellular communication through soluble mediators was maintained in this model and a pure population of either cell type could be recovered for further evaluation. As mentioned above, we have shown that KC produced IFN-α/β and PGE_2, which suppressed hepatic LAK cell induction.[82] Hepatocytes and their cytosolic extracts also inhibited LAK cell generation. More importantly, in addition to arginase and VLDL, two previously described hepatocyte-derived immunoinhibitors, we were able to isolate a novel immunoinhibitory protein from murine hepatocyte cytosol.[92] This 28-kDa protein was named liver-derived immunoinhibitory factor (LDIF) and inhibited the proliferation of lymphocytes and certain tumor cells. Since a specific assay is not available to us, it is not known whether LDIF is a normal secretory product like VLDL, or is released only after cell damage as in the case of arginase. We have observed that when experimental liver injury in mice was induced by D-galactosamine, serum levels of arginase and VLDL increased in parallel with the elevation of liver enzyme aspartate aminotransferase.[70] Concurrently, IL-2-induced LAK activity recovered from the liver and the spleen decreased significantly. We surmised that as a result of increased release of hepatocyte-derived immunoregulatory factors, both intrahepatic and systemic immune responses can be modulated following liver injury.

Recently, another effector pathway of activated macrophage has been reported.[93,94] L-Arginine was converted by activated macrophages into NO_2^-/NO_3^- and citrulline. Reactive nitrogen intermediates, in particular nitric oxide, generated during this process inhibited the growth of certain tumor cells. The inhibition was characterized by a reproducible pattern of metabolic changes, including mitochondrial iron loss, respiratory inhibition, and inhibition of DNA synthesis. The L-arginine analog, *N*^{G}-monomethyl-L-arginine, and arginase prevented the development of this effector mechanism, indicating that this pathway is arginine dependent. Billiar et al.[90] have demonstrated that rat KC activated by LPS utilized this arginine-dependent mechanism to inhibit hepatocyte protein synthesis. Interestingly, they observed that hepatocytes, in response to products of activated KC, were also capable of generating nitric oxide from arginine.[91] Therefore, this arginine-dependent effector pathway is employed by both KC and hepatocytes in modulating hepatocyte function and inhibiting tumor growth.

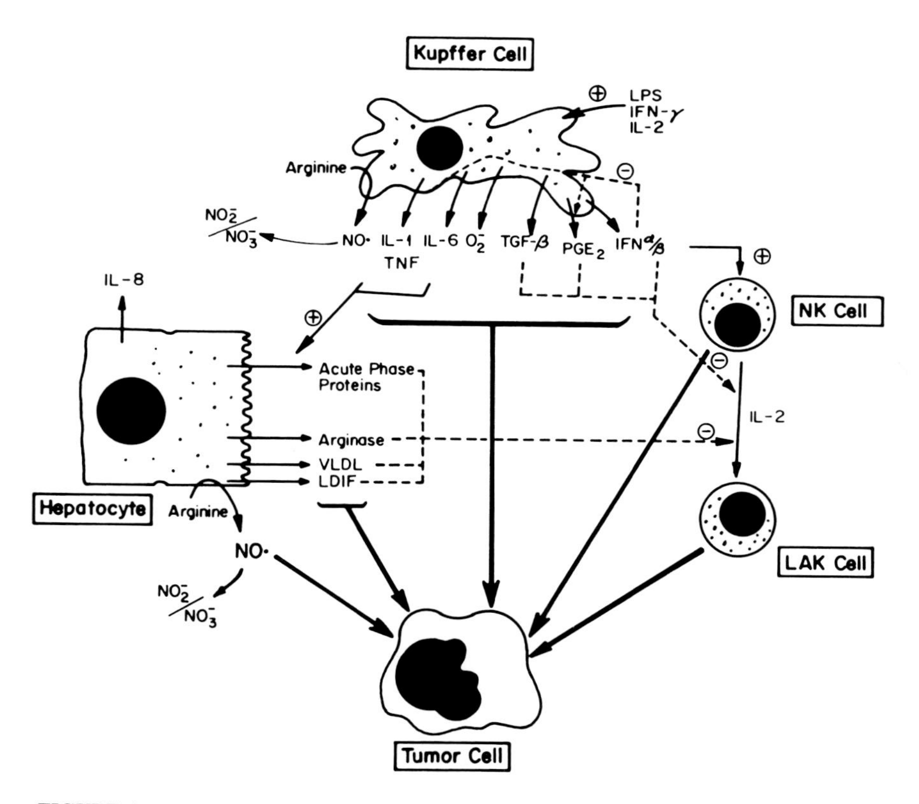

FIGURE 1. Hepatic antitumor responses involve interaction of different types of liver cells and a multitude of mediators.

V. SUMMARY

By considering experimental evidence discussed above, we have constructed a diagram to illustrate the complex cellular and cytokine circuits in the liver (Figure 1). The depiction is, of course, oversimplified and speculative, since many observations were reported only *in vitro*. In this diagram, KC and NK cells (including NK-derived LAK cells) are two antitumor effector cells. A number of factors produced by KC and hepatocytes inhibit the growth of tumor cells. KC plays a central regulatory role in that many of the KC-derived cytokines modulate the function of NK cells and hepatocytes. IFN-α/β augments NK while it suppresses LAK cell induction. IFN-α/β autoregulates KC function by dampening the production of several KC-derived cytokines, including IL-1, TNF, and PGE_2. IL-1, TNF, and IL-6 induce the secretion of acute-phase proteins, some of which are immunoinhibitory, e.g., CRP, α-fetoprotein, and α_1-acid glycoprotein. Hepatocytes also produce other immunoregulatory factors such as VLDL, arginase, and LDIF. The arginine-dependent pathway which generates nitric oxide is utilized by both activated KC and hepatocytes. Arginase can inhibit this mechanism by depletion of arginine. It is evident from this figure that one stimulus sets into motion a cascade of events that can further strengthen or offset other responses in the circuit. The ultimate outcome would be difficult to predict. For example, arginase by itself is inhibitory to tumor growth. However, the induction of LAK cells and the arginine-dependent effector mechanism are suppressed by arginase. *In vivo,* whether release of arginase is beneficial to the host or to the tumors remains unclear. To further complicate the situation, under various inflammatory conditions, extrahepatic recruitment of immune cells will certainly bring about an alteration in the hepatic environment.

Our knowledge of the cellular interactions pertaining to hepatic antitumor response is rudimentary. We are clearly far from realizing the exact processes of how tumor cells metastasize to the liver and grow, or instead are successfully eliminated from the liver. Our discussion summarizes many recent studies and serves to reveal the complex nature of this issue. Hopefully, continuing investigation will unravel the detailed mechanisms of hepatic antitumor response, the knowledge of which will lead to more efficient therapies against liver tumors, primary or metastatic.

REFERENCES

1. **Nolan, J. P. and Cohen, S. A.,** Interaction of endotoxin with sinusoidal cells of the liver, in *Sinusoidal Cells in Health and Disease,* P. Bioulac and C. Balabaut, Eds., Elsevier, Amsterdam, 1989, 231.
2. **Wisse, E.,** Ultrastructure and function of Kupffer cells and other sinusoidal cells in the liver, in *Kupffer Cells and Other Liver Sinusoidal Cells,* E. Wisse and D. L. Knook, Eds., Elsevier, Amsterdam, 1977, 33.
3. **Cady, B.,** Natural history of primary and secondary tumors of the liver, *Semin. Oncol.,* 86, 550, 1983.
4. **Foster, J. H. and Lundy, J.,** Liver metastases, *Curr. Prob. Surg.,* 18, 160, 1981.
5. **Wisse, E., De Zanger, R., and Jacobs, R.,** Scanning electromicroscopic observation on rat liver sinusoids relevant to microcirculation and transport processes, *J. Clin. Electron. Microsc.,* 16, 427, 1983.
6. **Riccardi, C., Santoni, A., Barlozzari, T., Puccetti, P., and Herberman, R. B.,** In vivo natural reactivity of mice against tumor cells, *Int. J. Cancer,* 25, 475, 1980.
7. **Cohen, S. A., Salazar, D., and Nolan, J. P.,** Natural cytotoxicity of isolated rat liver cells, *J. Immunol.,* 129, 495, 1982.
8. **Cohen, S. A., Salazar, D., von Muenchhausen, W., Werner-Wasik, M., and Nolan, J. P.,** Natural antitumor defense system of the murine liver, *J. Leuk. Biol.,* 37, 559, 1985.
9. **Leung, K., Salazar, D., Ip, M., and Cohen, S. A.,** Characterization of natural cytotoxic effector cells isolated from rat liver, *Nat. Immun. Cell Growth Reg.,* 6, 150, 1987.
10. **Kaneda, K., Dan, C., and Wake, K.,** Pit cells as natural killer cells, *Biomed. Res.,* 4, 567, 1983.
11. **Malter, M., Friedrich, E., and Suss, R.,** Liver as a tumor cell killing organ: Kupffer cells and natural killers, *Cancer Res.,* 46, 3055, 1986.
12. **Decker, T., Baccarini, M., and Lohmann-Matthes, M.,** Liver-associated macrophage precursors as natural cytotoxic effectors against Candida albicans and Yak-1 cells, *Eur. J. Immunol.,* 16, 693, 1986.
13. **Baccarini, M., Hao, L., Decker, T., and Lohmann-Matthes, M.,** Macrophage precursors as NK cells against tumor cells, *Natl. Immun. Cell Growth Reg.,* 7, 316, 1988.
14. **Crofton, R. W., Diesselhoff-Den Dulk, M. H. C., and van Furth, R.,** The origin, kinetics and characterization of the Kupffer Cell in the steady state, *J. Exp. Med.,* 22, 1, 1978.
15. **Portman, B., Schindler, A. M., Murray-Lyon, I. M., and Williams, R.,** Histological sexing of a reticulum cell sarcoma arising after liver transplantation, *Gastroenterology,* 70, 82, 1976.
16. **Bouwens, L., Baekeland, M., and Wisse, E.,** Cytokinetic analysis of the expanding Kupffer cell population in rat liver, *Cell Tissue Kinet.,* 19, 217, 1986.
17. **Bouwens, L., Knook, D. L., and Wisse, E.,** Local proliferation and extrahepatic recruitment of liver macrophages (Kupffer cells) in partially irradiated rats, *J. Leuk. Biol.,* 39, 687, 1986.

18. **Decker, T., Lohmann-Matthes, M., and Baccarini, M.,** Liver-associated macrophage precursor cells proliferate under impairment of regular hematopoiesis, *Eur. J. Immunol.,* 18, 697, 1988.
19. **Aterman, A.,** The Kupffer cell: a short historical view, in *Kupffer Cells and Other Liver Sinusoidal Cells,* E. Wisse and D. L. Knook, Eds., Elsevier, Amsterdam, 1977, 1.
20. **Roos, E. and Dingemans, K. P.,** Phagocytosis of tumor cells by Kupffer cell in vivo and in the perfused mouse liver, in *Kupffer Cells and Other Liver Sinusoidal Cells,* E. Wisse and D. L. Knook, Eds., Elsevier, Amsterdam, 1977, 184.
21. **Gardner, C. R., Wasserman, A. J., and Laskin, D. L.,** Differential sensitivity of tumor targets to liver macrophage-mediated cytotoxicity, *Cancer Res.,* 47, 6686, 1987.
22. **Decker, T., Kiderlen, A. F., and Lohmann-Matthes, M.,** Liver macrophages (Kupffer cells) as cytotoxic effector cells in extracellular and intracellular cytotoxicity, *Infect. Immun.,* 50, 358, 1985.
23. **Xu, Z. L., Bucana, C. D., and Fidler, I. J.,** In vitro activation of murine Kupffer cell by lymphokines or endotoxins to lyse syngeneic tumor cells, *Am. J. Pathol.,* 117, 372, 1984.
24. **Bodenheimer, H. C., Mulligan, J., and Charland, C.,** Cytotoxicity of resident rat Kupffer cell and peritoneal macrophage against syngeneic hepatocellular carcinoma in vitro, *Cells Hepatic Sinusoids,* 2, 235, 1988.
25. **Sherwood, E. R., Williams, D. L., McNamee, R. B., Jones, E. L., Browder, I. W., and DiLuzio, N. R.,** In vitro tumoricidal activity of resting and glucan-activated Kupffer cells, *J. Leuk. Biol.,* 42, 69, 1987.
26. **Decker, T., Lohmann-Matthes, M., Karck, U., Peters, T., and Decker, K.,** Comparative study of cytotoxicity, tumor necrosis factor and prostaglandin release after stimulation of rat Kupffer cells, murine Kupffer cells and murine inflammatory liver macrophages, *J. Leuk. Biol.,* 45, 139, 1989.
27. **Roh, M. S., Oyedeji, C., Wang, L., Curley, S. A., and Klostergaard, J.,** Human Kupffer cells are cytotoxic in vitro against human cancer, in *Proc. 5th Int. Symp. Cells Hepatic Sinusoids,* E. Wisse, T. Decker and R. S. McCusky, Eds., Kupffer Cell Foundation, Rijswijk, Netherlands, 1990, 117.
28. **Robb, R. J. and Rusk, C. M.,** High and low affinity receptors for interleukin 2: implications of pronase, phorbol ester, and cell membrane studies upon the basis for differential ligand affinities, *J. Immunol.,* 137, 142, 1986.
29. **Daeman, T., Veninga, A., Roerdink, F. H., and Scherphof, G. L.,** In vitro activation of rat liver macrophages to tumoricidal activity by free or liposome-encapsulated muramyl dipeptide, *Cancer Res.,* 46, 4330, 1986.
30. **Keller, F., Wild, M., and Kirn, A.,** In vitro cytostatic properties of unactivated rat Kupffer cells, *J. Leuk. Biol.,* 35, 467, 1984.
31. **Zang, S. R., Salup, R. R., Urias, P. E., Twilley, T. A., Talmadge, J. E., Herberman, R. B., and Wiltrout, R. H.,** Augmentation of natural killer activity and/or macrophage-mediated cytotoxicity in the liver by biological response modifiers including interleukin 2, *Cancer Immunol. Immunother.,* 21, 19, 1986.

32. **Xu, Z. and Fidler, I. J.,** The in situ activation of cytotoxic properties in murine Kupffer cells by the systemic administration of whole Mycobacterium bovis organisms or muramyl tripeptide, *Cancer Immunol. Immunother.*, 18, 118, 1984.
33. **Phillips, N. C. and Tsao, M. S.,** Inhibition of experimental liver tumor growth in mice by liposomes containing a lipophilic muramyl dipeptide, *Cancer Res.*, 49, 936, 1989.
34. **Ellens, H., Morselt, H. W. M., and Scherphof, G. L.,** In vivo fate of large unilamellar sphingomyelin-cholesterol liposomes after intraperitoneal and intravenous injection into rats, *Biochim. Biophys. Acta*, 674, 10, 1981.
35. **Mayhew, E. G., Goldrosen, M. H., Vaage, J., and Rustum, Y. M.,** Effects of liposome-entrapped doxorubicin on liver-metastasis of mouse colon carcinoma 26 and 38, *J. Natl. Cancer Inst.*, 78, 707, 1987.
36. **Shirahama, M., Ishibashi, H., Tsuchiya, Y., Kurokama, S., Hayashida, Y., Okumura, Y., and Niho, Y.,** Kupffer cells may autoregulate interleukin 1 production by producing interleukin 1 inhibitor and prostaglandin E_2, *Scand. J. Immunol.*, 28, 719, 1988.
37. **Decker, T.,** Hepatic mediators of inflammation, in *Cells of the Hepatic Sinusoids*, Vol. 2, E. Wisse, D. L. Knook, and K. Decker, Eds., Kupffer Cell Foundation, Rijswijk, Netherlands, 1988, 171.
38. **Andus, T., Geiger, T., Hirano, T., Kishimoto, T., Tran-Thi, T. A., Decker, T., and Heinrich, P. C.,** Regulation of synthesis and secretion of major rat acute phase proteins by recombinant IL-6 (BSA-2/IL-6) in hepatocyte culture, *Eur. J. Immunol.*, 173, 287, 1988.
39. **Kirn, A., Kowhren, F., and Steffan, A. M.,** Interferon synthesis in primary culture of Kupffer and endothelial cells from the rat liver, *Hepatology*, 2, 670, 1982.
40. **Werner-Wasik, M., von Muenchhausen, W., Nolan, J. P., and Cohen, S. A.,** Endogenous interferon α/β produced by murine Kupffer cells augments liver-associated natural killer activity, *Cancer Immunol. Immunother.*, 28, 107, 1989.
41. **Tzung, S. P. and Cohen, S. A.,** Endogenous interferon α/β produced by Kupffer cells inhibit interleukin-1 and tumor necrosis factor α production and interleukin-2-induced activation of nonparenchymal liver cells, *Cancer Immunol. Immunother.*, 34, 150, 1991.
42. **Laskin, D. L., Sirak, A. A., Pilaro, A. M., and Laskin, J. D.,** Functional and biochemical properties of rat Kupffer cells and peritoneal macrophages, *J. Leuk. Biol.*, 44, 71, 1988.
43. **Leser, H. G., Andreesen, R., and Gerok, W.,** Interleukin-1 production of differently stimulated Kupffer cells isolated from murine and human livers, in *Cells of the Hepatic Sinusoids*, Vol. 2, E. Wisse, D. L. Knook, and K. Decker, Eds., Kupffer Cell Foundation, Rijswijk, Netherlands, 1988, 179.
44. **Ramadori, G., van Damme, J., Rieder, H., and Meyer zum Buschenfelde, K. H.,** Interleukin 6, the third mediator of acute phase reaction, modulates hepatic protein synthesis in human and mouse. Comparison with IL-1 β and TNF α, *Eur. J. Immunol.*, 18, 1259, 1988.
45. **Ortaldo, J. R.,** Regulation of natural killer activity, *Cancer Metastasis Rev.*, 6, 637, 1987.

46. **Wisse, E., Van't Noordende, J. M., Van der Meulen, J., and Daems, W. T.,** The pit cells: description of a new type of cell occurring in rat liver sinusoids and peripheral blood, *Cell Tissue Res.*, 173, 432, 1976.
47. **Kaneda, K.,** Liver-associated large granular lymphocytes: morphological and functional aspects, *Arch. Histol. Cytol.*, 52, 447, 1989.
48. **Ito, H., Abo, T., Sugawara, S., Kanno, A., and Kumagai, K.,** Age-related variation in the proportion and cytotoxicity of murine liver natural killer cells and their cytotoxicity against regenerating hepatocytes, *J. Immunol.*, 141, 315, 1988.
49. **Bouwens, L., Jacobs, R., Remels, L., and Wisse, E.,** Natural cytotoxicity of rat hepatic natural killer cells and macrophages against a syngeneic colon adenocarcinoma, *Cancer Immunol. Immunother.*, 27, 137, 1988.
50. **Herberman, R. B.,** *NK Cells and Other Natural Effector Cells,* Academic Press, New York, 1982.
51. **Hanna, N. and Burton, R.,** Definitive evidence that natural killer cells inhibit experimental tumor metastasis in vivo, *J. Immunol.*, 121, 1754, 1981.
52. **Barlozzari, T., Leonhardt, J., Wiltrout, R. H., Herberman, R. B., and Reynolds, C. W.,** Direct evidence for the role of NK cells in the inhibition of tumor metastasis, *J. Immunol.*, 134, 2783, 1985.
53. **Wiltrout, R. H., Herberman, R. B., Zhang, S. R., Chirigos, M. A., Ortaldo, J. R., Green, K. M., and Talmadge, J. E.,** Role of organ-associated NK cells in decreasing formation of experimental metastases in lung and liver, *J. Immunol.*, 134, 4267, 1985.
54. **Kaneda, K., Natsuhara, Y., Oka, S., Uehara, K., Yano, I., and Wake, K.,** In vivo defensive role of pit cells in the liver, in *Cells of the Hepatic Sinusoids,* Vol. 2, E. Wisse, D. L. Knook, and K. Decker, Eds., Kupffer Cell Foundation, Rijswijk, Netherlands, 1988, 221.
55. **Talmadge, J. E., Herberman, R. B., Chirigos, M. A., Malwish, A. E., Schneider, M. A., Adams, J. S., Philips, H., Thurman, G. B., Varesio, L., Long, C., Oldman, R. K., and Wiltrout, R. H.,** Hyposensitivity to augmentation of murine NK cell activity in different anatomical compartments by multiple injections of various immunomodulators including recombinant interferons and interleukin 2, *J. Immunol.*, 135, 2483, 1985.
56. **Albini, B., Chen, C., Knoflach, P., Stinson, M. W., and Cohen, S. A.,** Hepatic nonparenchymal liver cells and immune deposits in a murine model of streptococcus-induced liver disease, in *Cells of the Hepatic Sinusoids,* Vol. 2, E. Wisse, D. L. Knook, and K. Decker, Eds., Kupffer Cell Foundation, Rijswijk, Netherlands, 1988, 339.
57. **Talmadge, J. E., Schneider, M., Collins, M., Phillips, H., Herberman, R. B., and Wiltrout, R. H.,** Augmentation of NK cell activity in tissue specific sites by liposomes incorporating MTP-PE, *J. Immunol.*, 135, 1477, 1985.
58. **Wiltrout, R. H., Mathieson, B. J., Talmadge, J. E., Reynolds, C. W., Zhang, S., Herberman, R. B., and Ortaldo, J. R.,** Augmentation of organ-associated NK activity by biological response modifiers, *J. Exp. Med.*, 160, 1431, 1984.
59. **Wiltrout, R. H., Pilaro, A. M., Gruys, M. E., Talmadge, J. E., Longo, D. L., Ortaldo, J. R., and Reynolds, C. W.,** Augmentation of mouse liver-associated NK activity by biological response modifiers occurs largely via rapid recruitment of large granular lymphocytes from the bone marrow, *J. Immunol.*, 143, 372, 1989.

60. **Blouin, A., Bolender, R. P., and Weibel, E. R.,** Distribution of organelles and membranes between hepatocytes and nonhepatocytes in the rat liver parenchyma, *J. Cell. Biol.,* 72, 441, 1977.
61. **Mavier, P., Preaux, A., Guigui, B., Lescs, M., Zaprani, E., and Dhumeaux, P.,** In vitro toxicity of polymorphonuclear neutrophils to rat hepatocytes: evidence for a proteinase-mediated mechanism, *Hepatology,* 8, 254, 1988.
62. **Ono, M., Tanaka, N., and Orita, K.,** Complete regression of mouse hepatoma transplanted after partial hepatectomy and the immunoregulatory mechanism of such regression, *Cancer Res.,* 46, 5049, 1986.
63. **Gately, M. K., Anderson, T., and Hayes, T. J.,** Role of Asialo-GM_1-positive lymphoid cells in mediating the toxic effects of IL-2 in mice, *J. Immunol.,* 141, 189, 1988.
64. **Chisari, F. V.,** Immunoregulatory properties of human plasma very low density lipoproteins, *J. Immunol.,* 119, 2129, 1977.
65. **Bennett, M. and Schmid, K.,** Immunosuppression by human plasma α_1-acid glycoprotein: importance of carbohydrate moiety, *Proc. Natl. Acad. Sci. U.S.A.,* 77, 6109, 1980.
66. **Murgita, R. A. and Tomasi, T. B., Jr.,** Suppression of the immune response by α-fetoprotein. II. The effect of α-fetoprotein on mixed lymphocyte reactivity and mitogen-induced lymphocyte transformation, *J. Exp. Med.,* 141, 440, 1974.
67. **Mortensen, R. F., Osmand, A. P., and Gewurz, H.,** Effects of C-reactive protein on the lymphoid system. I. Binding to thymus-dependent lymphocytes and alteration of their functions, *J. Exp. Med.,* 141, 821, 1974.
68. **Gautam, S. and Deohar, S.,** Generation of tumoricidal effector cells by human C-reactive protein and muramyl tripeptide: a comparative study, *J. Biol. Response Mod.,* 8, 560, 1989.
69. **Thornton, A. J., Strieter, R. M., Lindley, I., Baggiolini, M., and Kunkel, S. L.,** Cytokine-induced gene expression of a neutrophil chemotactic factor/IL-8 in human hepatocytes, *J. Immunol.,* 144, 2609, 1990.
70. **Tzung, S. P. and Cohen, S. A.,** The role of various liver-derived immunoregulatory proteins on IL-2-activated cytotoxicity of nonparenchymal liver cells, in *Cells of the Hepatic Sinusoids,* Vol. 3, E. Wisse, D. L. Knook, and R. S. McCusky, Eds., Kupffer Cell Foundation, Rijswijk, Netherlands, 1991, 315.
71. **Chisari, F. V., Nakamura, M., Milich, D. R., Han, K., Molden, D., and Leroux-Roels, C. G.,** Production of two distinct and independent hepatic immunoregulatory molecules by the perfused rat liver, *Hepatology,* 5, 735, 1985.
72. **Shinkai, K., Mukai, M., Komatsu, K., and Akedo, H.,** Factor from rat liver with antiinvasive potential on rat ascites hepatoma cells, *Cancer Res.,* 48, 3760, 1987.
73. **Gresser, I., Maury, C., Woodrow, D., Moss, J., Grutter, M. G., Vignaux, F., Belardelli, F., and Maunoury, M.,** Interferon treatment markedly inhibits the development of tumor metastases in the liver and spleen and increases survival time of mice after intravenous inoculation of Friend erythroleukemia cells, *Int. J. Cancer,* 41, 135, 1988.
74. **Goldrosen, M. H., Paolini, N., and Holyoke, D.,** Description of a murine model of exprimental hepatic metastases, *J. Natl. Cancer Inst.,* 77, 823, 1986.

75. **Cohen, S. A. and Goldrosen, M. H.,** Modulation of colon-derived experimental hepatic metastases by murine nonparenchymal liver cells, *Immunol. Invest.,* 18, 351, 1989.
76. **Lafreniere, R. and Rosenberg, S. A.,** A novel approach to the generation and identification of experimental hepatic metastases in a murine model, *J. Natl. Cancer Inst.,* 76, 309, 1986.
77. **Cohen, S. A., Tzung, S. P., Doerr, R. J., and Goldrosen, M. H.,** Asialo-GM_1 positive liver cells from athymic nude or polyinosinic-polycytidylic acid-treated mice in suppressing colon-derived experimental hepatic metastasis, *Cancer Res.,* 50, 1834, 1990.
78. **Doerr, R. J., Goldrosen, M. H., and Cohen, S. A.,** Antimetastatic role of the liver from beige mice against the MCA-38 colon-derived tumor, *Proc. Am. Assoc. Cancer Res.,* 31, 247, 1990.
79. **Cameron, R. B., McIntosh, J. K., and Rosenberg, S. A.,** Synergistic antitumor effects of combination immunotherapy with recombinant interleukin 2 and a recombinant hybrid α-interferon in the treatment of established murine hepatic metastases, *Cancer Res.,* 48, 5810, 1988.
80. **Mule, J. J., McIntosh, J. K., Jablons, D. M., and Rosenberg, S. A.,** Antitumor activity of recombinant interleukin 6 in mice, *J. Exp. Med.,* 171, 629, 1990.
81. **Cameron, R. B., Spiess, P. J., and Rosenberg, S. A.,** Synergistic antitumor activity of tumor-infiltrating lymphocytes, interleukin 2, and local tumor irradiation, *J. Exp. Med.,* 171, 249, 1990.
82. **Tzung, S. P., Gaines, K. C., Lance, P., Ehrke, M. J., and Cohen, S. A.,** Suppression of hepatic lymphokine-activated-killer cell induction by murine Kupffer cells and hepatocytes, *Hepatology,* 12, 644, 1990.
83. **Gresser, I., Maury, C., Vignaux, F., Haller, O., Belardelli, F., and Tovey, M. G.,** Antibody to mouse interferon α/β abrogates resistance to the multiplication of Friend erythroleukemia cells in the liver of allogeneic mice, *J. Exp. Med.,* 168, 1271, 1988.
84. **Rosenberg, S. A.,** Lymphokine-activated-killer cells: a new approach to the immunotherapy of cancer, *J. Natl. Cancer Inst.,* 75, 595, 1985.
85. **Ettinghausen, S. E., Lipford, E. H., Mule, J. J., and Rosenberg, S. A.,** Systemic administration of recombinant interleukin 2 stimulates in vivo lymphoid cell proliferation in tissues, *J. Immunol.,* 135, 1488, 1985.
86. **Puri, R. K., Travis, W., and Rosenberg, S. A.,** In vivo administration of interferon α and interleukin 2 induces proliferation of lymphoid cells in the organs of mice, *Cancer Res.,* 50, 5543, 1989.
87. **Goldstein, D. and Laszlo, J.,** Interferon therapy in cancer: from imagination to interferon, *Cancer Res.,* 46, 4315, 1986.
88. **Yasui, H., Proietti, D., Vignaux, F., Eid, P., and Gresser, I.,** Inhibition by mouse α/β-interferon of the multiplication of α/β-interferon-resistant Friend erythroleukemia cells cultured with mouse hepatocytes, *Cancer Res.,* 50, 3533, 1990.
89. **Curran, R. D., Billiar, T. R., Kispert, P. H., Bentz, B. G., May, M. T., and Simmons, R. L.,** Hepatocytes enhance Kupffer cell-mediated tumor cell cytostasis in vitro, *Surgery,* 106, 126, 1989.

90. **Billiar, T. R., Curran, R. D., Stuehr, D. J., Ferrari, F. K., and Simmons, R. L.,** Evidence that activation of Kupffer cells results in production of L-arginine metabolites that release cell-associated iron and inhibit hepatocyte protein synthesis, *Surgery,* 106, 364, 1989.
91. **Curran, R. D., Billiar, T. R., Stuehr, D. J., Hofmann, K., and Simmons, R. L.,** Hepatocytes produce nitrogen oxides from L-arginine in response to inflammatory products of Kupffer cells, *J. Exp. Med.,* 170, 1769, 1989.
92. **Tzung, S. P., Gaines, K. C., Henderson, M., Smith, T. J., and Cohen, S. A.,** Isolation and characterization of a novel liver-derived immunoinhibitory factor (LDIF), *Hepatology,* 14, 888, 1991.
93. **Hibbs, J. B., Varin, Z., and Tainter, R. R.,** L-Arginase is required for expression of the activated macrophage effector causing selective metabolic inhibition in target cells, *J. Immunol.,* 138, 550, 1987.
94. **Stuehr, D. J. and Nathan, C. F.,** Nitric oxide: a macrophage product responsible for cytostasis and respiration inhibition in tumor target cells, *J. Exp. Med.,* 169, 1543, 1989.
95. **Cohen, S., et al.,** manuscript in preparation.

Chapter 5

ALCOHOL-INDUCED LIVER DISEASE

S.N. Wickramasinghe

TABLE OF CONTENTS

ISBN 0-8493-6109-5

I. INTRODUCTION

Despite much research, the mechanisms underlying alcohol-related liver damage remain unclear. For many years, it has been considered likely that the hepatocytes are damaged as some consequence of their extensive capacity to metabolize ethanol. On the basis of the results of biochemical studies on liver cells, it was proposed that hepatocellular damage may result from: (1) alterations of the intrahepatocyte redox state as a consequence of ethanol metabolism[1] or (2) toxic effects of acetaldehyde or reactive oxygen metabolites generated during ethanol metabolism by hepatocytes.[2-6] It has also been suggested that such damage may eventually result in the exposure of neoantigens on the cell surface and the generation of antibodies, and possibly also lymphocytes, which react against hepatocytes.[7,8] Another current hypothesis is that one of the mediators of hepatocyte damage may be circulating acetaldehyde-modified albumin molecules.[9-11]

Until recently, the possibility that Kupffer cells play a role in the pathogenesis of liver disease in chronic alcoholics received little attention. However, the discovery that macrophages (including Kupffer cells) oxidize ethanol and consequently generate a cytotoxic activity *in vitro* suggested that alcohol-related tissue damage, including liver damage, may be partly caused by the effects of macrophage products on surrounding parenchymal cells. This chapter reviews the evidence that Kupffer cells metabolize ethanol and discusses some theoretical mechanisms by which these cells may initiate or aggravate alcohol-induced hepatocyte damage.

II. CAPACITY OF MACROPHAGES TO METABOLIZE ETHANOL

The first indication that macrophages are capable of oxidizing ethanol at significant rates came from studies into the cell types responsible for the modest ethanol-metabolizing capacity of suspensions of human bone marrow cells. It was observed that when such suspensions were depleted of cells capable of adhering to plastic, their ethanol-metabolizing activity (in incubation mixtures containing 1 mg ethanol per milliliter) was markedly reduced (Table 1).[12] Subsequent studies on human macrophage monolayers derived from blood monocytes, bone marrow, and spleen provided direct evidence that macrophages from these tissues metabolize ethanol to acetate at average rates between 2034 and 2745 nmol/10^7 cells per hour (Table 1). These values are about one fifth of the value of 11,790 nmol/10^7 cells per hour obtained for the rate of oxidation of ethanol (1 mg/ml) to acetate by isolated human hepatocyte suspensions prepared from a wedge biopsy of the liver.[13]

Similar results were obtained in mice. Thus, the rates of metabolism of ethanol to acetate by suspensions of cells prepared from mouse bone marrow, spleen, thymus, and testes were considerably reduced after removal of the

TABLE 1
Rates of Metabolism of [1-^{14}C]-Ethanol (1 mg/ml) to Acetate by Various Human Cell Preparations

Cell type	Number of experiments	Average rate of production of acetate ± S.D. (nmol/10^7 cells per hour)
Whole marrow cell suspension	4	399 ± 20
Adherent cell-depleted marrow cell suspension	4	12 ± 3
Bone marrow macrophages	6	2343 ± 317
Blood monocyte-derived macrophages	25	2142 ± 189
Splenic macrophages	3	2956 ± 266

TABLE 2
Effect of Removing Adhering Cells from Cell Suspensions Prepared from Various Murine Organs on the Rate of Oxidation of Ethanol (1 mg/ml) to Acetate

Organ	Number of expriments	Nature of cell suspension	Average rate of production of acetate ± S.D. (nmol/10^7 cells per hour)
Bone marrow	6	All cells	112 ± 35
		Adherent cell depleted	5 ± 2
Spleen	6	All cells	37 ± 5
		Adherent cell depleted	4 ± 2
Thymus	8	All cells	67 ± 17
		Adherent cell depleted	10 ± 3
Testis	6	All cells	19 ± 3
		Adherent cell depleted	4 ± 2

Reprinted with permission from *Alcohol Alcoholism,* 22, Wickramasinghe et al., The capacity of macrophages from different murine tissues to metabolize ethanol and generate an ethanol-dependent nondialyzable cytotoxic activity *in vitro,* 1987, Pergamon Press PLC.

adhering cells (Table 2).[14] Furthermore, the rates of oxidation of ethanol to acetate per 10^7 cells by macrophages derived from mouse bone marrow, spleen, and thymus were similar to those by mouse Kupffer cells (Table 3) and human blood monocyte-derived macrophages. Mouse Kupffer cells metabolized ethanol to acetate at one fifth of the rate shown by mouse hepatocytes (Table 3).[14]

The higher rate of metabolism of ethanol (1 mg/ml) to acetate in hepatocytes than in macrophages, when such rates are expressed per 10^7 cells per hour, is related to the fact that hepatocytes are much larger than macrophages.

TABLE 3
Rates of Metabolism of [1-^{14}C]-Ethanol (1 mg/ml) to Acetate by Various Cells Derived from C57 BL/6 Mice

Cell type	Number of experiments	Average rate of production of acetate ± S.D.	
		nmol/10^7 cells per hour	μmol/g wet weight per hour
Hepatocytes	11	9199 ± 1268	23 ± 3
Kupffer cells	35	1847 ± 443	45 ± 11
Bone marrow macrophages	27	2322 ± 465	66 ± 13
Splenic macrophages	25	1930 ± 298	60 ± 9
Thymic macrophages	26	1970 ± 632	59 ± 11

Reprinted with permission from *Alcohol Alcoholism,* 22, Wickramasinghe et al., The capacity of macrophages from different murine tissues to metabolize ethanol and generate an ethanol-dependent nondialyzable cytotoxic activity *in vitro,* 1987, Pergamon Press PLC.

When the data were expressed by unit of wet weight, mouse Kupffer cells gave higher rates than mouse hepatocytes, with values of 45 and 25 μmol/g wet weight per hour, respectively (Table 3).[14] Similarly, when the data were expressed per milligram of protein, human blood monocyte-derived macrophages gave slightly higher rates than two human hepatoma cell lines PLC/PRF/5 and Hep G2, with values of 621, 439, and 534 nmol/mg protein per hour, respectively.[15] Thus, it is evident that the ethanol-metabolizing capacity of macrophages is comparable to that of hepatocytes.

III. BIOCHEMICAL BASIS OF ETHANOL METABOLISM BY MACROPHAGES

The biochemical pathways involved in ethanol metabolism by macrophages and hepatocytes have been studied by investigating the effects of various metabolic inhibitors on the ability of these cells to convert ethanol to acetate. The inhibitors employed were pyrazole (an inhibitor of alcohol dehydrogenase [ADH]), 4-iodopyrazole (an inhibitor of π-ADH), 3-amino-1,2,4-triazole (an inhibitor of catalase), and three inhibitors of cytochrome P-450, namely carbon monoxide, metyrapone, and tetrahydrofurane. Such studies have shown that, unlike hepatocytes, which metabolize ethanol mainly via ADH, macrophages do so mainly via ADH-independent, catalase-independent, and cytochrome-P-450-dependent pathways. This applies to human macrophages derived from blood monocytes, the bone marrow, or spleen[16,17] as well as to mouse Kupffer cells and mouse macrophages derived from the bone marrow, spleen, and thymus.[14] The effects of various inhibitors on the oxidation of ethanol (1 mg/ml) to acetate by mouse hepatocytes and Kupffer cells are compared in Table 4.

TABLE 4
Effect of Various Inhibitors on the Metabolism of Ethanol (1 mg/ml) to Acetate by Murine Hepatocytes and Kupffer Cells

Cell type	Additive	Number of experiments	Averge rate of production of acetate ± S.D. (nmol/10^7 cells per hour)		Change with additive (%)
			Without additive	With additive	
Hepatocytes	Pyrazole, 100 m*M*	11	9,199 ± 1,268	103 ± 30	−99
	Pyrazole, 10 m*M*	8	8,348 ± 149	339 ± 96	−96
	Iodopyrazole, 10 m*M*	7	8,390 ± 151	526 ± 123	−84
	Aminotriazole, 12 m*M*	5	8,371 ± 178	7,717 ± 308	−8
	CO, 1 min	10	8,601 ± 1,886	7,728 ± 1,365	−10
	Metyrapone, 10 m*M*	5	10,330 ± 718	8,824 ± 253	−15
	Tetrahydrofurane, 10 m*M*	6	8,314 ± 160	7,504 ± 565	−10
Kupffer cells	Pyrazole, 100 m*M*	10	2,163 ± 231	2,002 ± 238	−7
	Pyrazole, 10 m*M*	5	1,836 ± 46	1,748 ± 33	−5
	Iodopyrazole, 10 m*M*	4	1,834 ± 201	1,786 ± 196	−3
	Aminotriazole, 12 m*M*	4	2,036 ± 206	1,939 ± 178	−5
	CO, 1 min	4	1,691 ± 43	772 ± 92	−54
	Metyrapone, 10 m*M*	6	2,035 ± 58	1,005 ± 13	−51
	Tetrahydrofurane, 10 m*M*	8	2,050 ± 197	967 ± 67	−53

Reprinted with permission from *Alcohol Alcoholism,* 22, Wickramasinghe et al., The capacity of macrophages from different murine tissues to metabolize ethanol and generate an ethanol-dependent nondialyzable cytotoxic activity *in vitro,* 1987, Pergamon Press, PLC.

Further studies with human blood monocyte-derived macrophages have shown that ethanol metabolism by these cells is also partly dependent on the generation of superoxide anion radicals (i.e., is partly inhibited by superoxide dismutase).[18]

IV. RELEASE OF ACETALDEHYDE BY MACROPHAGES

After ethanol (2 mg/ml) is added to cultures of human macrophages derived from blood monocytes[9,19] and of murine macrophages derived from the liver, bone marrow, spleen, and thymus,[14] and the cultures are incubated for 3 d, the culture supernatants contain a nondialyzable cytotoxic activity which causes detachment of A9 cells. In the case of human blood monocyte-derived macrophages, it was shown that: (1) the degree of cytotoxic activity increased progressively in cultures containing 0.1, 1, and 2 mg ethanol per milliliter; (2) the cytotoxic activity was largely abolished by heating to 56°C for 10 min; and, (3) the cytotoxic activity not only caused detachment of A9 cells, but also reduction of ^{3}H-thymidine incorporation by K562 cells.[19]

When supernatants from cultures of human blood monocyte-derived macrophages which had been incubated with 2 mg ethanol per milliliter for 72 h were subjected to Sephacryl® S-300 gel filtration and the fractions so obtained were tested for cytotoxicity against A9 cells, the cytotoxic activity coeluted with some of the fractions containing albumin molecules. Circumstantial evidence that the cytotoxic molecules represented acetaldehyde-modified albumin molecules came from the observations that when macrophage culture medium alone (i.e., without macrophages) was incubated with ^{14}C-acetaldehyde for 72 h and the medium was subjected to gel filtration, both radioactivity and cytotoxic activity appeared in the fractions containing albumin.[9] Furthermore, pure albumin obtained from supernatants of ethanol-containing macrophage cultures by affinity chromatography on blue Sepharose® CL-6B[10] shows cytotoxic activity,[15] proving that at least some of the cytotoxicity was associated with albumin molecules rather than with some other molecule coeluting with albumin. Finally, after treatment with the reducing agent sodium borohydride at a pH of 9.5,[10] the cytotoxicity of macrophage culture supernatants was markedly reduced (Table 5),[20] suggesting that the cytotoxic activity resided in the unstable Schiff bases formed during the first stage of reaction between acetaldehyde and proteins. Such bases result from a reaction between the C=O groups of acetaldehyde and the NH_2 groups of proteins and thus contain C=N bonds. They are stabilized and, consequently, made nontoxic by addition across the C=N bond of either hydrogen or a thiol group.[21]

Investigations have also been performed into the nature of the cytotoxicity in supernatants from cultures of mouse Kupffer cells and mouse bone marrow-

TABLE 5
Effects of Various Treatments on the Cytotoxicity Against A9 Cells of Dialyzed Supernatants from Cultures of Human Blood Monocyte-Derived Macrophges[a] which were Incubated with 0 or 2 mg Ethanol per milliliter for 72 hour

Treatment[10]	Average number of residual adherent A9 cells per well[b] ± S.D. ($\times 10^{-5}$)	
	Ethanol-free cultures	Ethanol-containing cultures
None	1.24 ± 0.78	0.71[c] ± 0.47
6.4 m*M* NaOH for 1.5 h	1.21 ± 0.66	0.67 ± 0.35
6.4 m*M* NaOH + 1.55 m*M* $NaBH_4$ for 1.5 h	1.23 ± 0.77	1.03[d] ± 0.59

[a] Monocytes from five different healthy donors were used.
[b] Wells contained 50% (v/v) of the solution under study.
[c] Significance of difference from value obtained with untreated supernatants from ethanol-free cultures, calculated using a paired *t* test: $p < 0.05$, N = 5.
[d] Significance of difference from value obtained after treatment with NaOH alone, calculated using a paired *t* test: $p < 0.05$, N = 5.

From Wickramasinghe, S. N., unpublished data.

derived macrophages which were prepared and incubated with 2 mg ethanol per milliliter for 3 d as described by Wickramasinghe et al.[14] The results were generally similar to those from corresponding studies on supernatants from ethanol-containing cultures of human blood monocyte-derived macrophages. Thus, the albumin-containing fractions obtained by Sephacryl® S-300 gel filtration caused greater detachment of A9 cells when such fractions were derived from ethanol-containing Kupffer cell or bone marrow-derived macrophage cultures than from ethanol-free control cultures (Figures 1 and 2).[20] In addition, when pure albumin (Pool B) was obtained by affinity chromatography on blue Sepharose® CL-6B and tested for cytotoxicity against A9 cells,[10] fewer residual A9 cells were found in wells containing albumin from ethanol-containing Kupffer cell cultures than in those containing albumin from ethanol-free control cultures (Table 6).[20]

The data summarized above demonstrate that during the process of metabolizing ethanol, macrophages release substantial quantities of acetaldehyde molecules extracellularly, some of which form unstable cytotoxic complexes with albumin present in the macrophage growth medium. Other acetaldehyde molecules are presumed to form nontoxic stable complexes with proteins. The cytotoxic activity of the unstable acetaldehyde-albumin complexes results from the release of acetaldehyde from such complexes and the binding of the released acetaldehyde to target cells.[10]

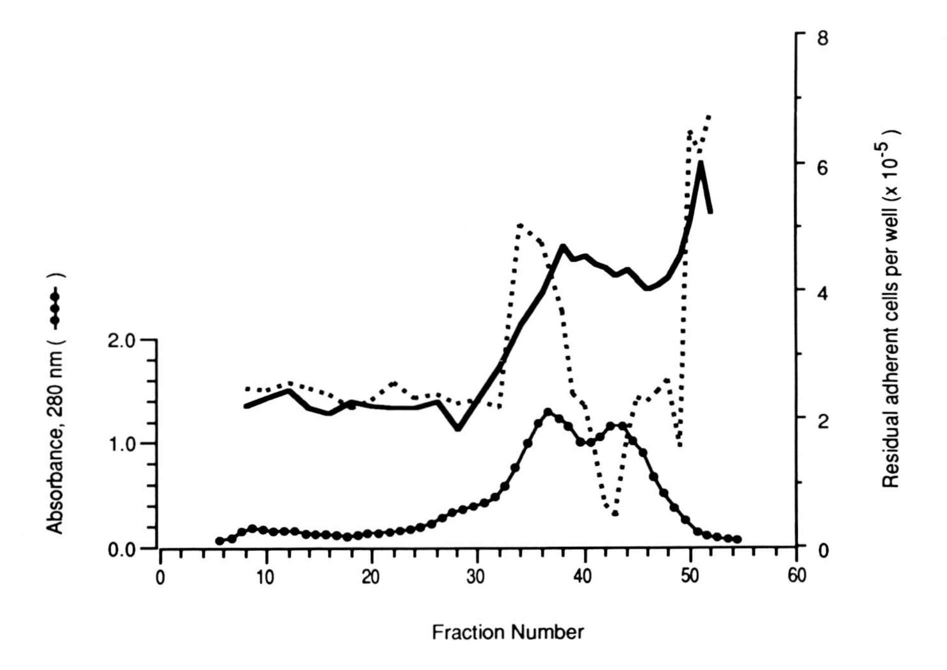

FIGURE 1. Sephacryl® S-300 gel filtration of growth medium from flasks containing murine Kupffer cells which were cultured for 72 h with 0 mg ethanol per milliliter (solid line) or 2 mg ethanol per milliliter (broken line).[20] The graph shows the relationship between the absorption at 280 nm of individual fractions and their cytotoxic activity. The latter was considered to be inversely proportional to the number of residual adherent A9 cells per well after incubating cells adherent to the walls of a multiwell culture plate with 2 ml of each fraction per well at 37°C for 16 h; damaged cells become detached.[19]

V. COMPARISON BETWEEN THE EXTENT OF EXTRACELLULAR RELEASE OF ACETALDEHYDE BY MACROPHAGES AND HEPATOCYTES

When 2×10^6 mouse hepatocytes (estimated wet weight 73 mg) in 10 ml macrophage growth medium (RPMI 1640 with antibiotics plus 25% newborn calf serum) are incubated at 37°C for 12 h in the presence of 0 or 2 mg ethanol per milliliter, the supernatants from the ethanol-containing suspensions develop a nondialyzable cytotoxic activity against A9 cells (Figure 3).[20] The extent of cytotoxic activity generated was slightly less than that in cultures of mouse Kupffer cells containing 6.3×10^5 cells (estimated wet weight 2.6 mg) which were incubated with 2 mg ethanol per milliliter for 3 d (i.e., six times longer than the hepatocytes) (Figure 3).[20] As the shorter incubation period in the case of the hepatocytes should have been more than compensated for by the fact that the average wet weight of hepatocytes was 28 times greater than that of the Kupffer cells, these data suggest that, per unit wet weight,

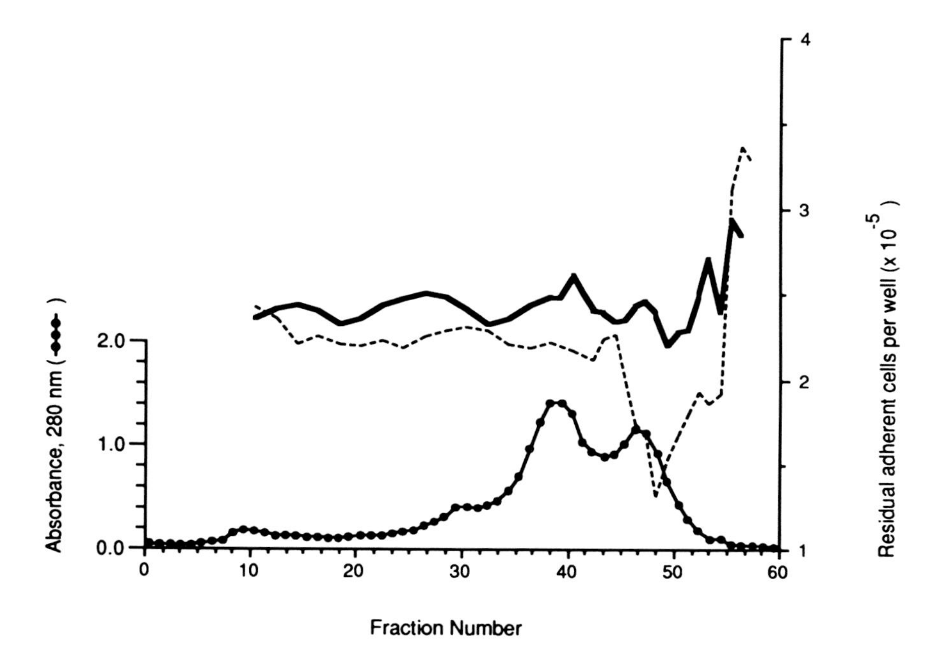

FIGURE 2. Sephacryl® S-300 gel filtration of supernatants from cultures of murine bone marrow-derived macrophages which were grown without ethanol (solid line) and with 2 mg ethanol per milliliter (broken line) for 72 h.[20] The graph shows the relationship between the absorption at 280 nm of individual fractions and their cytotoxic activity against A9 cells; cytotoxicity is inversely proportional to the number of residual adherent A9 cells per well.

mouse hepatocytes may generate much less cytotoxic activity than mouse Kupffer cells. In another study, cultures of human blood monocyte-derived macrophages and of the two human hepatoma cell lines, PLC/PRF/5 and Hep G2, containing similar quantities of cellular protein per flask and converting roughly similar quantities of ethanol to acetate per flask, were incubated with 0 or 2 mg ethanol per milliliter for 72 h and the culture supernatants subjected to affinity chromatography on blue Sepharose® CL-6B. When pure albumin preparations obtained in this way were adjusted to the same optical density and tested for cytotoxicity against A9 cells, albumin obtained from ethanol-containing macrophage cultures was considerably more toxic than that obtained from the ethanol-containing cultures of PLC/PRF/5 and Hep G2 cells.[15] It therefore appears that for a given amount of ethanol metabolized, considerably more acetaldehyde is released extracellularly by macrophages than by the two hepatoma cell lines. It is possible that, unlike hepatocytes which are known to be rich in acetaldehyde dehydrogenase, macrophages contain lower levels of this enzyme and are therefore not able to oxidize intracellular acetaldehyde to acetate as effectively as hepatocytes.

In order to obtain precise information on the rate of extracellular release of acetaldehyde by macrophages and hepatocytes, it would be necessary to

TABLE 6
Cytotoxicity of Fractions Obtained by Affinity Chromatography on Blue Sepharose® CL-6B of Supernatants from C57 BL/6 Mouse Kupffer Cell Cultures which were Incubated with and without 2 mg Ethanol per milliliter for 3 days

		No. of residual adherent A9 cells per well[b] ($\times 10^{-5}$)	
Nature of culture	Pool[a]	Mean[c]	S.D.
Without ethanol	A	2.157	0.065
	B	2.257	0.081
With ethanol	A	1.975	0.023
	B	1.251[d]	0.093

[a] Pool A = mixture of fractions containing unbound proteins; pool B = mixture of fractions containing eluted albumin.[10]
[b] Wells contained 100% (v/v) pool A or B.
[c] From four experiments
[d] Significance of difference from value obtained with pool B from cultures without ethanol, calculated using a paired *t* test: $p < 0.001$.

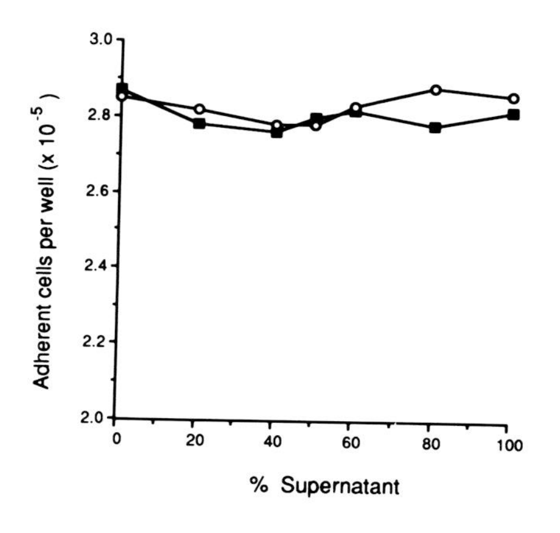

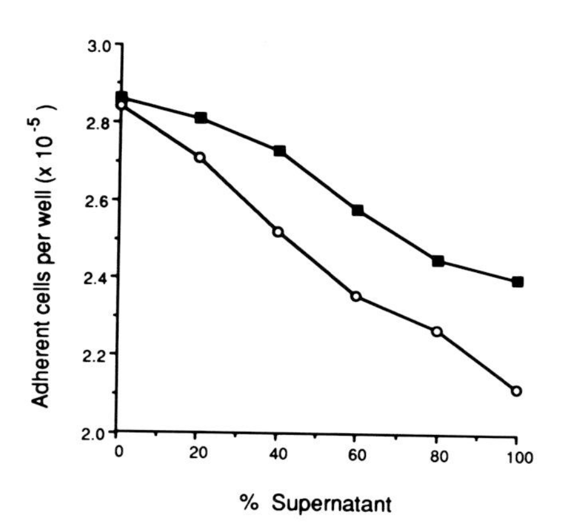

FIGURE 3. Number of adherent A9 cells remaining after incubation for 16 h in medium containing various percentages of dialyzed supernatants from cultures of murine hepatocytes (■) or Kupffer cells (○) containing 0 mg ethanol per milliliter (graph left) or 2 mg ethanol per milliliter (graph right).[20] The duration of culture, with and without ethanol, was 12 h for the hepatocytes and 72 h for the Kupffer cells. The data represent average values obtained from three sets of experiments. The average number of cells per culture was 2×10^6 for hepatocytes and 6.3×10^5 for Kupffer cells.

TABLE 7
Rates of Conversion of Ethanol to Acetate by Various Cell Types from Control C57 BL/6 Mice and Mice Continuously Exposed to Ethanol Vapor for 20 to 37 d

	Rate of production of acetate (nmol/10^7 cells per hour)[14]						
	Ethanol concentration	Control			Alcohol treated[22]		
Cell type	(mg/ml)	N[a]	Mean	S.D.	N	Mean	S.D.
Hepatocytes	0.1	12	1,196	344	15	1,487	367
	1.0	17	9,118	747	16	11,833[b]	1,521
Bone marrow macrophages	0.1	5	206	15	10	203	16
	1.0	7	2,091	146	14	2,165	138
Splenic macrophages	0.1	7	180	6	13	194	19
	1.0	9	1,877	36	14	1,920	69
Kupffer cells	0.1	6	174	10	10	174	7
	1.0	7	1,802	171	10	1,857	61

[a] N = Number of C57 BL/6 mice studied.
[b] Significance of difference from mean value in control animals, $p < 0.001$.

make quantitative biochemical measurements of all the released acetaldehyde (i.e., the free acetaldehyde and the acetaldehyde in both the unstable acetaldehyde-protein complexes and the stable acetaldehyde-protein complexes). Unfortunately, such biochemical estimations are not yet possible.

VI. EFFECT OF CHRONIC ETHANOL ADMINISTRATION ON THE METABOLISM OF ETHANOL BY MACROPHAGES AND HEPATOCYTES

When mice were exposed to ethanol vapor (15 to 20 mg/l of inhaled air) for 20 to 37 d as described by Malik and Wickramasinghe,[22] the hepatocytes, but not the Kupffer cells or other tissue macrophages, showed an increased rate of ethanol metabolism (Table 7);[20] the macrophages were concentrated by allowing them to adhere for 2 h[14] and were studied immediately afterwards. The effects of various metabolic inhibitors on the oxidation of ethanol by Kupffer cells and hepatocyte from alcohol-exposed mice were similar to those from control mice (Table 8). However, the difference between the rates of ethanol metabolism by hepatocytes in pyrazole-containing and corresponding pyrazole-free incubation mixtures was greater when the hepatocytes were derived from alcohol-exposed animals than from control animals ($p < 0.005$). Thus, under the experimental conditions of the study, hepatocytes, but not Kupffer cells, showed metabolic adaptation and the higher rate of ethanol

TABLE 8
Effects of Various Additives on the Rates of Metabolism of Ethanol (1 mg/ml) to Acetate by Hepatocytes and Kupffer Cells[14] from Control C57 BL/6 Mice and Mice Continuously Exposed to Ethanol Vapor for 20 to 37 d[20]

Cells	Animals[22]	Additive	Number of experiments	Rate of production of acetate (nmol/10^7 cells per hour)				Change with additive (%)
				Without additive		With additive		
				Mean	S.D.	Mean	S.D.	
Hepatocytes	Controls	Pyruvate, 10 m*M*	8	9065	997	9,799[a]	666	+8.1
		Pyrazole, 10 m*M*	9	9,076	933	679[b]	169	−92.5
		Iodopyrazole, 10 m*M*	3	9,344	97	649[b]	57	−93.0
		Aminotriazole, 12 m*M*	8	9,099	978	8,464[c]	772	−7.0
		Tetrahydrofurane, 10 m*M*	8	9,106	974	8,302[c]	822	−8.8
	Alcohol treated	Pyruvate, 10 m*M*	8	11,405	1,865	12,407[c]	1,223	+8.8
		Pyrazole, 10 m*M*	10	11,873	1,659	758[b]	144	−93.6
		Iodopyrazole, 10 m*M*	6	11,913	2,037	918[b]	74	−92.3
		Aminotriazole, 12 m*M*	9	12,313	1,656	10,874[d]	2,201	−11.7
		Tetrahydrofurane, 10 m*M*	9	11,935	1,747	10,402[d]	2,075	−12.8
Kupffer cells	Controls	Pyrazole, 10 m*M*	3	1,867	195	1,765[c]	195	−5.5
		Tetrahydrofurane, 10 m*M*	3	1,757	200	677[c]	103	−61.5
	Alcohol treated	Pyrazole, 10 m*M*	4	1,835	79	1,774	113	−3.3
		Aminotriazole, 12 m*M*	4	1,855	83	1,792	132	−3.4
		Tetrahydrofurane, 10 m*M*	5	1,856	72	806[c]	105	−56.6

Note: Significance of difference from values in cultures without the additive, calculated using a paired t test.

[a] = $p < 0.05$, [b] = $p < 0.001$, [c] = $p < 0.005$, [d] = $p < 0.025$.

metabolism by hepatocytes from ethanol-exposed animals appeared to be mainly due to increased ADH-dependent rather than ADH-independent metabolism.

VII. BIOLOGICAL SIGNIFICANCE OF THE RELEASE OF ACETALDEHYDE

The capacity of all types of tissue macrophages, including Kupffer cells, to oxidize substantial quantities of ethanol to acetaldehyde and acetate and to release large amounts of acetaldehyde extracellularly is likely to be of considerable importance in the pathogenesis of ethanol-related tissue damage. One suggestion has been that the high concentrations of acetaldehyde generated at the surface of macrophages following ethanol consumption may react with and damage surrounding parenchymal cells.[12,19,23] It is probable that little if any damage is caused by acetaldehyde generated and retained within normal hepatocytes as these cells rapidly oxidize intracellular acetaldehyde molecules and render them ''nontoxic'' via the action of acetaldehyde dehydrogenase. On the other hand, hepatocytes may be vulnerable to damage by extracellular acetaldehyde molecules, much if not most of which could be generated by Kupffer cells. Acetaldehyde generated by Kupffer cells may also disturb the function of neighboring lipocytes.

Acetaldehyde is a highly reactive substance, and there are abundant data from *in vitro* studies that micromolar concentrations of acetaldehyde are toxic to various cell types, including hepatocytes. Thus, at concentrations of 50 to 200 μM, 30 to 60 μM, 100 to 580 μM, 10 to 50 μM, and 45 to 360 μM, respectively, acetaldehyde has been shown to impair rat liver mitochondrial function,[24,25] inhibit guineapig cardiac microsomal protein synthesis,[26] depress protein synthesis and glycoprotein secretion by rat liver slices,[27,28] impair PHA- or concanavalin A-stimulated human lymphocyte transformation,[29] and prolong the doubling time and increase the modal volume of human cell lines.[30] There are also data indicating that ^{14}C-acetaldehyde derived from the oxidation of ^{14}C-ethanol by rat liver slices binds to preformed hepatocyte protein;[31] since liver slices contain Kupffer cells, the bound acetaldehyde may have been generated by Kupffer cells, hepatocytes, or both. The binding of acetaldehyde to cellular proteins (e.g., enzymes) may impair their normal function and, in the case of surface proteins, may result in the formation of neoantigens and the mounting of an immune response which may damage affected hepatocytes.[7,8] Acetaldehyde may also inactivate antioxidants and thereby promote free radical-mediated cellular damage.

The measurement of acetaldehyde levels in the blood is fraught with many technical difficulties, and over the years investigators have progressively revised downwards the levels considered to be found after ethanol consumption.[32,33] According to some current estimates, the blood acetaldehyde concentration after ethanol ingestion by healthy volunteers is only <1 to 6 μM, and it may therefore be argued that many of the effects of acetaldehyde outlined in the preceding paragraph which were observed at higher acetaldehyde concentrations are biologically irrelevant. However, the newer assays for acetaldehyde probably measure only free acetaldehyde and not the substantial quantities of reversibly protein-bound and potentially cytotoxic acetaldehyde which are present in blood. Furthermore, the concentration of acetaldehyde immediately around tissue macrophages may easily be 10- to 100-fold higher than blood acetaldehyde levels.

VIII. OTHER POSSIBLE KUPFFER CELL-DEPENDENT HEPATOTOXIC MECHANISMS IN ALCOHOLISM

Alcoholism may induce functional changes in Kupffer cells which result in damage to adjacent hepatocytes. These changes may be caused either by a direct effect of ethanol on Kupffer cells (e.g., as some consequence of ethanol metabolism by these cells) or by an indirect effect of chronic alcoholism.

A. RELEASE OF CYTOTOXIC SUBSTANCES OTHER THAN ACETALDEHYDE

Macrophages are known to release a number of potentially cytotoxic substances when appropriately stimulated. These include tumor necrosis factor (TNF), components of complement, neutral proteases, prostaglandin E, nitric oxide, and reactive oxygen metabolites (hydrogen peroxide, superoxide anion radicals, and hydroxyl radicals).[34-37] It is therefore theoretically possible that toxic substances other than acetaldehyde are released by the macrophages of chronic alcoholics and that these damage surrounding parenchymal cells.

Several stimuli such as lipopolysaccharide (endotoxin), immune complexes, complement components, some T cell products (e.g., interferon-gamma or GM-CSF) and even some macrophage products (e.g., TNF) can activate macrophages. There is, however, only limited information as to whether or not macrophages are activated by excessive alcohol intake. In one study, the peritoneal macrophages of mice which were given daily intraperitoneal injections of ethanol for 4 d showed increased phagocytosis of microspheres when compared with cells from control mice injected with saline.[38] In another study, peritoneal, alveolar, and splenic macrophages from rats fed alcohol chronically for 12 weeks showed increased expression of Fc and C3b surface receptors, but a decreased capacity for Fc- and C3b-mediated phagocytosis, compared with cells from control animals.[39] Since macrophage activation is not always associated with an increase of all effector functions, studies of the direct effects of ethanol and acetaldehyde on the different effector functions of Kupffer cells and other tissue macrophages are needed. Ethanol may also activate Kupffer cells indirectly by promoting the absorption from the gut of lipopolysaccharides,[40] which are potent activators of macrophages.

The previously reported finding that the cytotoxicity of dialyzed macrophage culture supernatants is not significantly different from that of undialyzed supernatants[19] only excludes an important cytotoxic role for stable substances with a molecular weight of less than 10,000, but does not exclude such a role for unstable and highly reactive substances such as nitric oxide or oxygen-derived free radicals.

1. Tumor Necrosis Factor

The observation that the cytotoxicity of supernatants from ethanol-containing cultures of human blood monocyte-derived macrophages was largely lost after heating to 56°C for 10 min suggests that TNF-α may not contribute substantially to the cytotoxicity of such supernatants,[19] since TNF-α is reported to retain 90% of its activity at this temperature.[41,42] Nevertheless, the possibility that Kupffer cell-generated TNF plays a role in alcohol-related liver disease should continue to receive serious attention, since plasma TNF levels were found to be high in patients with acute alcoholic hepatitis, normal in chronic alcoholics without liver disease, and slightly elevated in patients with inactive alcoholic cirrhosis. Furthermore, in the patients with alcoholic hepatitis, TNF levels correlated positively with serum bilirubin ($r = 0.74$) and serum creatinine ($r = 0.81$), and with prognosis.[43] In another study, blood monocytes from patients with alcoholic hepatitis were shown to release increased quantities of TNF *in vitro,* both spontaneously and after stimulation with lipopolysaccharide, when compared with monocytes from healthy control subjects.[44]

2. Nitric Oxide

A recent study has shown that lipopolysaccharide-stimulated rat Kupffer cells inhibit protein synthesis in cocultured rat hepatocytes by a mechanism dependent on the metabolism of L-arginine to L-citrulline, nitrites, and nitrates via reactive nitrogen intermediates (including nitric oxide); the authors considered it likely, but did not prove that the hepatotoxic agent was nitric oxide.[45] In another investigation, prestimulation of Kupffer cells with interleukin 2 resulted in a significant reduction of the concentration of lipopolysaccharide necessary to cause inhibition of hepatocyte protein synthesis.[46] Subsequently, Billiar et al.[47] directly demonstrated that the addition of lipopolysaccharide to Kupffer cell:hepatocyte cocultures results in the release of more nitric oxide into the culture supernatant than the addition of lipopolysaccharide to Kupffer cell cultures, and suggested that the Kupffer cell and hepatocyte together formed a functional unit capable of generating large quantities of nitric oxide. More recently, it has been shown that ethanol stimulates nitric oxide production *in vitro* by human Kupffer cells, human hepatocytes and mixtures of both cell types. Interestingly, the ethanol-containing cocultures generated more nitric oxide than expected on the basis of the number of each cell type present and this effect appeared to be mediated by the release of one or more nondialyzable soluble factors by both cell types.[57] Whether ethanol also increases the production of nitric oxide by Kupffer cells indirectly either by reducing the threshold for lipopolysaccharide-dependent nitric oxide production or by increasing absorption of lipopolysaccharides from the gut remains to be determined.

3. Superoxide Anion Radicals

The findings that ethanol metabolism in unstimulated cultures of human blood monocyte-derived macrophages is partly inhibited by superoxide dismutase, that phorbol myristate acetate (PMA) considerably increases ethanol metabolism by cultures of such macrophages, and that this increase is prevented by superoxide dismutase indicate that some superoxide anion radicals are generated both in unstimulated and PMA-stimulated cultures.[18] Since the dismutase is a large molecule, which is unlikely to enter macrophages other than by pinocytosis, its main effect was probably to prevent the extracellular oxidation of ethanol by superoxide anion radicals. This implies that some superoxide anion radicals are released extracellularly, particularly by activated macrophages. Whether or not sufficient numbers of radicals are released to react both with ethanol and with neighboring cells is unclear. Furthermore, results obtained with cultured human blood monocyte-derived macrophages may not apply to resident human tissue macrophages. Studies in mice have shown that Kupffer cells isolated by a collagenase perfusion technique release negligible quantities of H_2O_2 and superoxide anion radicals even after stimulation by substances (PMA, opsonized zymosan, or interferon-gamma) which trigger a respiratory burst in peritoneal macrophages.[48] Similarly, resident mouse bone marrow macrophages have been shown to generate little or no reactive oxygen intermediates.[49] If these data in mice also apply to healthy humans and chronic alcoholics, then Kupffer cell-derived reactive oxygen metabolites may not play a role in ethanol-related hepatocyte damage.

B. INTERFERENCE WITH THE PHAGOCYTIC FUNCTION OF KUPFFER CELLS

The possibility also exists that the decreased phagocytic activity of blood monocytes[50] and macrophages[51] reported in chronic alcoholics may play some role in the pathogenesis of alcohol-related liver disease.

Nolan and Camara[52] have suggested that ethanol may impair the ability of Kupffer cells to clear and detoxify gut-derived lipopolysaccharide present in portal blood and thereby permit such substances to reach and damage hepatocytes.

C. INTERFERENCE WITH THE HEPATOCYTE-SUPPORT FUNCTION OF KUPFFER CELLS

There is increasing experimental evidence that hepatocytes, lipocytes (Ito cells), and Kupffer cells are functionally interdependent. For example, experiments with cocultures of rat hepatocytes and Kupffer cells have shown that Kupffer cells regulate the types and amounts of individual proteins synthesized by hepatocytes[53] and that hepatocytes increase Kupffer cell-mediated tumor cell cytostasis.[54] Furthermore, when lipocytes from rat liver are exposed to Kupffer cell-conditioned medium, they show cytoplasmic and nuclear enlargement, and increased synthesis of collagen and total protein. They also show increased proliferation due to an increased expression of receptors for platelet-derived growth factor.[55] A theoretical mechanism by which ethanol might induce hepatic damage may therefore be by affecting one or more such hepatocyte- or lipocyte-support functions of Kupffer cells.

IX. CONCLUSIONS

There is good evidence that Kupffer cells and other tissue macrophages have a substantial capacity to metabolize ethanol to acetate. When expressed per milligram of protein, the rate of ethanol metabolism by these cells is comparable to that by hepatocytes. Unlike hepatocytes, which metabolize ethanol predominantly via the enzyme ADH, Kupffer cells do so by ADH-independent, cytochrome P-450-dependent, and, in the case of human blood monocyte-derived macrophages, superoxide-dependent pathways. The metabolism of ethanol by Kupffer cells and other tissue macrophages is associated with the release of acetaldehyde extracellularly, and for a given quantity of ethanol that is converted to acetate, blood monocyte-derived macrophages release more acetaldehyde than human hepatocyte cell lines. These data suggest a new mechanism for alcohol-induced liver damage, namely damage via reaction with acetaldehyde molecules formed and released by Kupffer cells. Since the concentration of free acetaldehyde would be highest immediately around Kupffer cells, hepatocytes and lipocytes lying in contact with or close to Kupffer cells would be expected to be most affected. Whether or not other nonparenchymal liver cells behave like Kupffer cells with respect to acetaldehyde release is not known. However, it is likely that the endothelial cells of hepatic sinusoids metabolize ethanol and release acetaldehyde to a much smaller extent than Kupffer cells, as the rate of metabolism of ethanol (1 mg/ml) to acetate[16] by cultures of human umbilical cord-derived endothelial cells[56] was only 125 nmol/10^7 cells per hour (N = 6, SD = 14 nmol/10^7 cells per hour),[20] which is 6.8% of the value given by murine Kupffer cells. Interestingly, this small metabolic activity appeared to be mainly ADH independent, as it was reduced only by 26.8% to 91 nmol/10^7 cells per hour (N = 4, SD = 6 nmol/10^7 cells per hour) in the presence of 10 m*M* pyrazole.[20]

Recent investigations also point to the theoretical possibility that Kupffer cells may mediate alcohol-related hepatocyte damage by additional mechanisms. For example, it is possible, though not proven, that in chronic alcoholics, Kupffer cells: (1) release TNF or reactive nitrogen intermediates which damage surrounding cells; (2) suffer from a defect in phagocytic activity and, consequently, an impaired ability to protect the hepatocyte from the toxic effects of lipopolysaccharide or other substances absorbed through the gut;

or (3) fail to perform some important hepatocyte or lipocyte support function normally.

Further studies are required to determine whether in chronic alcoholics the generation of acetaldehyde or other substances by Kupffer cells or portal tract macrophages or the failure of some normal macrophage function is involved in the transformation of lipocytes to myofibroblasts and the formation of increased quantities of collagen by the latter.

REFERENCES

1. **Lieber, C. S.,** Alcohol, protein metabolism and liver injury, *Gastroenterology,* 79, 373, 1980.
2. **Thomas, M. and Peters, T. J.,** Acetaldehyde, its role in alcoholic toxicity and dependence, *Br. J. Addict.,* 76, 375, 1981.
3. **Eriksson, C. J. P.,** The role of acetaldehyde in drinking behavior and tissue damage, *Br. J. Alc. Alcoholism,* 17, 57, 1982.
4. **Lewis, K. O. and Paton, A.,** Could superoxide cause cirrhosis?, *Lancet,* i, 188, 1982.
5. **Winston, G. W. and Cederbaum, A. I.,** NADPH-dependent production of oxyradicals by purified components of the rat liver mixed function oxidase system. II. Role of microsomal oxidation of ethanol, *J. Biol. Chem.,* 258, 1514, 1983.
6. **Feierman, D. E., Winston, G. W., and Cederbaum, A. I.,** Ethanol oxidation by hydroxyl radicals, role of iron chelates, superoxide, and hydrogen peroxide, *Alcoholism, Clin. Exp. Res.,* 9, 95, 1985.
7. **MacSween, R. N. M. and Anthony, R. S.,** Immune mechanisms in alcoholic liver disease, in *Alcoholic Liver Disease,* P. Hall, Ed., Edward Arnold, London, 1985, 69.
8. **Eddleston, A. L. W. F. and Vento, S.,** Relevance of immune mediated mechanisms in progressive alcoholic liver injury, *Mol. Aspects Med.,* 10, 169, 1988.
9. **Wickramasinghe, S. N., Gardner, B., and Barden, G.,** Cytotoxic protein molecules generated as a consequence of ethanol metabolism *in vitro* and *in vivo, Lancet,* ii, 823, 1986.
10. **Wickramasinghe, S. N., Gardner, B., and Barden, G.,** Circulating cytotoxic protein generated after ethanol consumption: identification and mechanisms of reaction with cells, *Lancet,* ii, 122, 1987.
11. **Wickramasinghe, S. N., Marjot, D. H., Rosalki, S. B., and Fink, R. S.,** Correlations between serum proteins modified by acetaldehyde and biochemical variables in heavy drinkers, *J. Clin. Pathol.,* 42, 295, 1989.
12. **Wickramasinghe, S. N.,** Rates of metabolism of ethanol to acetate by human neutrophil precursors and macrophages, *Alcohol Alcoholism,* 20, 299, 1985.
13. **Wickramasinghe, S. N.,** Neuroglial and neuroblastoma cell lines are capable of metabolizing ethanol via an alcohol-dehydrogenase-independent pathway, *Alcoholism, Clin. Exp. Res.,* 11, 234, 1987.

14. **Wickramasinghe, S. N., Barden, G., and Levy, L.,** The capacity of macrophages from different murine tissues to metabolize ethanol and generate an ethanol-dependent nondialyzable cytotoxic activity *in vitro, Alcohol Alcoholism,* 22, 31, 1987.
15. **Wickramasinghe, S. N. and Hasan, R.,** Comparison of the generation of albumin-associated cytotoxic activity in supernatants from ethanol-containing cultures of human blood monocyte-derived macrophages and of two human hepatoma cell lines, *Alcohol Alcoholism,* 26, 147, 1991.
16. **Wickramasinghe, S. N.,** Observations on the biochemical basis of ethanol metabolism by human macrophages, *Alcohol Alcoholism,* 21, 57, 1986.
17. **Wickramasinghe, S. N.,** Oxidation of ethanol by human macrophages derived from the bone marrow and spleen, *Hematol. Rev. Commun.,* 2, 179, 1988.
18. **Wickramasinghe, S. N.,** Role of superoxide anion radicals in ethanol metabolism by blood monocyte-derived human macrophages, *J. Exp. Med.,* 169, 755, 1989.
19. **Wickramasinghe, S. N.,** Supernatants from ethanol-containing macrophage cultures have cytotoxic activity, *Alcohol Alcoholism,* 21, 263, 1986.
20. **Wickramasinghe, S. N.,** unpublished data.
21. **Tuma, D. J. and Sorrell, P.,** Covalent binding of acetaldehyde to hepatic proteins: role in alcoholic liver injury, in *Aldehyde Adducts in Alcoholism,* M. A. Collins, Ed., Alan R. Liss, New York, 1985, 3.
22. **Malik, F. and Wickramasinghe, S. N.,** Haematological abnormalities in mice continuously exposed to ethanol vapor, *Br. J. Exp. Pathol.,* 67, 831, 1986.
23. **Wickramasinghe, S. N.,** Role of macrophages in the pathogenesis of alcohol-induced tissue damage, *Br. Med. J.,* 294, 1137, 1987.
24. **Orrego, H., Israel, Y., and Blendis, L. M.,** Alcoholic liver disease, information in search of knowledge?, *Hepatology,* 1, 267, 1981.
25. **Burke, J. P. and Rubin, E.,** The effects of ethanol and acetaldehyde on the products of protein synthesis by liver mitochondria, *Lab. Invest.,* 41, 393, 1979.
26. **Schreiber, S. S., Oratz, M., Rothschild, M. A., Reff, F., and Evans, C.,** Alcoholic cardiomyopathy. II. The inhibition of cardiac microsomal protein synthesis by acetaldehyde, *J. Mol. Cell Cardiol.,* 6, 207, 1974.
27. **Perin, A., Scalabrino, G., Sessa, A., and Arnaboldi, A.,** *In vitro* inhibition of protein synthesis in rat liver as a consequence of ethanol metabolism, *Biochim. Biophys. Acta,* 366, 101, 1974.
28. **Tuma, D. J., Zetterman, R. K., and Sorrell, M. F.,** Inhibition of glycoprotein secretion by ethanol and acetaldehyde in rat liver slices, *Biochem. Pharmacol.,* 29, 35, 1980.
29. **Roselle, G. A. and Mendenhall, C. L.,** Alteration of *in vitro* human lymphocyte function by ethanol, acetaldehyde, and acetate, *J. Clin. Lab. Immunol.,* 9, 33, 1982.
30. **Wickramasinghe, S. N. and Malik, F.,** Acetaldehyde causes a prolongation of the doubling time and an increase in the modal volume of cells in culture, *Alcoholism, Clin. Exp. Res.,* 10, 350, 1986.
31. **Medina, V. A., Donohue, T. M., Jr., Sorrell, M. F., and Tuma, D. J.,** Covalent binding of acetaldehyde to hepatic proteins during ethanol oxidation, *J. Lab. Clin. Med.,* 105, 5, 1985.

32. **De Master, E. G., Redfern, B., Weir, E. K., Pierpont, G. L., and Crouse, L. J.,** Elimination of artifactual acetaldehyde in the measurement of human blood acetaldehyde by the use of polyethylene glycol and sodium azide, normal blood acetaldehyde levels in the dog and human after ethanol, *Alcoholism, Clin. Exp. Res.*, 7, 436, 1983.
33. **Peters, T. J., Ward, R. J., Rideout, J., and Lim, C. K.,** Blood acetaldehyde and ethanol levels in alcoholism, *Genet. Alcoholism, Progr. Clin. Biol. Res.*, 241, 215, 1987.
34. **Reiko, T. and Werb, Z.,** Secretory products of macrophages and their physiological functions, *Am. J. Physiol.*, 246, C1, 1984.
35. **Nathan, C. F.,** Secretory products of macrophages, *J. Clin. Invest.*, 79, 319, 1987.
36. **Beutler, B. and Cerami, A.,** Cachectin: more than a tumor necrosis factor, *N. Engl. J. Med.*, 316, 379, 1987.
37. **Marletta, M. A.,** Nitric oxide, biosynthesis and biological significance, *Trends Biochem. Sci.*, 14, 488, 1989.
38. **Yamamoto, H., Mori, H., Okano, K., Sassa, R., and Uchida, T.,** The effect of alcohol on the activity of macrophages: assessment of macrophage activity by determination of intracellular reactive oxygen, *Jpn. J. Alc. Drug Depend.*, 21, 183, 1986.
39. **Bagasra, O., Howeedy, A., and Kajdacsy-Balla, A.,** Macrophage function in chronic experimental alcoholism. I. Modulation of surface receptors and phagocytosis, *Immunology,* 65, 405, 1988.
40. **Bode, C., Kugler, V., and Bode, J. C.,** Endotoxaemia in patients with alcoholic and non-alcoholic cirrhosis and in subjects with no evidence of chronic liver disease following acute alcohol excess, *J. Hepatol.*, 4, 8, 1987.
41. **Matthews, N. and Watkins, J. F.,** Tumor-necrosis factor from the rabbit. I. Mode of action, specificity and physicochemical properties, *Br. J. Cancer,* 38, 302, 1978.
42. **Kull, F. C., Jr. and Cuatrecasas, P.,** Preliminary characterization of the tumor cell cytotoxin in tumor cell necrosis serum, *J. Immunol.*, 126, 1279, 1981.
43. **Bird, G. L. A., Sheron, N., Goka, A. K. J., Alexander, G. J., and Williams, R. S.,** Increased plasma tumor necrosis factor in severe alcoholic hepatitis, *Ann. Intern. Med.*, 112, 917, 1990.
44. **McClain, C. J. and Cohen, D. A.,** Increased tumor necrosis factor production by monocytes in alcoholic hepatitis, *Hepatology,* 9, 349, 1989.
45. **Billiar, T. R., Curran, R. D., Stuehr, D. J., Ferrari, F. K., and Simmons, R. L.,** Evidence that activation of Kupffer cells results in production of L-arginine metabolites that release cell-associated iron and inhibit hepatocyte protein synthesis, *Surgery,* 106, 364, 1989.
46. **Curran, R. D., Billiar, T. R., West, M. A., Bentz, B. G., and Simmons, R. L.,** Effect of interleukin-2 on Kupffer cell activation: interleukin-2 primes and activates Kupffer cells to suppress hepatocyte protein synthesis *in vitro, Arch. Surg.*, 123, 1373, 1988.
47. **Billiar, T. R., Curran, R. D., Ferrari, F. K., Williams, D. L., and Simmons, R. L.,** Kupffer cell:hepatocyte cocultures release nitric oxide in response to bacterial endotoxin, *J. Surg. Res.*, 48, 349, 1990.

48. **Lepay, D. A., Nathan, C. F., Steinman, R. M., Murray, H. W., and Cohn, Z. A.,** Murine Kupffer cells: mononuclear phagocytes deficient in the generation of reactive oxygen intermediates, *J. Exp. Med.*, 161, 1079, 1985.
49. **Crocker, P. R. and Gordon, S.,** Isolation and characterization of resident stromal macrophages and hematopoietic cell clusters from mouse bone marrow, *J. Exp. Med.*, 162, 993, 1985.
50. **Morland, H., Johnsen, J., Bjorneboe, G.-E. Aa., Drevon, C. A., Morland, J., and Morland, B.,** Reduced IgG Fc-receptor-mediated phagocytosis in human monocytes isolated from alcoholics, *Alcoholism, Clin. Exp. Res.*, 12, 755, 1988.
51. **Liu, Y. K.,** Phagocytic function of RES in alcoholics, *J. Reticuloend. Soc.*, 25, 605, 1979.
52. **Nolan, J. P. and Camara, D. S.,** Endotoxin and liver disease, in *Sinusoidal Liver Cells,* D. L. Knook and E. Wisse, Eds., Elsevier, Amsterdam, 1982, 377.
53. **West, M. A., Billiar, T. R., Mazuski, J. E., Curran, R. D., Cerra, F. B., and Simmons, R. L.,** Endotoxin modulation of hepatocyte secretory and cellular protein synthesis is mediated by Kupffer cells, *Arch. Surg.*, 123, 1400, 1988.
54. **Curran, R. D., Billiar, T. R., Kispert, P. H., Bentz, B. G., May, M. T., and Simmons, R. L.,** Hepatocytes enhance Kupffer cell-mediated tumor cell cytostasis *in vitro, Surgery,* 106, 126, 1989.
55. **Friedman, S. L. and Arthur, M. J. P.,** Activation of cultured rat hepatic lipocytes by Kupffer cell conditioned medium: direct enhancement of matrix synthesis and stimulation of cell proliferation via induction of platelet-derived growth factor receptors, *J. Clin. Invest.*, 84, 1780, 1989.
56. **Zuhrie, S. R., Pearson, J. D., and Wickramasinghe, S. N.,** Haemoglobin synthesis in K562 erythroleukaemia cells is affected by intimate contact with monolayers of various human cell types, *Leuk. Res.*, 12, 567, 1988.
57. **Goldin, R. D., Clark, J. D., Hunt, N. C. A., and Wickramasinghe, S. N.,** The effect of ethanol on the production of nitric oxide by human hepatocytes and Kupffer cells, *Alcoholism, Clin. Exp. Res.*, in press.

Chapter 6

VIRAL INFECTIONS AND HEPATITIS

D. Vetter, A. Kirn, and J.-P. Gut

TABLE OF CONTENTS

ISBN 0-8493-6109-5

I. INTRODUCTION

Viruses may infect the body by different routes. Often, systemic infection occurs with generalization of the virus via the bloodstream. The liver may thus be the main target organ for virion multiplication. It may also be a reservoir serving to establish a second viremia which conveys newly synthesized viral particles to other sites such as the skin or the central nervous system. This situation is well illustrated by ectromelia in mice.[1]

Whether the parenchymal cells are the final site of infection or merely an intermediate, the interactions between Kupffer cells and hepatocyte, or the consequences of their interactions, are fundamental. It is a fact that before reaching the hepatocytes the circulating virus particles must gain access to the space of Disse. This may occur either by their passage through the fenestrae of the endothelial cells, or by their uptake by Kupffer cells (and/or endothelial cells) followed by their emergence into the space of Disse; phagocytosis of virus particles by Kupffer cells is probably the most frequent situation.[2,3]

To date, the role of nonparenchymal cells, and particularly of Kupffer cells, in the physiopathology of viral infections of the liver has not been fully estimated. Our knowledge is mainly based on the study of animal models such as mouse virus hepatitis,[4] Rift Valley fever,[5] ectromelia in mice,[6] infectious canine hepatitis,[7] and yellow fever in the monkey.[8] However, misinterpretation of the architecture of the sinusoid has often resulted in erroneous explanations of the experimental results.

Theoretically, Kupffer cells may be involved in the pathogenesis of viral hepatitis in at least three ways: (1) direct uptake of the virus with protective consequences; (2) transmission of antigenic information from the virus itself as well as from infected hepatocytes to immunocompetent cells; and (3) production of cytokines or other mediators that would further modulate the immune response to the virus and/or the infected cells and, on the other hand, participate in hepatocellular or sinusoidal damage, noteworthy in fulminant hepatitis. The pathogenic implication of Kupffer cells in various types of human virus-induced liver diseases is still far from being fully understood because of the lack of any experimental model suitable for the concomitant study of the immune system and the infected hepatocytes. Furthermore, the complex interacting network of the human hepatic sinusoid is beyond our current experimental knowledge. Nevertheless, on the basis of some histologic data and extrapolations concerning small rodents with rival or toxin-induced hepatic damage, speculations and questions may be anticipated.

In this chapter, we first describe two experimental models of viral hepatitis which related the interaction between Kupffer cells and hepatocytes. In the first model, hepatocytolysis involves the previous destruction of Kupffer cells by toxic viral proteins; in the second, the viral replication in the Kupffer cells represents a limiting factor for the secondary infection of the hepatocytes. We then discuss the role of Kupffer cells in human viral infections, which is much less documented and which mainly relays histopathological or histochemical examination of liver specimens.

II. EXPERIMENTAL MODELS OF VIRAL HEPATITIS INVOLVING THE INFECTION OF KUPFFER CELLS

A. PATHOGENICITY RELATED TO THE DESTRUCTION OF THE KUPFFER CELLS BY TOXIC VIRAL PROTEINS: FROG VIRUS 3 (FV 3)-INDUCED HEPATITIS

1. Features of FV 3 Hepatitis

FV 3 is an icosahedral deoxyribovirus isolated from *Rana pipiens*.[9] It multiplies in the cytoplasm of amphibian, bird, and mammalian cells at a temperature of 28°C. At 37°C its multiplication is completely inhibited, but it still prevents cellular macromolecular metabolism.[10] A parenteral inoculation of a suspension of purified virus into mice,[11] rats,[12] and Saimiri monkeys[102] kills the animals within 18 to 36 h. This lethal effect is specific, since animals immunized against FV 3 become resistant.[13] Pathological examination, 30 h after inoculation of one 100% lethal dose (LD 100) into rats reveals acute degenerative lesions in the liver, the other organs showing only slight congestion.[14,15] The hepatotropism of FV 3 is confirmed by the patterns of systemic distribution of radioactive-labeled virus. More than two thirds of the radioactivity is found in the liver; however, per gram of organ, the splenic tissue is the richest in virus particles.[12] Thus, there is a divergence between the organ distribution of the virus and the localization of the lesions, which suggests that nonviral factors may be involved in parenchymal damage.

The pathogenicity of FV 3 is related to structural viral proteins which can be solubilized from the virus particles. Inoculation of viral proteins into mice produces an acute hepatitis, which kills the animals within 24 to 48 h.[16]

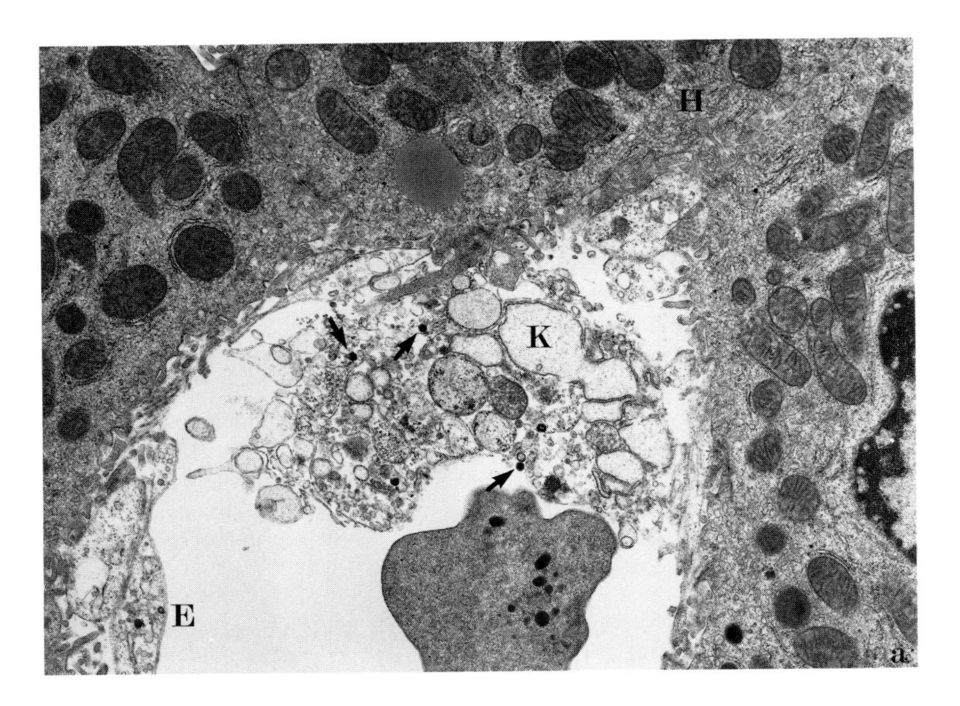

FIGURE 1A. Ultrastructural lesions observed in wistar AG rats 1.5 h after an intravenous inoculation of one 100% lethal dose of FV 3 (morphological observation, Dr. A. Bingen, Virology Institute). In a sinusoid, lined by intact endothelial cells (E), a completely disorganized Kupffer cell (K), containing some FV 3 viral particles (→), can be observed. (Original magnification × 12,000.)

2. Morphological Lesions of the Liver

The hepatocellular changes are characterized by granulo-vacuolar degeneration of the cytoplasm at the periportal and mediolobular areas and by disorganization of the lobular architecture; the sinusoid capillary structures disappear.[15]

Ultrastructural examinations of the liver specimens at different intervals after infection reveal that the first lesions appear in the Kupffer cells.[17] Parenchymal cell damage occurs in two stages, the earlier stage affecting the nuclei[18,19] and the later one involving cytoplasmic degeneration[20] (Figure 1).

Numerous virus particles are found inside vacuoles in the cytoplasm of Kupffer cells 15 min after intravenous inoculation of FV 3. The virus penetrates by phagocytosis, pinocytosis, and fusion.[21] Kupffer cell damage occurs as soon as 1 h after administration of one LD 100 and is characterized by many dense bodies, multiple small vesicles, vacuoles, and altered mitochondria. Within 2 h, the Kupffer cells are destroyed and gaps appear in the endothelial lining; finally, the endothelial cells have disappeared 4 to 6 h after infection.

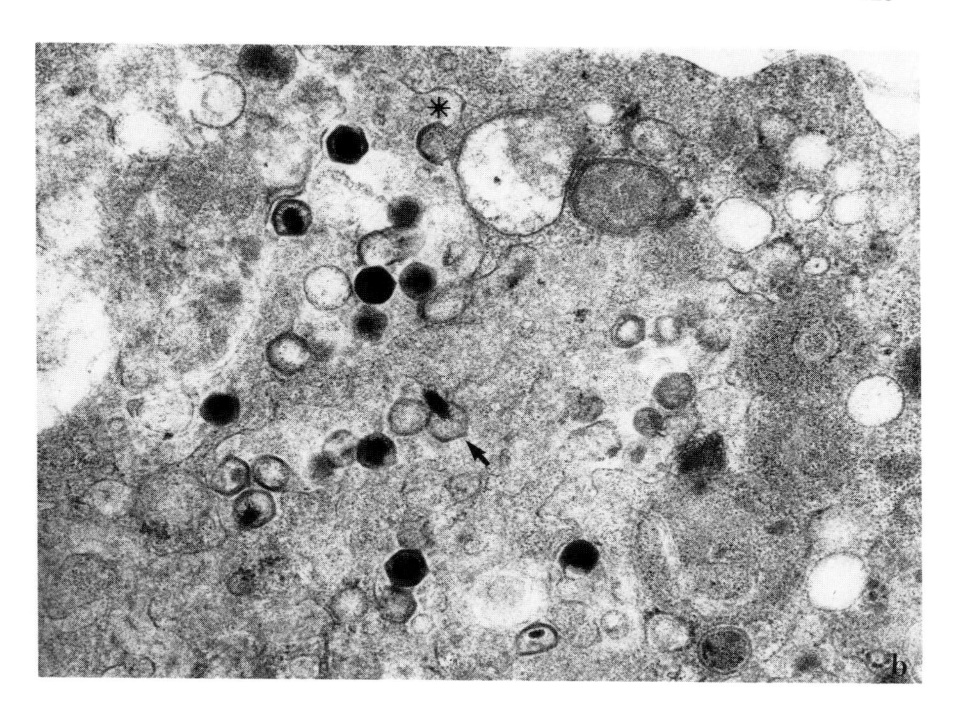

FIGURE 1B. Ultrastructural lesions observed in wistar AG rats 1.5 h after an intravenous inoculation of one 100% lethal dose of FV 3 (morphological observation, Dr. A. Bingen, Virology Institute). Another lysed Kupffer cell, at a higher magnification, showing several viral particles at different stages of maturation. Worthy of note: a viral particle liberating its nuceloprotein content (→) and another one which has fused with the phagocytic vacuole membrane (*). (Original magnification × 54,000).

An *in vivo* microscopic study[22] describing the dynamic events occurring in FV 3 hepatitis shows that within the first hours of infection there is a fragmentation of the Kupffer cells which embolize the sinusoids and the formation of platelet aggregates which adhere to the sinusoid walls, particularly to the injured Kupffer cells.

When Kupffer cell lesions occur, hepatocytes show the following ultrastructural nuclear lesions: condensation of chromatin and accumulation of interchromatin granules in dense clusters at the center of the nucleus. Dose-response experiments reveal a direct correlation between the number of damaged nuclei and the amount of virus inoculated.[12]

It should be emphasized that once the nuclear lesions are complete (3 to 5 h postinfection), the cytoplasm remains uninvolved. Obvious cytoplasmic lesions occur in a few hepatocytes 8 h after inoculation of one LD 100 in rats and become generalized 14 h after infection. These lesions are characterized by damage to the endoplasmic reticulum. It is worth noting that, at this stage, the chondrium is well preserved; mitochondria lysis occurs later

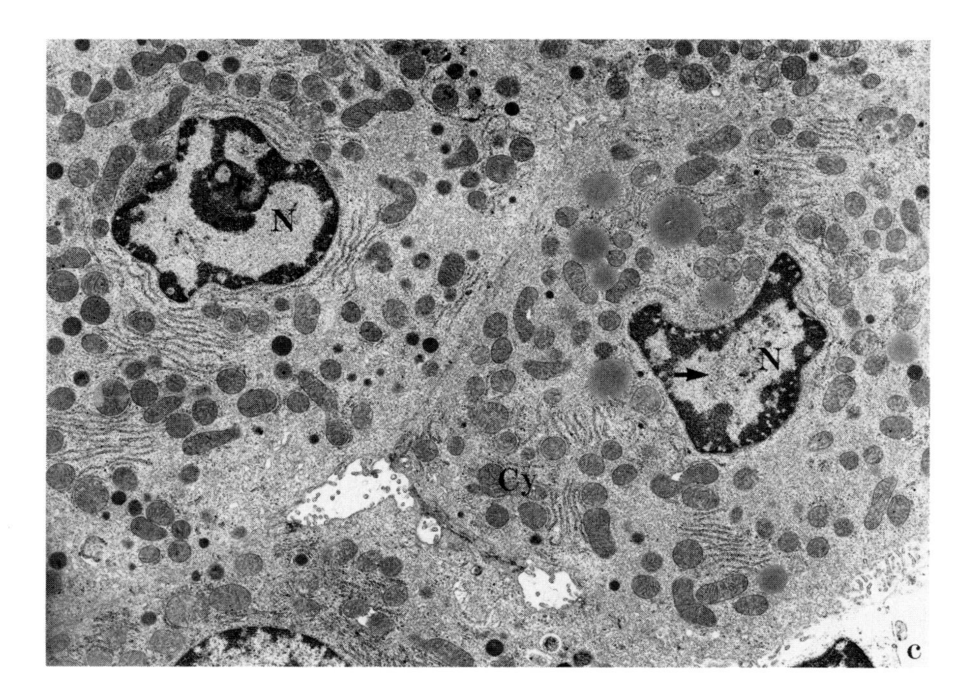

FIGURE 1C. Ultrastructural lesions observed in wistar AG rats 1.5 h after an intravenous inoculation of one 100% lethal dose of FV 3 (morphological observation, Dr. A. Bingen, Virology Institute). Alterations in the hepatocyte nuclei (N) include changes in shape, condensation of the perinuclear and perinucleolar chromatin, and an increase in the number of interchromatin granules (→). The surrounding cytoplasm (Cy) is free of lesions, but there is glycogen depletion. (Original magnification × 6,500.)

when cell necrosis takes place. Whole viral particles are never observed in the hepatocytes, although autoradiographic studies reveal viral proteins, but not DNA, inside parenchymal cells.[23]

3. Biological Changes in the Liver

In animals infected with FV 3, there is a rapid inhibition in DNA, RNA, and, to a lesser extent, protein synthesis in the liver a few hours after infection.[24-26] For instance, a 50% inhibition in the synthesis of RNA occurs within 4 h after inoculation of one LD 100 into rats, and there is a positive correlation between the number of damaged nuclei and RNA synthesis inhibition.[12]

However, there is no direct relationship between metabolic failure and hepatocellular necrosis. Following the inoculation of one LD 100 into two strains of rats, one of them sensitive and the other one more resistant, inhibition in RNA synthesis in the livers of sensitive animals was less marked than that observed in the resistant strain. Likewise, there was no apparent correlation

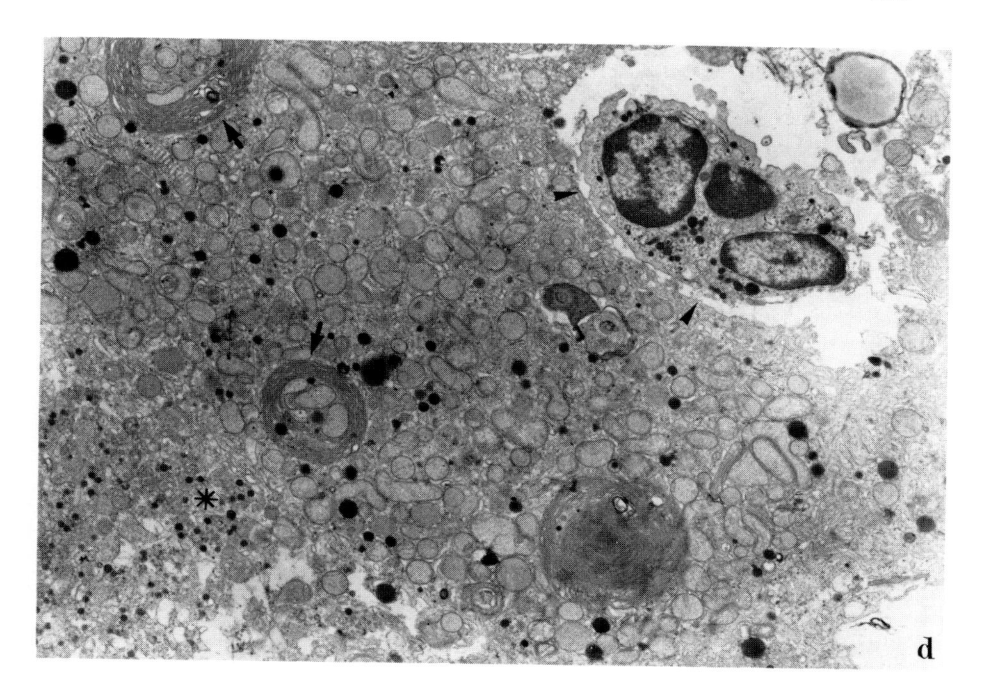

FIGURE 1D. Ultrastructural lesions observed in wistar AG rats 35 h after an intravenous inoculation of one 100% lethal dose of FV 3 (morphological observation, Dr. A. Bingen, Virology Institute). Cytoplasm of a hepatocyte showing concentric lamellar formations (fingerprints) encircling some mitochondria (→). The rough endoplasmic reticulum is no longer visible. Many small dense bodies are apparent, particularly with the lysed organelles lying probably in a sinusoid (*). On the opposite side, the sinusoidal plasma membrane is still evident (→), but no longer lined with endothelial cells. (Original magnification × 9,000.)

in the more resistant strain between the appearance of damaged nuclei and the intensity of hepatocellular necrosis.[12]

FV 3 hepatitis is characterized by cytolysis evolving in two steps (Figure 2). An increase in serum activity of cathepsin D, a lysosomal hydrolase mainly contained in Kupffer cells, takes place after the second hour; a peak equivalent to four times the physiological value is reached 3 h after FV 3 inoculation and the activity returns progressively to normal by the eighth hour. The serum activity of transaminases increases later and reaches a maximum just before death; for aminotransferase ALT, the titer is 188 times greater than that found in uninfected animals. The biological syndrome of cytolysis thus bears out the morphological findings.

4. Consequences of Sinusoidal Necrosis

Kupffer and endothelial cell destruction causes the hepatocytes and the blood in the sinusoids to come into direct contact. This permits material in

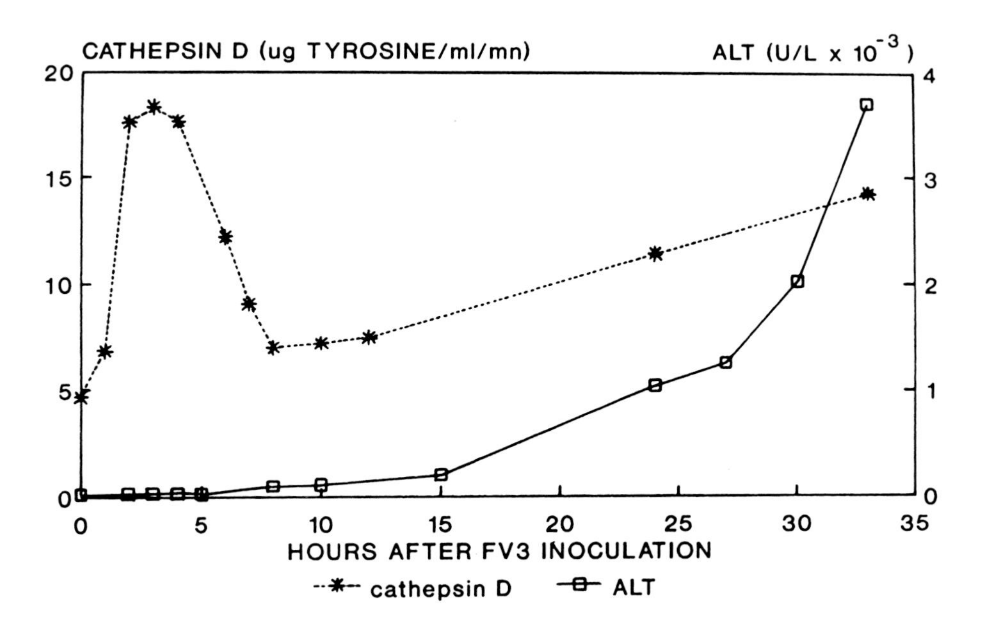

FIGURE 2. Concentration of cathepsin D and ALT in sera of wistar AG rats after intravenous inoculation of one 100% lethal dose of FV 3. (Reprinted with permission from Gut, J. P., Anton, M., Bingen, A., Vetter, J. M., and Kirn, A., *Lab. Invest.*, 45, 218, 1981. © Canadian Academy of Pathology.)

the blood which, under normal conditions, does not reach the hepatocytes, to be taken up by the parenchymal cells. Accordingly, in normal mice colloidal carbon inoculated intravenously is observed mainly in Kupffer cells, but it is also found in the cytoplasm of hepatocytes when the mice had previously received a sublethal dose of FV 3, 4 h prior to carbon inoculation.[27] At this time, the plasma membrane of the hepatocytes is intact and the cytoplasm does not show any alteration.

Not only do the hepatocytes become capable of taking up colloidal carbon, but they also phagocytose larger particles, such as latex grains or nonhepatotrophic viruses. Vaccinia or herpes viruses do not produce hepatitis in mice, principally because they are taken up by Kupffer cells in which they cannot multiply and are therefore prevented from reaching the parenchymal cells. In animals in which the sinusoidal cells have been previously destroyed by a nonlethal dose of FV 3, these viruses are capable of infecting the hepatocytes and multiplying within the parenchymal cells, thus leading to fatal hepatitis. However, if the delay between inoculation of FV 3 and the vaccinia virus is greater than 72 h, a sublethal dose of FV 3 no longer causes mortality. This results from regeneration of the sinusoidal cells, as demonstrated by the clearance of carbon particles from the blood[27] (Figure 3).

Activation of the complement system at the beginning of infection might be another consequence of the sinusoidal cell necrosis.[12] Excessive comple-

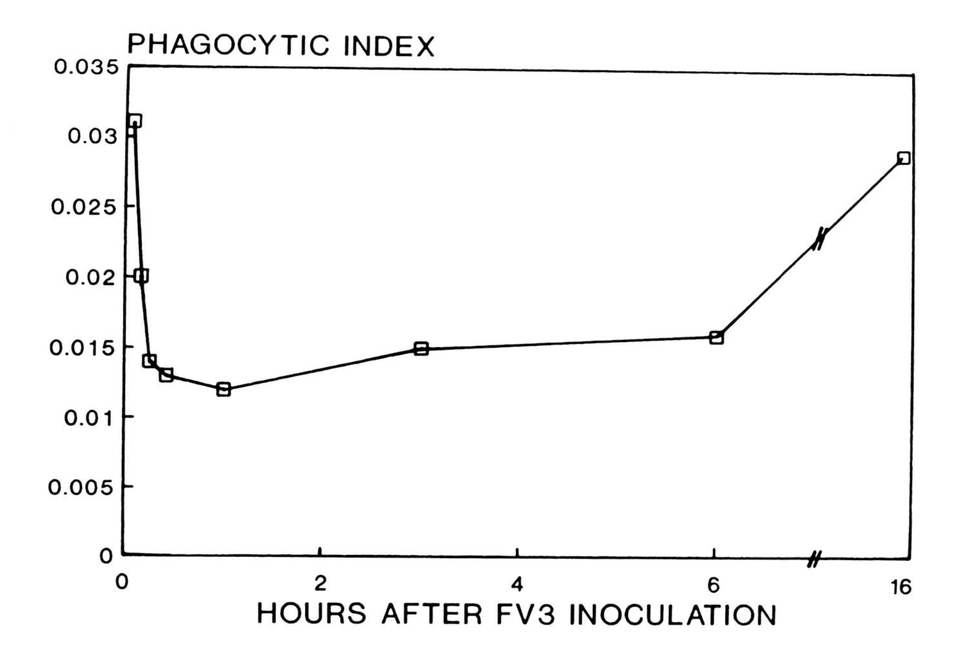

FIGURE 3. Phagocytic index in SWR/J mice after intravenous inoculation of one 100% lethal dose of FV 3. (Gut, J. P., personal results.)

ment consumption may be caused by viral proteins and lysosomal enzymes, e.g., cathepsin, which are released from lysed Kupffer cells during the first few hours.[12]

5. Role of Bacterial Endotoxin in the Pathogenesis of FV 3 Hepatitis

The question as to how cellular necrosis occurs is intriguing. Kupffer cell damage may be directly related to the virus particles which they take up in large amounts. It is known that these cells are particularly sensitive to any metabolic inhibition.[29,30] However, since FV 3 seems unable to reach the hepatocytes, one may wonder how parenchymal cell degeneration occurs. Cessation of macromolecular cell metabolism, which is responsible for the nuclear damage, cannot be the cause of the hepatocellular necrosis because there is no direct relationship between metabolic failure and hepatonecrosis; the inoculation of actinomycin D into mice at a dose which inhibits liver RNA either does not kill or does so only at a late stage.[31] Because the cellular lesions are localized in the liver, although FV 3 is found in great amounts in the spleen and lung, the effect of a second etiopathogenic factor, specific to the liver, must be considered. Bacterial endotoxins appear as the best candidates.

Presently, it is generally accepted that endotoxins released by the Gram-negative bacteria of the digestive tract are continuously absorbed[32] and constitute a physiological component of the portal blood.[33] Since the Kupffer cells are damaged early on in FV 3 hepatitis, endotoxins may no longer be cleared by the reticuloendothelial system and may exert their hepatotoxic potentialities.[34] Several arguments support this hypothesis:

1. An increase in the pool of endotoxins by inoculation of exogenous endotoxin after FV 3 worsens the development of the hepatitis. Accordingly, mice inoculated with a sublethal dose of virus are 20 times more sensitive to endotoxins than control animals.[35]
2. Decrease or suppression of endogenous endotoxin strengthens the animals' resistance. Germfree mice are twice as resistant to FV 3,[36] and colectomy protects rats against a lethal dose of FV 3. However, inoculation of 0.01 LD 100 of endotoxin into colectomized rats suppresses their resistance to the virus.[37]
3. Inhibition of certain physiopathological effects of endotoxin by treating mice with polymyxin B or indomethacin also attenuates hepatitis. Moreover, the sensitivity of mice to FV 3 is decreased in animals rendered endotoxin tolerant or which are genetically resistant to lipopolysaccharide (LPS).[38]

It is not known if endotoxins act directly on the hepatic parenchyma or through a secondary effect, since the biological events caused by these toxins are numerous and various. Leukotrienes, known to play a role in the inflammatory response, may be the mediators of hepatic injury. Recent results show that leukotrienes are produced *in vitro* by Kupffer cells and that their bile concentration increases during the course of FV 3 hepatitis.[39] On the other hand, blockade of leukotriene biosynthesis by lipooxygenase inhibitors leads to a reduction in hepatocellular injury after FV 3 infection.[40] Other mediators such as tumor necrosis factor alpha (TNF-α) or interleukin 6 (IL-6), which play a role in human toxin-induced liver disease, could also be relevant for explaining parenchymal cell damage.[41]

Finally, the organic lesions caused by the endotoxin become irreversible, since the inhibition in macromolecular synthesis produced by viral proteins inhibits any regeneration of the hepatocytes.[25]

In conclusion, FV 3 hepatitis revealed itself to be an interesting model as it brought to light new data concerning the role of Kupffer cells in the pathogenesis of liver disease:

1. The destruction of the Kupffer cells (and the endothelial cells) may have dramatic consequences for the parenchymal cells which can be infected by nonhepatotropic viruses or become able to take up toxic substances directly from the blood. The recent finding that primary cultures of Kupffer cells can produce interferon stresses the importance of the intactness of the sinusoidal cells in protecting the liver.

2. When detoxification in Kupffer cells is impaired, endotoxins may exert a hepatotoxic effect. However, additional factors such as complement cleavage product, lysosomal enzymes, leukotrienes, TNF-α, and IL-6 may also play a role in hepatocellular necrosis.

B. PATHOGENICITY RELATED TO THE MULTIPLICATION OF VIRUSES IN THE KUPFFER CELLS: MOUSE HEPATITIS VIRUS TYPE 3 (MHV 3)-INDUCED HEPATITIS

Another example of Kupffer cell involvement in the pathogenicity of the viral infection of the liver is that of MHV 3. In this case, the Kupffer cells constitute a reservoir in which viral multiplication takes place and from which dissemination into parenchymal cells occurs.

1. Features of MHV 3 Hepatitis

MHV 3 is a member of the *Coronavirus* family, which produces focal hepatic necrosis,[42] the severity of this necrosis varying according to the mouse strain.[43] The A/J strain develops a small number of very well-defined necrotic foci in the liver which disappear 8 to 12 d later. On the other hand, most mouse strains, including Balb/c, show a large number of necrotic foci 2 to 3 d after the infection and the mice die of acute hepatitis 3 to 4 d later. Animals from the C3H strain over 3 months in age have an intermediate susceptibility to MHV 3; most of them survive the acute stage of the infection, but become chronic virus carriers with an evolving disease accompanied by signs of neurological involvement.

2. Role of the Kupffer Cell

Although the hepatocytes constitute the main site of MHV 3 replication, the virus particles carried by the blood have to cross the sinusoidal barrier before they infect the parenchymal cells. A direct relationship has been demonstrated between the resistance of A/J mice and the ability of their peritoneal macrophages[4,44,45] or hepatocytes[46] to partially resist MHV 3 replication. However, the interaction of MHV 3 with Kupffer cells is also of importance in the outcome of the disease. After infection of the primary culture of Kupffer cells with MHV 3, the final virus yield was similar in cell cultures from resistant (A/J)) and susceptible (Balb/c) mice, but the maximum titer was always observed 24 to 36 h later in the cells from the resistant mice (Figure 4).[47] In cell cultures from susceptible mice, cell fusion spreads rapidly over the entire monolayer, whereas in cultures from resistant mice there were reduced numbers of cells and only limited spreading.

The longer time lapse needed for MHV 3 replication in Kupffer cells from resistant mice, together with that observed for hepatocytes, may be of crucial importance in their resistance to the infection. When viruses present in the blood reach the sinusoids, they first interact with the Kupffer and/or the endothelial cells of the hepatic sinusoid and only afterwards with the

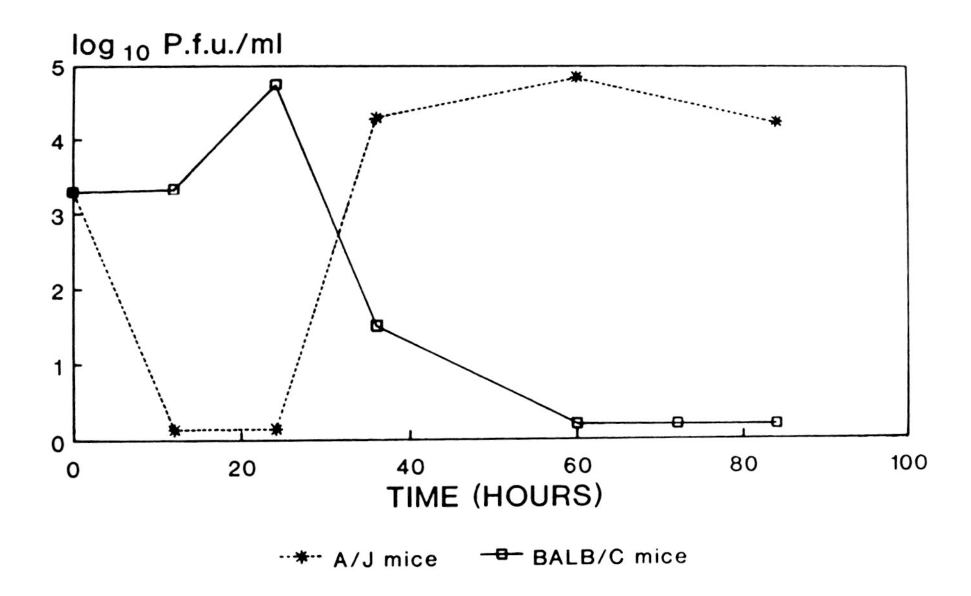

FIGURE 4. Kinetics of MHV 3 multiplication in primary cultures of Kupffer cells. (Reprinted from Pereira, C. A., Steffan, A. M., and Kirn, A., *J. Gen. Virol.*, 65, 1618, 1984. With permission.)

hepatocytes. It may be speculated that *in vivo* the prolongation of the lag phase in these cells allows the specific and nonspecific mechanisms to clear a large amount of virus particles, thus suppressing virus infection in numerous hepatocytes. On the other hand, in susceptible mice where the virus is produced very rapidly, the immune system as well as the nonspecific defenses, if stimulated, would be overwhelmed and would not be able to clear the organism of the virus particles.

Another indication of the essential role of Kupffer cells in the pathogenesis of MHV 3 is given by the following experiment. When A/J mice resistant to MHV 3 were preinfected with sublethal doses of FV 3 in order to destroy their Kupffer cells, they became highly susceptible to MHV 3 hepatitis (Figures 5 and 6).[48]

The resistance of A/J mice to MHV 3 hepatitis may be broken down by inducing a nutritional hypercholesterolemia.[49] The animals, which were fed with the hypercholesterolemic diet for 15 to 60 d before the inoculation of MHV 3, developed an acute hepatitis which led to high levels of mortality. Kupffer cells of hypercholesterolemic mice were shown to exhibit an impairment of several of their functions, such as their ability to take up C3-coated immunoglobulin M (IgM)-opsonized sheep red blood cells or ^{3}H-thymidine-labeled *Escherichia coli* after activation with LPS. On the other hand, these cells were less susceptible to interferon during the induction of an antiviral state.

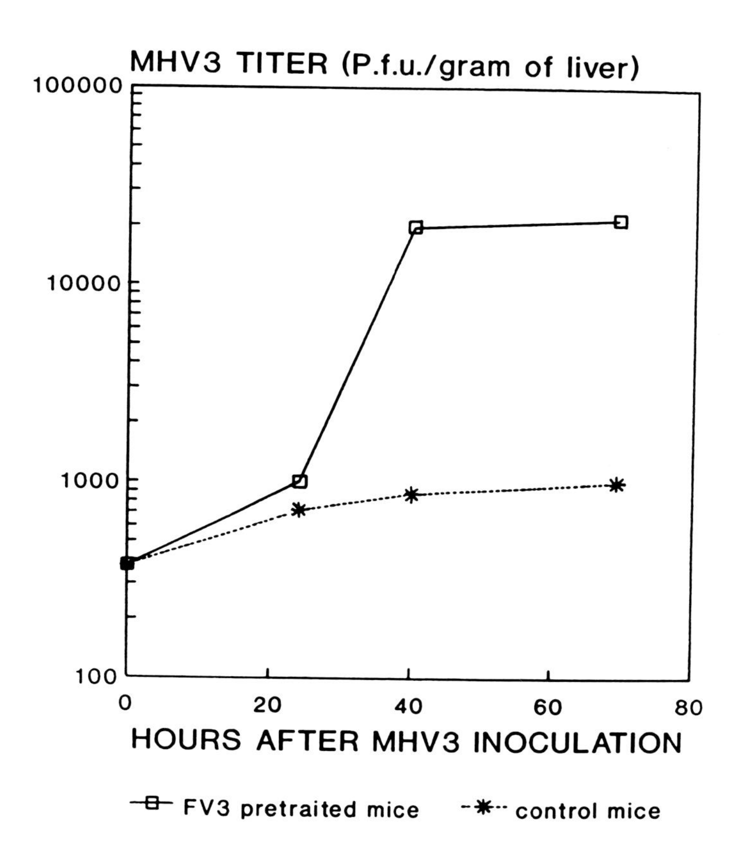

FIGURE 5. Kinetics of virus multiplication in livers of control and FV 3-pretreated mice. Mice were pretreated with a sublethal dose of FV 3 6 h before infection with MHV 3. (Reprinted from Pereira, C. A., Steffan, A. M., and Kirn, A., *Virus Res.*, 1, 561, 1984. With permission.)

Finally, it has been postulated that the susceptibility or resistance to MHV 3 is genetically linked to the monocyte procoagulant activity induced in the course of the viral infection.[50] Accordingly, severe abnormalities in liver microcirculation are observed early in the course of MHV 3 infection in Balb/c mice, whereas normal streamlined blood flow occurs in the livers of the resistant A/J animals.

On the basis of these data and of previous results concerning the role of liver phagocytic cells in resistance to MHV 3, experiments were carried out in order to evaluate the effect of infection on two of the Kupffer cell functions which may be involved in the resistance displayed. Accordingly, the capacity of endotoxin to induce restriction of MHV 3 replication was compared to the synthesis of IL-1 in infected Kupffer cells of both A/J and Balb/c strains of mice. In addition, the accessory cell function of Ia(+) Kupffer cells in the presence of MHV 3 antigen has been studied.[51] The data obtained showed

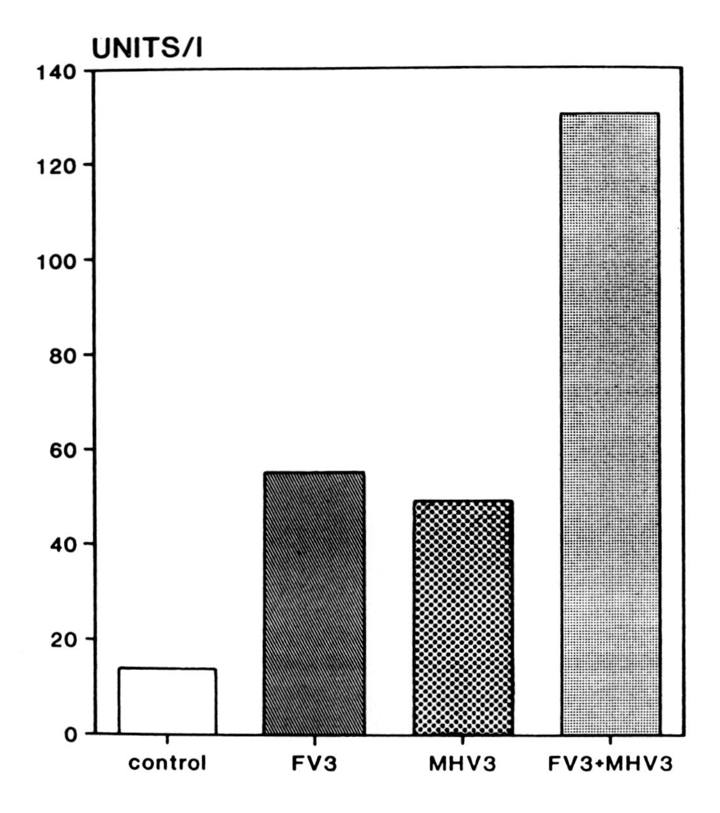

FIGURE 6. Serum ALT activity in control and FV 3-pretreated mice (one sublethal dose of FV 3 h before infection with MHV 3) after 72-h MHV 3 infection. (Reprinted from Pereira, C. A., Steffan, A. M., and Kirn, A., *Virus Res.*, 1, 561, 1984. With permission.)

that Kupffer cell activation leads to a decrease in the multiplication of MHV 3 in A/J, but not in Balb/c mice. The data also showed that MHV 3 infection induces in the two strains of mice the synthesis of IL-1 by LPS-activated Kupffer cells in a dose-dependent manner. Indomethacin treatment, which blocks the synthesis of prostaglandin E_2, an inhibitor of IL-1 activity, alters the innate resistance of A/J mice. Additionally, it was elsewhere demonstrated that pretreatment by prostaglandin E_2 prevents fulminant hepatitis in susceptible Balb/c mice.[52] These findings suggest that in the two strains of mice, the Kupffer cells are perfectly able to process the MHV 3 antigen in a way which efficiently triggers splenic lymphocyte proliferation. While in A/J resistant mice the delayed replication of the virus enables the protective immune response to develop, this response appears too late in susceptible mice to be effective.

Thus, resistance to acute liver infection appears to depend to a large extent on the replication potential of MHV 3 in the Kupffer cells.

III. KUPFFER CELLS IN HEPADNAVIRUS INFECTIONS AND OTHER HUMAN VIRUS-INDUCED HEPATITIS TYPES

A. HEPADNAVIRUS INFECTIONS

1. Histopathologic Changes Affecting Kupffer Cells During the Course of Acute Disease Due to Hepatitis B Virus (HBV) and Hepadnaviruses

When a liver biopsy is performed at the beginning of the clinical phase of the disease, an activation of the Kupffer cells is usually reported. The Kupffer cells appear hyperplastic in the space of Disse or even in the sinusoid itself, mingled with circulating macrophages and surrounding affected hepatocytes or cell debris which undergo phagocytosis and digestion.[53-56] No current data are available for histological alterations during the preclinical or early phase of the disease in humans. Studies in experimental infection have been carried out with woodchuck hepatitis virus (WHV), one of the hepadnaviruses that is the best analogous animal model for HBV infection.[57] At the onset of symptoms, 6 weeks after inoculation, mild lymphocytic and neutrophilic infiltration of the sinusoid was the only histological finding; portal tracts and parenchyma appeared unaffected. At 8 weeks postinoculation, there was a mild hyperplasia of the Kupffer cells adjacent to small foci of degenerative hepatocytes surrounded by lymphocytes, neutrophils, and eosinophils. WHV Ag, as detected by victoria blue, was found in Kupffer cells between 6 and 8 weeks, whereas it was completely undetectable in hepatocytes at that time. In this model, therefore, Kupffer cells appeared as the first hepatic cells infected by the virus. It could be speculated that the sinusoidal inflammation and focal hepatocytic damage are the consequences of Kupffer cell infection and a specific activation, as it has been suggested for about 30 years for various human and animal viral infections.[58] At 8 weeks after inoculation, WHV Ag was detectable in the hepatocytes and the classical histological pattern with inflammatory infiltrate and hepatocyte necrosis appeared.

As for the other viruses, the question arises concerning how Kupffer cells may internalize HBV, as no specific receptor has been as yet identified. Indeed, uptake of various viruses by macrophagic cells is an ancient question.[58] The direct uptake of virions by macrophagic cells is well established for many viruses[59-61] and it seems to be the case for HBV. Other mechanisms are likely to be involved for other viruses, such as phagocytosis of infected cells and immune complexes containing the virus. Direct evidence of human hepatotropic virus uptake by monocytes and Kupffer cells is lacking, and this point remains under debate.

The first protective consequence of Kupffer cell activation is the production of interferon-alpha, which exerts a local inhibition of virus replication and induces the expression by other cells, including hepatocytes,[62] of the major histocompatibility complex class I molecules. The concomitant recruitment and activation of T helper lymphocytes, with which the Kupffer cells interact in a positive feedback, further allows the T cytotoxic eradication of HBV-infected hepatocytes and the production of protective antibodies. The latter seems to play an important role in the early limitation of the infection with antipre-S antibodies that inhibit the attachment of HBV to hepatocytes. Lastly, production of anti-HBV antibodies prevents reinfection.[62] However, it has been recently shown that, during the replicative phase of chronic HBV infection, functional alterations of the monocytes occur, leading to a defect in the initial steps of the immune response.[63] These alterations seem to be a direct consequence of monocyte infection by HBV and could allow in part for the passage to chronicity.

2. Acute Benign vs. Fulminant Hepatitis

Kupffer cells are not the only cells to be infected by hepadnavirus early on: this occurs also for peripheral and splenic mononuclear cells, including macrophages.[64] However, this point remains controversial for the latter cells, as other authors did not find WHV or HBV DNA and/or RNA in these cells after animal transmission, while they were detected in lymphocytes.[64,65] This supports the hypothesis that macrophages and probably Kupffer cells are not permissive for the replication of these viruses, although they can take them up and undergo functional activation or alterations. Thus, the significance of the very early infection of Kupffer cells by the virus remains to be explained.

It could occur as the first step of virus recognition, Kupffer cells being involved as a part of the immune system that would rapidly transmit antigenic information to the lymphocytes. It is now clearly demonstrated that recovery from acute hepatitis B is due to an adequate early immune response.[62] Some striking human observations have shown that this response and all its regulatory mechanisms appear in the first days following inoculation, T lymphocytes being sensitized to pre-S2 Ag, then to HBc and HBs Ag 30 d before the cytolytic clinical phase that occurs when the hepatocytes are infected.[66]

On the other hand, excessive Kupffer cell activation may be relevant to explain liver damage in fulminant hepatitis, by the directed production of TNF and, maybe, IL-1 and IL-6, the former being directly hepatotoxic, as demonstrated in human alcoholic hepatitis.[67] Leukotrienes are another important product of activated monocytes and Kupffer cells that have been shown to be able to cause severe hepatocytic damage in various toxic- or virus-induce liver injuries in animal and human experiments,[68-70] by inducing oxidative free-radical production. Lastly, procoagulant factors are secreted by activated Kupffer cells and may play an important role in the formation of platelet aggregates, inducing *in situ* ischemic sinusoidal damage.[71] Furthermore, production of TNF and leukotrienes is enhanced by endotoxinemia, caused by Kupffer cell inactivation, so that a vicious circle could occur.[68,69,71,72] It appears likely that this plays a role in human viral hepatitis, but to what extent remains to be assessed. Actually TNF or procoagulant activity production by macrophages occurs in most animal models of virus- or toxic-induced acute hepatic failure.[62] It is in accordance with the histological pattern of fulminant hepatitis, in which massive necrosis evokes ischemic or toxic mechanisms rather than immune-mediated cell destruction.[62] This point has not been experimentally demonstrated for HBV of hepadnaviruses, but it supports the current attempts to treat severe viral hepatitis with prostaglandins. Contrarily, monocytes and Kupffer cells themselves are able to secrete prostaglandins that could exert a protective action upon toxic- or immune-mediated hepatocytic damage.[52,62,68,71,72]

Furthermore, activated Kupffer cells could create focal breaches in the endothelial barrier, which would allow parenchymal damage by intestinal endotoxins.[71] Actually, pathological interactions between Kupffer cells and the other sinusoidal cells are likely to be of major importance, but no data concerning humans are available.

3. Kupffer Cells in Chronic Hepatitis B

One of the prominent functions of the Kupffer cells and other cells of the same type is antigen presentation to the lymphocytes, using the major histocompatibility complex class II molecules. During the late integrative phase of chronic hepatitis B, the Kupffer cells strongly express these antigens, though so do the hepatocytes themselves to a comparable extent.[62] An antigenic information can therefore bypass the antigen-presenting cells without its physiological regulation. As a consequence, a protracted immunological liver damage involving nonviral hepatocytic antigens such as the so-called ''liver-specific protein'' complex seems to occur.[73] Additionally, the partial loss of the protective function of the Kupffer cells against bacteria and endotoxins which come from the portal blood could participate in parenchymal damage. However, there is no correlation between the decrease of isotope uptake by the liver and that of the number of Kupffer cells during chronic active human hepatitis;[74] this suggests that intrahepatic shunts play a predominant role in this abnormality. Lastly, the activation of Kupffer cells is able to enhance hepatic fibrosis, as monocyte production of IL-1 is strongly correlated with liver fibrosis, but independent of hepatocyte necrosis.[75] So far, no specific Kupffer cell involvement has been reported in hepatitis D occurring as a coinfection as well as a superinfection with HBV.

B. KUPFFER CELL INVOLVEMENT IN OTHER HUMAN VIRUS-INDUCED LIVER DISEASES

1. Hepatitis A Virus (HAV) Infection

During the acute phase of hepatitis A, histological examination provides findings similar to those for hepatitis B.[53-56,76] Furthermore, the presence of HAV in Kupffer cells has been demonstrated by immunoperoxydase staining and ultrastructural examination during acute hepatitis A in man.[76] Experimental parenteral transmission to the marmoset[77] or to the chimpanzee[78] showed that HAV antigens were found in hepatocytes as well as in sinusoidal macrophages and Kupffer cells. In the proximity of infected Kupffer cells, inflammatory changes were predominant. Among Kupffer cells, IgM deposits were abundant, while they were undetectable on hepatocytes. This strongly suggests that the uptake by the Kupffer cells of immune complexes, constituted by anti-HAV IgM and the virus itself, could have led to their activation and the beginning of the liver disease.[79] As HAV infection is via the enteric route, the Kupffer cells appear as a component of the front-line protective barrier together with the other sinusoidal cell types, including endothelial cells and especially pit cells, the natural killer resident lymphocyte of the liver. To what extent these cells could interact with Kupffer cells remains to be assessed. Since HAV-induced hepatitis is now regarded as immune mediated rather than due to a cytopathic effect of the virus,[80] speculations similar to those evoked for HBV may be drawn concerning the further evolution of the disease in the case of recovery as well as for fulminant hepatitis.

2. Non-A, Non-B Hepatitis, Type E

Epidemic non-A, non-B hepatitis invades the liver like HAV, after enteric contamination. Histological findings at the onset of symptoms are closely similar to those of hepatitis A, with mild Kupffer cell hyperplasia, polymorphonuclear leukocytes, and few lymphocytes in the sinusoid.[81-83] Close to these inflammatory areas appears focal hepatocytic necrosis, followed by widespread infection and destruction of the parenchyma. Whether it is due to a cytopathic effect of the virus or to an immunological mechanism remains to be ascertained. Experimental transmission to the marmoset has provided similar data.[84] However, the likelihood of developing a fulminant hepatitis is vastly higher than for hepatitis A.[84] It has been clearly suggested that a directed consequence of the early Kupffer cell damage is the release of toxic cytokines or the crossing of the endothelial barrier by enterotoxins.[84] This hypothesis is in accordance with the putative role imputed to Kupffer cells as a component of the front-line defense of the liver against enterically transmitted viruses.[84] However, this point of view does not take into account the role of the other sinusoidal cells, especially the endothelial ones, for which information is completely missing.

3. Non-A, Non-B Hepatitis, Type C

Hepatitis C virus is transmitted mainly via the bloodstream. Histological examination of the liver in this disease[81,85] as well as in experimental inoculation of the chimpanzee[81] shows concomitant alterations of the sinusoid and of the parenchyma, but no specific data about Kupffer cells are yet available.

4. Yellow Fever

Experimental inoculation of the rhesus monkey with the flavivirus of yellow fever, which is directly cytopathic, has shown that the earliest event was virus uptake by the Kupffer cells, which underwent degeneration and death before the hepatocytes themselves underwent the same fate.[8]

5. Cytomegalovirus (CMV) Hepatitis

During the course of adult acquired CMV infection, hepatitis is common and usually benign; Kupffer cells displaying mild hyperplasia are often seen, but may be difficult to distinguish from atypical infected lymphocytes.[86-88] On the other hand, a more severe disease occurs in immunocompromised patients, and major problems arise in liver-transplant recipients.[89] The hepatocytes are the predominant target cells for CMV infection,[87] whereas Kupffer cell involvement appears to be a marginal event in the general disease that requires a prompt and strong T cell response for recovery. The situation is quite different in hepatitis occurring in immunocompromised hosts as well as in congenital infection in which giant cell granulomas are often encountered, while the virus has been found in the Kupffer cells themselves.[87] The true pathogenic implications and consequences of such an involvement, as well as the significance of the granulomatous hepatitis that would involve the macrophagic systems remains unclear.

6. Epstein-Barr Virus (EBV) Hepatitis

During infectious mononucleosis, symptoms of mild hepatitis are frequent and mostly resolve without consequences in healthy adults.[90] In these benign forms, a sinusoidal lymphocytic infiltrate and a marked hyperplasia of the Kupffer cells, which seem to contain the virus, are the most common findings.[90,91] A mild lobular hepatitis is usually found, but, as EBV does not infect the hepatocytes, all these abnormalities are thought to be a consequence of an aberrant immune response due to infection of B and T lymphocytes that would further interact with parenchymal cells and Kupffer cells. Fatal hepatitis occurs in patients immunocompromised or bearing an X-linked lymphoproliferative syndrome,[92] or, very seldom, in healthy adults. Liver examination shows a massive infiltrate with atypical lymphocytes and histiocytes, but a general systemic failure is usually associated. This implies the destruction of the whole monomacrophagic system, and death is often due to uncoercible bacterial or fungal infection.[92]

7. Herpes Simplex Hepatitis

As for CMV, herpes simplex 1 and 2 are systemic infections which occasionally involve the liver. Severe or fatal hepatitis can occur in pregnancy, immunocompromised subjects, or in the newborn.[93,94] Whatever the case, the virus, that is directly cytopathic, has been found in the hepatocytes. There is no evidence of any Kupffer cell infection, but macrophages are known to be able to limit the replication of the virus in adjacent cells, namely by interferon production.[95,96]

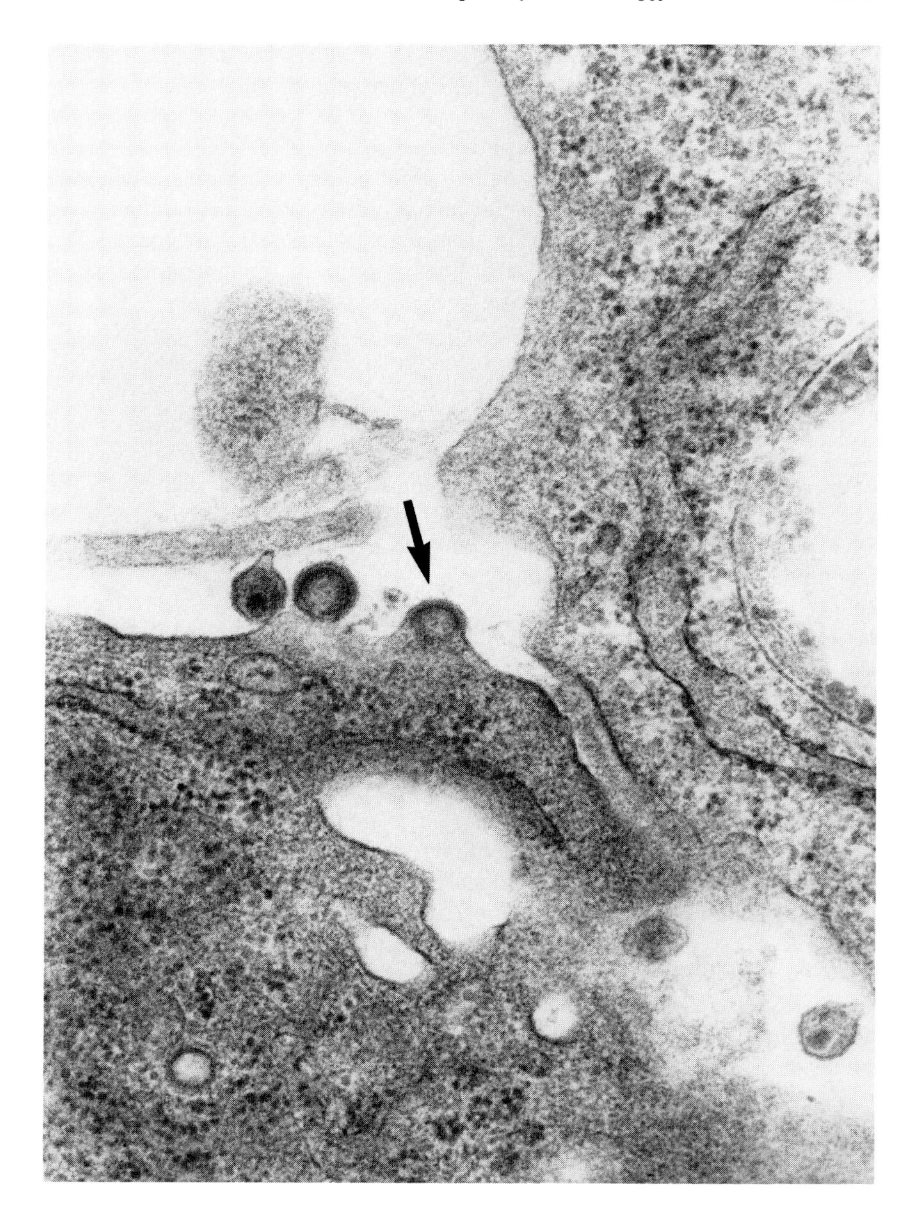

FIGURE 7. Production of HIV_1 particles by a human Kupffer cell in culture examined by transmission electron microscopy. One virion is budding at the plasma membrane (→); two others have already been released. (Original magnification × 72,500.) (Morphological observation, Dr. J. L. Gendrault, Virology Institute.)

8. AIDS

Although the liver does not constitute a target organ in AIDS, several arguments allow HIV-infected Kupffer cells to be suspected of playing a role in the pathogenesis of the disease. First, sinusoidal cell damage has been reported by several authors to occur in AIDS patients. The histological changes concerned are mainly hyperplasia of Kupffer cells[97] and peliosis hepatitis.[98] Furthermore, in several samples obtained from patients with AIDS or AIDS-related complex, HIV-specific antigens (P24 gag protein) could be found in Kupffer cells and other undetermined sinusoidal cells.[99] Finally, it has been shown recently that human Kupffer cell cultures (Figure 7) do support the multiplication of HIV-1 as well as that of other primate lentiviruses.[100,101] The fact that these cells are relatively resistant to the infection means that they are likely to constitute a reservoir from which the infection may disseminate to other cells, especially to lymphocytes. On the other hand, infected Kupffer cells expressing viral antigens at their surface may play a role in the depletion of uninfected T4 lymphocytes.

IV. CONCLUSION

During the course of most common human virus-induced liver diseases, early alterations of the Kupffer cells are usually reported. For hepatitis A, V, and E, this could appear as the first step of the infection in which these viruses could exert a cytopathic effect in nonpermissive cells, followed by either recovery or invasion of the liver. Nothing is known about the subclinical forms of these diseases that are, numerically speaking, the most frequent. The Kupffer cells may play a prominent role in promptly limiting the infection, either by themselves or through the rapid recruitment of immunocompetent cells, transmitting in all cases the antigenic information that allows the production of protective antibodies. Contrarily, a weaker or aberrant initial response of the Kupffer cells is likely to be responsible for certain fulminant hepatitis forms which they may aggravate by disrupting the sinusoidal barrier and/or by the direct production of toxic cytokines.

REFERENCES

1. **Fenner, F.,** The clinical features and pathogenesis mouse-pox (infectious ectromelia of mice), *J. Pathol. Bacteriol.,* 60, 529, 1948.
2. **Kirn, A., Gut, J. P., and Gendrault, J. L.,** Interaction of viruses with sinusoidal cells, in *Progress in Liver Diseases,* Vol. 7, H. Popper and F. S. Schaffner, Eds., Grune & Stratton, New York, 1982, 377.
3. **Wake, K., Decker, K., Kirn, A., Knook, D. L., McCuskey, R. S., Bouwens, L., and Wisse, E.,** Cell biology and kinetics of Kupffer cells in the liver, *Int. Rev. Cytol.,* 118, 173, 1990.

4. **Bang, F. B. and Warwick, A.,** Macrophages and mouse hepatitis, *Virology,* 9, 715, 1959.
5. **Mims, C. A.,** Rift Valley fever virus in mice. II. Adsorption and multiplication of virus, *Br. J. Exp. Pathol.,* 37, 110, 1956.
6. **Mims, C. A.,** The response of mice to large intravenous injections of ectromelia virus. II. The growth of virus in the liver, *Br. J. Exp. Pathol.,* 40, 543, 1959.
7. **Coffin, D. L., Coons, A. H., and Cabasso, V. J.,** A histological study of infectious canine hepatitis by means of fluorescent antibody, *J. Exp. Med.,* 98, 13, 1953.
8. **Tigertt, W. D., Berge, T. O., Gouchenour, W. S., Gleiser, C. A., Eveland, W. C., Vorderbruegge, C., and Smetana, H. F.,** Experimental yellow fever, *Trans. N.Y. Acad. Sci.,* 22, 323, 1960.
9. **Granoff, A., Came, P. E., and Breeze, D. C.,** Viruses and renal carcinoma of Rana pipiens. I. The isolation and properties of virus from normal and tumor tissue, *Virology,* 29, 133, 1966.
10. **Maes, R. and Granoff, A.,** Viruses and renal carcinoma of Rana pipiens. IV. Nucleic acid synthesis in Frog Virus 3-infected BHK 21/13 cells, *Virology,* 33, 491, 1967.
11. **Kirn, A.,** Pouvoir lethal pour la souris du virus 3 de la grenouille (FV 3), *CR Acad. Sci. Ser. D (Paris),* 272, 2504, 1971.
12. **Gut, J. P., Anton, M., Bingen, A., Vetter, J. M., and Kirn, A.,** Frog Virus 3 induces a fatal hepatitis in rats, *Lab. Invest.,* 45, 218, 1981.
13. **Kirn, A., Gut, J. P., Steffan, A. M., and Gendrault, J. L.,** Immunization of mice against the toxic hepatitis produced by FV 3: inhibition of virus penetration into the liver, *Intervirology,* 2, 244, 1973/74.
14. **Kirn, A., Gut, J. P., Bingen, A., and Hirth, C.,** Acute hepatitis produced by Frog Virus 3 in mice, *Arch. Gesamte Virusforsch.,* 35, 394, 1971.
15. **Bingen-Brendel, A., Batzenschlager, A., Gut, J. P., Hirth, C., Vetter, J. M., and Kirn, A.,** Etude histologique et virologique de l'hepatite degenerative aigue provoquee par le FV 3 (Frog Virus 3) chez la souris, *Ann. Inst. Pasteur, Paris,* 122, 125, 1972.
16. **Aubertin, A. M., Anton, M., and Bingen, A.,** Solubilized viral proteins produce fatal hepatitis in mice, *Nature (London),* 265, 456, 1977.
17. **Gendrault, J. L., Steffan, A. M., and Bingen, A.,** Interaction of Frog Virus 3 with sinusoidal cells, in *Kupffer Cells and Other Liver Sinusoidal Cells,* E. Wisse and D. L. Knook, Eds., Elsevier, Amsterdam, 1977, 223.
18. **Bingen, A. and Kirn, A.,** Modifications ultrastructurales precoces des noyaux des hepatocytes de souris aucours de l'hepatite degenerative aigue provoquee par le FV3 (Frog Virus 3), *J. Ultrastruct. Res.,* 45, 343, 1973.
19. **Bingen, A. and Kirn, A.,** Fibrillar bodies in hepatocyte nuclei during the course of the toxic hepatitis produced by Frog Virus 3 in mice, *J. Ultrastruct. Res.,* 50, 167, 1975.
20. **Bingen, A. and Kirn, A.,** Hepatocellular necrosis during Frog Virus 3-induced hepatitis of mice. An electron microscopic study, *Exp. Mol. Pathol.,* 27, 68, 1977.
21. **Gendrault, J. L., Steffan, A. M., Bingen, A., and Kirn, A.,** Penetration and uncoating of Frog Virus 3 (FV 3) in cultured rat Kupffer cells, *Virology,* 112, 375, 1981.

22. **McCuskey, R. S., McCuskey, P. A., Ditter, B., Ditter, B., and Kirn, A.,** "In vivo" microscopic study of dynamic events occurring in hepatic sinusoids following FV 3 virus infection, *Hepatology,* 3, 844, 1983.
23. **Bingen, A., Bouteille, M., and Kirn, A.,** Localization by autoradiography of viral proteins in the parenchymal cells of the liver during Frog Virus 3-induced hepatitis of mice, *J. Submicrosc. Cytol.,* 14, 55, 1982.
24. **Elharrar, M., Hirth, C., and Blanc, J.,** Pathogenie de l'hepatite toxique de la souris provoquee par le FV 3 (Frog Virus 3), inhibition de la synthese des macromolecules de foie, *Biochim. Biophys. Acta,* 319, 91, 1973.
25. **Elharrar, M. and Kirn, A.,** Effect of Frog Virus 3 infection on protein synthesis activity of mouse liver ribosomes, *FEBS Lett.,* 1, 13, 1977.
26. **Elharrar, M. and Kirn, A.,** Inhibition of DNA synthesis by isolated liver nuclei from Frog Virus 3-infected mice, *Biochem. Biophys. Res. Commun.,* 57, 801, 1974.
27. **Kirn, A., Steffan, A. M., and Anton, M.,** Phagocytic properties displayed by mouse hepatocytes after virus-induced damage of the sinusoidal lining, *Biomedicine,* 29, 25, 1978.
28. **Steffan, A. M. and Kirn, A.,** Multiplication of vaccinia virus in the livers of mice after Frog Virus 3-induced damage to sinusoidal cells, *J. Reticuloendo. Soc.,* 26, 531, 1979.
29. **Brinton, M. A. and Plagemann, P. G. W.,** Actinomycin D cytotoxicity for mouse peritoneal macrophages and effect on lactate dehydrogenase-elevating virus replication, *Intervirology,* 12, 349, 1979.
30. **Skilleter, D. N., Paine, A. J., and Stirpe, F.,** A comparison of the accumulation of ricin by hepatic parenchymal and non-parenchymal cells and its inhibition of protein synthesis, *Biochim. Biophys. Acta,* 677, 495, 1981.
31. **Farber, J.,** *Pathology of Transcription and Translation,* Marcel Dekker, New York, 1972, 55.
32. **Gans, H. and Matsumo, K.,** The escape of endotoxin from the intestine, *Surg. Gynecol. Obstet.,* 139, 395, 1974.
33. **Jacob, A. I., Goldberg, P. K., and Bloom, N.,** Endotoxin and bacteria in portal blood, *Gastroenterology,* 72, 1268, 1977.
34. **Nolan, J. P.,** Endotoxin, reticuloendothelial function, and liver injury, *Hepatology,* 1, 458, 1981.
35. **Gut, J. P., Steffan, A. M., Anton, M., and Kirn, A.,** Kupffer cell functions and Frog Virus 3 hepatitis in mice and rats, in *The Reticuloendothelial System and the Pathogenesis of Liver Disease,* H. Liehr and M. Grun, Eds., Elsevier/North-Holland, Amsterdam, 1980, 211.
36. **Gut, J. P., Anton, M., Bingen, A., Schmitt, S., and Kirn, A.,** Further indications for the role of enteric endotoxins in the pathogenesis of FV 3 hepatitis, in *Sinusoidal Liver Cells,* D. L. Knook and E. Wisse, Eds., Elsevier/North-Holland, Amsterdam, 1982, 413.
37. **Gut, J. P., Schmitt, S., and Bingen, A.,** Protective effect of colectomy in Frog Virus 3 (FV 3) hepatitis of rats. Possible role of endotoxins, *J. Infect. Dis.,* 146, 594, 1982.
38. **Gut, J. P., Schmitt, S., Bingen, A., Anton, M., and Kirn, A.,** Probable role of endogenous endotoxins in hepatocytolysis during murine hepatitis caused by Frog Virus 3, *J. Infect. Dis.,* 149, 621, 1984.

39. **Hagmann, W., Kirn, A., and Keppler, D.,** Role of leukotrienes in acute inflammatory liver disease, in *Modulation of Liver Cell Expression,* W. Reuter, I. M. Arias, L. Bianchi, T. C. Heinrich, D. Keppler and L. Landmann, Eds., MTP Press, Lancaster, England, 1987, 423.
40. **Hagmann, W., Steffan, A. M., Kirn, A., and Keppler, D.,** Leukotrienes as mediators in Frog Virus 3-induced hepatitis in rats, *Hepatology,* 7, 732, 1987.
41. **Abecassis, M., Falk, J. A., Makowka, L., Dindzans, V. J., Falk, R. E., and Levy, G. A.,** 16-16 Dimethyl prostaglandin E2 prevents the development of fulminant hepatitis and blocks the induction of monocyte/macrophage procoagulant activity after murine hepatitis virus strain 3 infection, *J. Clin. Invest.,* 80, 881, 1987.
42. **Piazza, M., Panet, G., and de Ritis, F.,** The fate of MHV 3 after intravenous injection into susceptible mice, *Arch. Gesamte Virusforsch.,* 22, 472, 1967.
43. **Levy-Leblond, E., Orth, D., and Dupuy, J. M.,** Genetic study of mouse sensitivity to MHV 3 infection: influence of the H-2 complex, *J. Immunol.,* 122, 1359, 1979.
44. **Virelizier, J. L. and Allison, A. C.,** Correlation of persistent mouse hepatitis virus 3 (MHV 3) infection with its effect on mouse macrophage cultures, *Arch. Virol.,* 50, 279, 1976.
45. **Bang, F. B. and Warwick, A.,** Mouse macrophages as host cells for the mouse hepatitis virus and the genetic basis of their susceptibilities, *Proc. Natl. Acad. Sci. U.S.A.,* 46, 1065, 1960.
46. **Arnheiter, H., Thomas, B., and Haller, O.,** Adult mouse hepatocytes in primary monolayer culture express genetic resistance to mouse hepatitis virus 3, *J. Immunol.,* 129, 1275, 1982.
47. **Pereira, C. A., Steffan, A. M., and Kirn, A.,** Interaction between mouse hepatitis viruses and primary cultures of Kupffer and endothelial liver cells from resistant and susceptible inbred mouse strains, *J. Gen. Virol.,* 65, 1617, 1984.
48. **Pereira, C. A., Steffan, A. M., and Kirn, A.,** Kupffer and endothelial liver cell damage renders A/J mice susceptible to mouse hepatitis virus type 3, *Virus Res.,* 1, 557, 1984.
49. **Pereira, C. A., Steffan, A. M., Koehren, F., Douglas, D. R., and Kirn, A.,** Increased susceptibility of mice to MHV 3 infection induced by hypercholesterolemic diet: impairment of Kupffer cell function, *Immunobiology,* 174, 253, 1987.
50. **Levy, G. A., Leibowitz, J. L., and Edgington, T. S.,** Induction of monocyte procoagulant activity by murine hepatitis virus type 3 parallels disease susceptibility in mice, *J. Exp. Med.,* 154, 1150, 1981.
51. **Keller, F., Schmitt, D., and Kirn, A.,** Interaction of mouse hepatitis virus 3 with Kupffer cells explanted from susceptible and resistant mouse strains. Antiviral activity, interleukin-1 synthesis, *FEMS Microbiol. Immunol.,* 47, 87, 1988.
52. **Abecassis, M., Falk, J., Dindzans, V., Lopatin, W., Makowka, L., Levy, G., and Falk, R.,** Prostaglandin E2 prevents fulminant hepatitis and the induction of procoagulant activity in susceptible animals, *Transplant. Proc.,* 19, 1103, 1987.

53. Morphological criteria in viral hepatitis. Review by an international group, *Lancet,* 1, 333, 1971.
54. **Dienstag, J. L., Popper, H., and Purcell, R. H.,** The pathology of viral hepatitis types A and B: a comparison, *Am. J. Pathol.,* 85, 131, 1976.
55. **Ishak, K. G.,** Light microscopic morphology of viral hepatitis, *Am. J. Clin. Pathol.,* 65, 787, 1976.
56. **McSween, R. N. M.,** Pathology of viral hepatitis and its sequelae, *Clin. Gastroenterol.,* 9, 23, 1986.
57. **Tyler, G. V., Snyder, R. L., and Summer, J.,** Experimental infection of the woodchuck (marmotta monax monax) with woodchuck hepatitis virus, *Lab. Invest.,* 55, 51, 1986.
58. **Mims, C. A.,** Tumor necrosis factor, the acute phase response and the pathogenesis of alcoholic liver disease (editorial), *Hepatology,* 9, 497, 1989.
59. **Morahan, P. S., Connor, J. R., and Leay, K. R.,** Viruses and the versatile macrophage, *Br. Med. Bull.,* 41, 15, 1985.
60. **Mims, C. A.,** Interactions of the viruses within the immune system, *Clin. Exp. Immunol.,* 66, 1, 1986.
61. **Rouse, B. T. and Horohow, D. W.,** Immunosuppression in viral infections, *Rev. Infect. Dis.,* 8, 850, 1986.
62. **Vetter, D., Doffoel, M., and Bockel, R.,** Aspects immunologiques de la physiopathologie des hepatites virales B, *Gastroenterol. Clin. Biol.,* 13, 916, 1989.
63. **Prieto, J., Castilla, A., Subira, M. L., Serrano, M., Morte, S., and Civeira, M. P.,** Cytoskeletal organization and functional changes in monocytes from patients with chronic hepatitis B: relationship with viral replication, *Hepatology,* 9, 720, 1989.
64. **Harrison, T. J.,** Hepatitis B virus DNA in peripheral leukocytes: a brief review, *J. Med. Virol.,* 31, 33, 1990.
65. **Korba, B. E., Wells, F., Tennant, B. C., Yoakum, G. H., Purcell, R. H., and Gein, J. L.,** Hepadnavirus infection of peripheral blood lymphocytes in vivo: woodchuck and chimpanzee models of viral hepatitis, *J. Virol.,* 58, 1, 1986.
66. **Vento, S., Ranieri, S., Williams, R., Rondanell, E. G., O'Brien, C. J., and Eddleston, A. L. W. F.,** Prospective study of cellular immunity to hepatitis B virus antigens from the early incubation phase of acute hepatitis B, *Lancet,* 2, 119, 1987.
67. **Thiele, D. L.,** Tumor necrosis factor, the acute phase response and the pathogenesis of alcoholic liver disease (editorial), *Hepatology,* 9, 497, 1989.
68. **Shiratori, Y., Tanaka, M., Hai, K., Kawase, T., Shiina, S., and Sugimoto, T.,** Role of endotoxin-responsive macrophages in hepatic injury, *Hepatology,* 11, 183, 1990.
69. **Deviere, J., Content, J., Denys, D., Vandenbussche, P., Schandene, L., Wybran, J., and Dupont, E.,** Excessive in vitro bacterial lipopolysaccharide-induced production of monokines in cirrhosis, *Hepatology,* 11, 628, 1990.
70. **Huber, M. and Keppler, D.,** Eicosanoids and the liver, in *Progress in Liver Disease,* H. Popper and F. Schaffner, Eds., W.B. Saunders, Philadelphia, 9, 117, 1990.
71. **Nolan, J. P.,** Intestinal endotoxins as mediators of hepatic injury. An idea whose time has come again (editorial), *Hepatology,* 10, 887, 1989.

72. **Shiratori, Y., Kawase, T., Shiina, S., Okano, K., Sugimoto, T., Teraoka, M., Matano, S., Matsumoto, K., and Kamii, K.,** Modulation of hepatotoxicity by macrophages in the liver, *Hepatology,* 8, 815, 1988.
73. **Eddleston, A. L. W. F., Mondelli, M., Mieli-Vergani, G., and Williams, R.,** Lymphocyte cytotoxicity against autologous hepatoytes in chronic hepatitis B virus infection, *Hepatology,* 2, 122s, 1982.
74. **Triger, D. R., Bialas, M. C., Segasdby, C. A., and Underwood, J. C. E.,** Hepatic reticuloendothelial function: a correlation of radioisotopic and immunohistochemical assessment, *Liver,* 9, 86, 1989.
75. **Anastassakos, C. H., Alexander, G. J. M., Woltenscroft, R. A., Dumonde, D. C., Eddleston, A. L. W. F., and Williams, R.,** Interleukin-1 and interleukin-2 activity in chronic hepatitis B virus infection, *Gastroenterology,* 94, 999, 1988.
76. **Takinawa, K., Sata, M., Setoyama, H., and Abe, H.,** Changes of the Kupffer cell and clinical manifestations in acute hepatitis type A, in *Cells of the Hepatic Sinusoid,* A. Kirn, D. C. Knook, and E. Wisse, Eds., The Kupffer Cell Foundation, Rijswijk, Netherlands, 1, 371, 1985.
77. **Mathiesen, L. R., Drucker, J., Lorenz, D., Wagner, J. A., Gerety, R. J., and Purcell, R. H.,** Localization of hepatitis A in marmoset organs during infection with hepatitis A virus, *J. Infect. Dis.,* 138, 369, 1978.
78. **Margolis, H. S., Nainan, O. V., Krawczynski, K., Bradley, D. W., Ebert, J. W., Spelbring, J., Fields, H. A., and Maynard, J. E.,** Appearance of immune complexes during experimental hepatitis A infection in chimpanzees, *J. Med. Virol.,* 26, 31, 1988.
79. **Margolis, H. S. and Nainan, O. V.,** Identification of virus components in circulating immune complexes isolated during hepatitis A virus infection, *Hepatology,* 11, 31, 1990.
80. **Valbracht, A., Gabriel, P., Maier, K., Hartmann, F., Steinhardt, H. J., Muller, C., Wolf, A., Mannke, K. H., and Flehming, B.,** Cell-mediated cytotoxicity in hepatitis A virus infection, *Hepatology,* 6, 1308, 1986.
81. **Schmid, M., Pirovino, M., Altofer, J., Gudat, F., and Bianchi, L.,** Acute hepatitis Non A, Non B; are there any specific light microscopic features?, *Liver,* 2, 61, 1982.
82. **Purcell, R. H.,** Enterically transmitted Non A, Non B hepatitis, in *Progress in Liver Diseases,* H. Popper and F. Schaffner, Eds., W.B. Saunders, Philadelphia, 9, 497, 1990.
83. **Ramalingaswami, V. and Purcell, R. H.,** Waterborne Non A, Non B hepatitis, *Lancet,* 1, 571, 1988.
84. **Inoue, O., Nagataki, S., Itakura, H., Iida, F., Shimizu, Y., Yano, M., Shikata, T., and Tandon, B. N.,** Liver morphology in marmoset infected with epidemic Non A, Non B hepatitis in India, *Liver,* 6, 178, 1986.
85. **Dienes, H. P., Popper, H., Arnold, W., and Lobeck, H.,** Histologic observations in human hepatitis Non A, Non B, *Hepatology,* 2, 562, 1982.
86. **Ten Napel, C. H. H. and Houthoff, J. F.,** The T.H. cytomegalovirus hepatitis in normal and immune compromised hosts, *Liver,* 4, 184, 1984.
87. **Friffith, P. D.,** Cytomegalovirus and the liver, in *Seminars in Liver Disease,* Thieme-Stratton, New York, 4, 307, 1984.

88. **Snover, D. C. and Horwitz, C. A.,** Liver disease in cytomegalovirus mononucleosis: a light microscopical and immunoperoxidase study of six cases, *Hepatology,* 4, 408, 1984.
89. **Bronsther, O., Makowka, L., Jaffe, R., Demetris, A. J., Breinig, M. K., Ho, M., Esquivel, C. W., Gordon, R. D., Iwatsuki, S., Tzakis, A., Marsh, J. W., Jr., Mazzaferro, V., Van Thiel, D., and Starzl, T. E.,** Occurrence of cytomegalovirus hepatitis in liver transplant patients, *J. Med. Virol.,* 24, 423, 1988.
90. **Jacobson, I. M., Gang, D. L., and Schapiro, R. H.,** Epstein-Barr viral hepatitis: an unusual case and review of the literature, *Am. J. Gastroenterol.,* 79, 628, 1984.
91. **Gowing, N. F. C.,** Infectious mononucleosis. Histopathologic aspects, *Pathol. Annu.,* 10, 1, 1975.
92. **Markins, R. E., Linder, J., Zuerlein, K., Mroczek, E., Grierson, H. L., Brichacek, B., and Purtilo, D. T.,** Hepatitis in fatal mononucleosis, *Gastroenterology,* 93, 1210, 1987.
93. **Marrie, T. J., McDonald, A. T. J., Conen, P. E., and Boudreau, S. F. J.,** Herpes simplex hepatitis. Use of immunoperoxidase to demonstrate the viral antigen in hepatocytes, *Gastroenterology,* 82, 71, 1982.
94. **Raga, J., Chrystal, V., and Coovadia, H. M.,** Usefulness of clinical features and liver biopsy in diagnosis of disseminated herpes simplex infection, *Arch. Dis. Child.,* 59, 820, 1984.
95. **Stevens, T. G. and Cook, M. L.,** Restriction of herpes virus infection by macrophages, *J. Exp. Med.,* 133, 19, 1971.
96. **Moahan, P. S., Glasgow, L. A., Crane, J. R., and Kern, E. R.,** Comparison of antiviral and antitumor activity of activated macrophages, *Cell. Immunol.,* 28, 404, 1977.
97. **Czapar, C. A., Weldon-Linne, C. M., Moore, D. M., and Rhone, D. P.,** Peliosis hepatitis in the acquired immunodeficiency syndrome, *Arch. Pathol. Lab. Med.,* 110, 611, 1986.
98. **Scoazec, J. Y., Marche, C., Girard, P. M., Houtmann, J., Durand-Schneider, A. M., Saimot, A. G., Benhamou, J. P., and Feldmann, G.,** Peliosis hepatitis and sinusoidal dilation during infection by the human immunodeficiency virus (HIV), an ultrastructural study, *Am. J. Pathol.,* 131, 38, 1988.
99. **Housset, C., Boucher, O., Girard, P. M., Leibowitch, J., Saimot, A. G., Brechot, C., and Marche, C.,** Immunohistochemical evidence for human immunodeficiency virus-1 infection of liver Kupffer cells, *Hum. Pathol.,* 21, 404, 1989.
100. **Schmitt, M. P., Steffan, A. M., Gendrault, J. L., Jaeck, D., Royer, C., Schweitzer, C., Beyer, C., Schmitt, C., Aubertin, A. M., and Kirn, A.,** Multiplication of human immunodeficiency virus in primary cultures of human Kupffer cells, possible role of liver macrophage infection in the physiopathology of AIDS, *Res. Virol.,* 141, 143, 1990.
101. **Schmitt, M. P., Gendrault, J. L., Steffan, A. M., Beyer, C., Jaeck, D., Kirn, A., and Aubertin, A. M.,** Human Kupffer cell cultures support the multiplication of primate AIDS-inducing lentiviruses, in *Cells of the Hepatic Sinusoid,* E. Wisse, D. L. Knook, and R. S. McCuskey, Eds., The Kupffer Cell Foundation, Rijswijk, Netherlands, 3, 1991.
102. **Gut, J. P.,** unpublished observations.

Chapter 7

ROLE OF MACROPHAGES AND ENDOTHELIAL CELLS IN HEPATOTOXICITY

D.L. Laskin

TABLE OF CONTENTS

ISBN 0-8493-6109-5

I. INTRODUCTION

Approximately 30 to 35% of the cells in the liver are nonparenchymal cells. The majority of these cells, which include Kupffer cells, endothelial cells, fat-storing cells, and pit cells, reside within the hepatic sinusoids. Epithelial cells and fibroblasts are also found in the liver. The epithelial cells form the walls of the bile ducts, whereas fibroblasts are distributed throughout the liver lobule. Nonparenchymal cells have been largely ignored in studies aimed at elucidating cellular mechanisms of hepatotoxicity, both because of their small size relative to hepatocytes and the small volume they occupy in the liver. However, following exposure to hepatotoxic chemicals, nonparenchymal cells can become ''activated''. They release large quantities of highly reactive mediators such as superoxide anion, hydrogen peroxide, nitric oxide, eicosinoids, and proteolytic enzymes which can damage hepatic tissue. These ''activated'' cells may thus contribute to toxicity.

Our laboratory has been interested in studying the potential role of activated macrophages and endothelial cells in hepatotoxicity. For our studies, we have focused on acetaminophen and lipopolysaccharide as model hepatotoxicants. Acetaminophen is a mild analgesic and antipyretic agent which is safe and effective when taken in low doses. Ingestion of high doses, however, leads to acute liver failure accompanied by centrilobular degeneration and necrosis in the liver in both man and experimental animals.[1-5] Lipopolysaccharide is a toxic cell wall component of Gram-negative bacteria. It is present in large quantities in the intestine as a result of bacterial death and is released during bacterial growth. Small amounts of lipopolysaccharide are rapidly cleared from the portal circulation by Kupffer cells.[6,7] Impairment of Kupffer cell function or exposure to large amounts of lipopolysaccharide, however, is associated with hepatocellular necrosis.[8,9] A number of studies have suggested that the toxicity of acetaminophen and lipopolysaccharide is not limited to a direct effect of these agents on hepatocytes.[9-14] Other cell types, in particular hepatic nonparenchymal cells and inflammatory leukocytes, may participate in this process. For example, our laboratory has found that treatment of rats with hepatotoxic doses of acetaminophen or lipopolysaccharide results in a significant increase in the number of macrophages and endothelial cells in the liver (Table 1). These cells are known to release reactive mediators which may be inflammatory, cytotoxic, and/or vasoactive[15] and have been implicated in liver injury.[16] When animals are treated with agents that activate hepatic macrophages and endothelial cells, such as lipopolysaccharide or poly I:C, acetaminophen hepatotoxicity is enhanced.[17] In contrast, a hepatoprotective effect is observed following treatment of animals with gadolinium chloride or dextran sulfate, which block Kupffer cell function.[17] Thus, in this example, macrophages and endothelial cells appear to play a critical role in liver parenchymal cell damage. This cytotoxic process most likely involves extracellular mediators such as reactive oxygen and

TABLE 1
Effects of Lipopolysaccharide (LPS) and Acetaminophen Treatment of Rats on Hepatic Macrophage and Endothelial Cell Number

Treatment	Cells per gram perfused liver × 10^6	
	Macrophages	Endothelial cells
None	2.3 ± 0.3^a	12.7 ± 1.8
LPS	6.2 ± 0.8^b	19.9 ± 2.8^b
Acetaminophen	5.7 ± 0.8^b	16.0 ± 2.0^b

Note: Macrophages and endothelial cells were isolated from female Sprague-Dawley rats (200 to 250 gm) by *in situ* perfusion followed by differential centrifugation and centrifugal elutriation as previously described.[11,13,84,110] Rats were treated with 5 mg/kg *Escherichia coli* lipopolysaccharide (LPS, i.v., 48 h) or 1.2 g/kg acetaminophen (p.o., 24 h) prior to nonparenchymal cell isolation.

[a] Mean ± S.E. from six to eight separate experiments.
[b] Significant difference ($p \leq 0.05$) between control and treated rats.

(Reprinted from Laskin, D. L., *Semin. Liver Dis.*, 10, 293, 1990. With permission.)

nitrogen intermediates, lysosomal enzymes, leukotrienes, tumor necrosis factor (TNF), and interleukins (IL) released by hepatic macrophages and endothelial cells. The present chapter summarizes the experimental data from our laboratory implicating these hepatic nonparenchymal cells in the toxicity of lipopolysaccharide and acetaminophen.

II. EFFECTS OF LIPOPOLYSACCHARIDE AND ACETAMINOPHEN ON LIVER HISTOLOGY

Treatment of rats with toxic doses of acetaminophen or lipopolysaccharide is associated with infiltration of macrophages into the liver (Figure 1). With both of these agents, accumulation is rapid, typically occurring within 24 to 48 h. However, the localization of the macrophages within the liver varies with the chemical agent. Thus, treatment of rats with acetaminophen results in the accumulation of macrophages in centrilobular regions of the liver (Figure 1A), presumably due to high levels of cytochrome P-450 mixed-function oxidases in these regions, which metabolically activate acetaminophen to toxic intermediates.[4,5,18] In contrast, macrophages that accumulate in the liver following lipopolysaccharide treatment of rats are scattered throughout the liver lobule (Figure 1B). These patterns of localization of macrophages appear to be correlated with areas of the liver that subsequently become necrotic.[4,5,19,20] We hypothesized that the macrophages that accumulate in the liver following hepatotoxicant exposure become "activated" and release me-

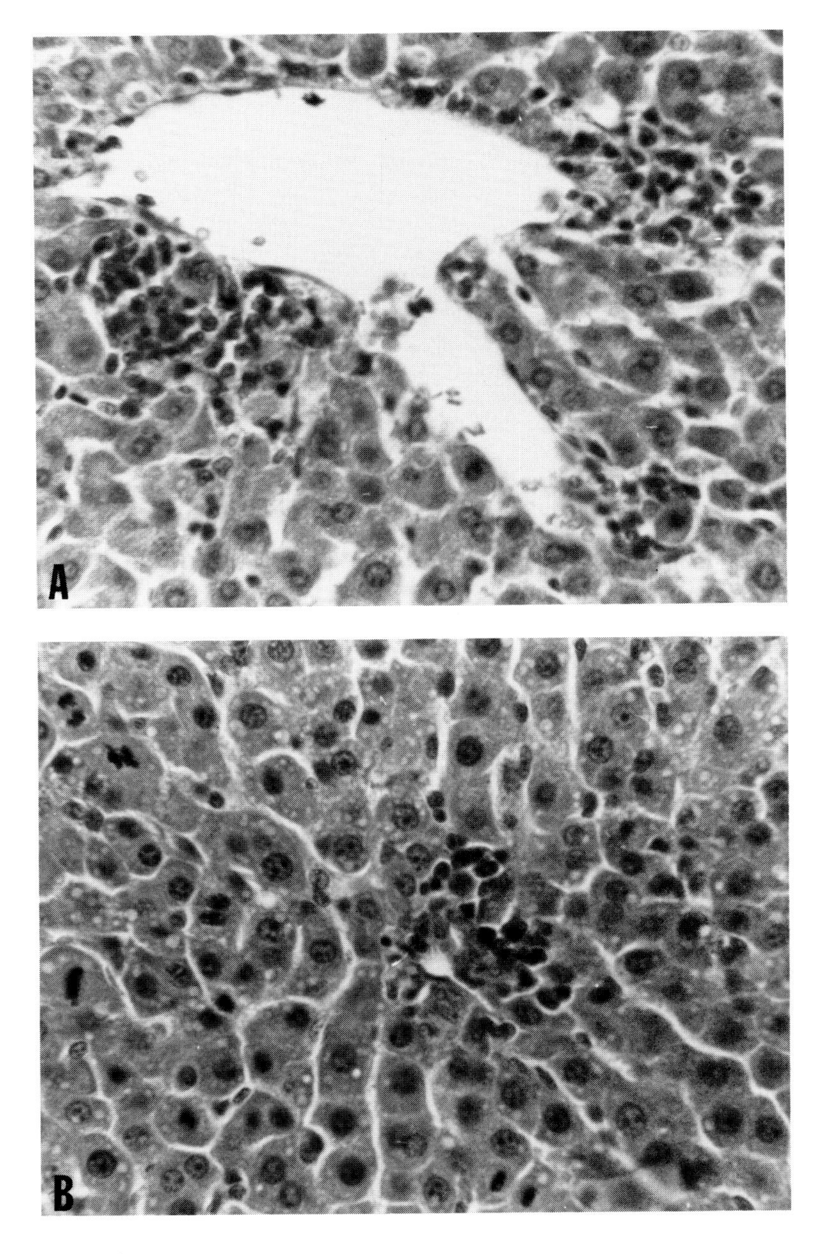

FIGURE 1. Histologic sections of perfused liver from rats treated with acetaminophen (1.2 g/kg) (A) and lipopolysaccharide (5 mg/kg) (B). (Magnification × 400.)

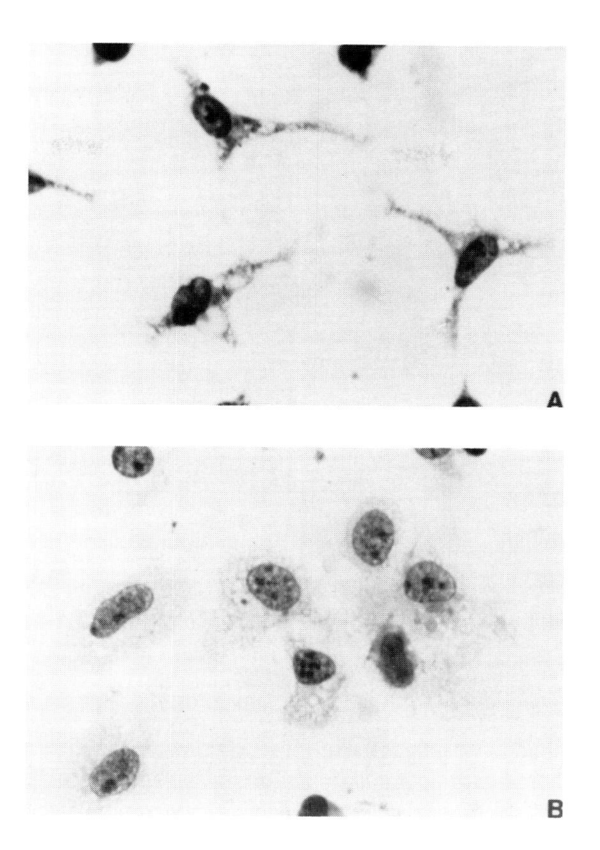

FIGURE 2. Photomicrographs of macrophages from livers of control (A) and acetaminophen-treated (B) rats. Macrophages, cultured for 24 h, were stained with Giemsa and photographed under oil immersion. (Magnification × 1000.) (Reprinted from Laskin, D. L. and Pilaro, A. M., *Toxicol. Appl. Pharmacol.*, 86, 204, 1986. With permission.)

diators that contribute to toxicity.[11,13] To test this hypothesis, we compared the morphologic and functional properties of macrophages isolated from livers of control and acetaminophen- or lipopolysaccharide-treated rats.

III. ACTIVATION OF LIVER MACROPHAGES FOLLOWING HEPATOTOXICANT EXPOSURE

Activated macrophages can be distinguished from resident macrophages morphologically.[11,21] When macrophages were isolated from livers of lipopolysaccharide- or acetaminophen-treated rats and examined microscopically, we found that they displayed very distinct morphology when compared to resident Kupffer cells. In general, macrophages from lipopolysaccharide-[11] or acetaminophen-treated rats (Figure 2B) were larger than resident Kupffer cells, highly vacuolated, and displayed an increased cytoplasmic to nuclear

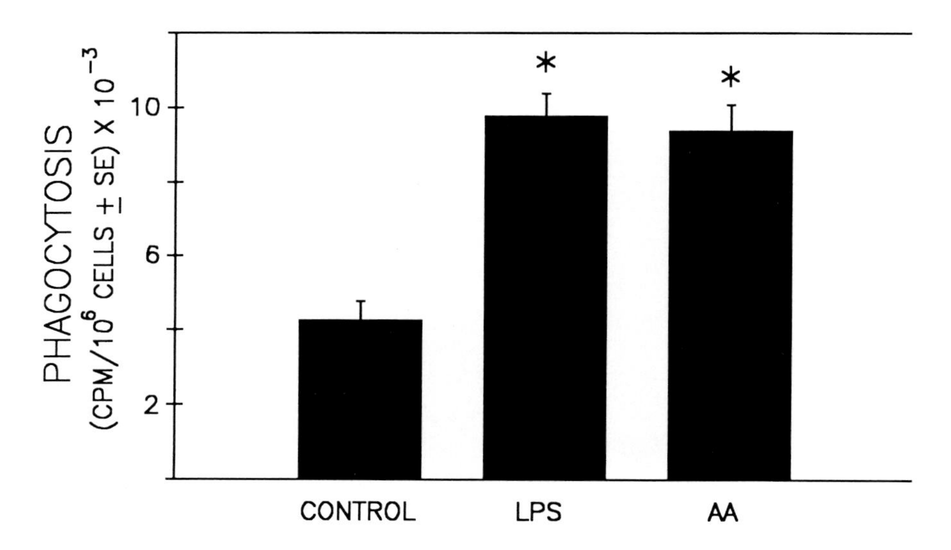

FIGURE 3. Comparison of phagocytosis of opsonized ^{51}Cr-sheep erythrocytes by macrophages from livers of control, lipopolysaccharide (LPS)-, or acetaminophen (AA)-treated rats. Macrophage phagocytosis was quantified as previously described.[11,13] Each bar represents the mean of four samples ± S.E. Statistically significant differences ($p \leq 0.05$) between macrophages from control and treated rats are shown by asterisks.

ratio. In addition, macrophages from lipopolysaccharide- or acetaminophen-treated rats adhered to and spread on culture dishes more rapidly than resident Kupffer cells. In contrast, resident Kupffer cells were stellate with long, thin processes (Figure 2A). These cells were also more refractile than macrophages from treated rats. These properties are characteristic of morphologically activated macrophages.

Since activated macrophages also display enhanced functional capacity, we next compared the ability of the two macrophage cell types to phagocytize sheep erythrocytes, respond to chemotactic stimuli, produce reactive and immune mediators, and kill tumor cells. We found that macrophages from both lipopolysaccharide- and acetaminophen-treated rats exhibited enhanced phagocytic, chemotactic, and cytotoxic activity when compared to resident Kupffer cells (Figures 3 to 5), as well as increased release of superoxide anion, hydrogen peroxide, nitric oxide, IL-1, IL-6, and TNF (Tables 2 and 3, not all shown). These results demonstrate that liver macrophages from hepatotoxicant-treated rats are functionally "activated". It has been proposed that these activated cells promote damage induced by hepatotoxicants such as lipopolysaccharide or acetaminophen through the release of secretory products such as reactive oxygen and nitrogen intermediates, hydrolytic enzymes, and cytokines.[12,20,22-28]

Reactive oxygen intermediates such as superoxide anion and hydrogen peroxide have been implicated as primary mediators of macrophage-induced

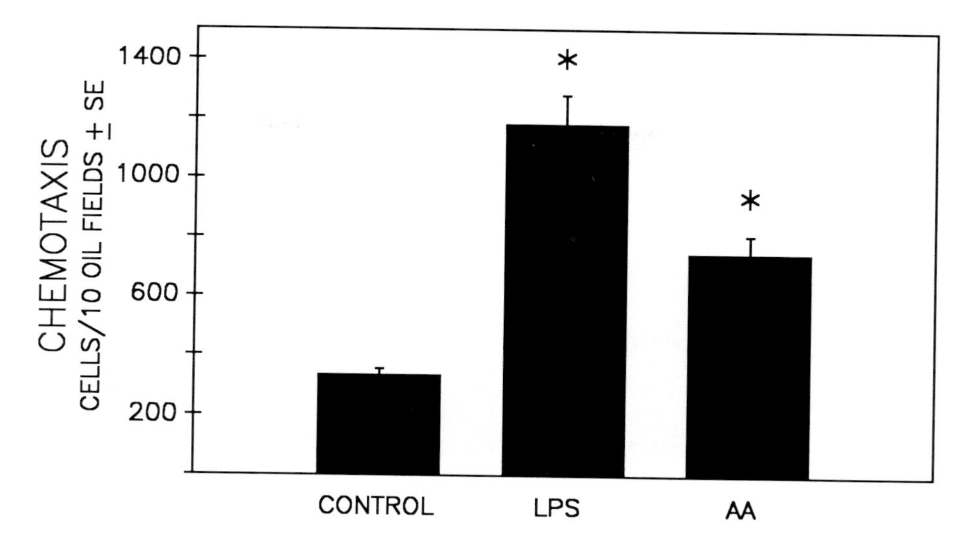

FIGURE 4. Comparison of the chemotactic activity of macrophages from livers of control, lipopolysaccharide (LPS)-, or acetaminophen (AA)-treated rats. Macrophage chemotaxis toward phorbol myristate acetate was quantified by the modified Boyden chamber technique as previously described.[11,13] Each bar represents the average of six samples ± S.E. Statistically significant differences ($p \leq 0.05$) between macrophages from control and treated rats are shown by asterisks.

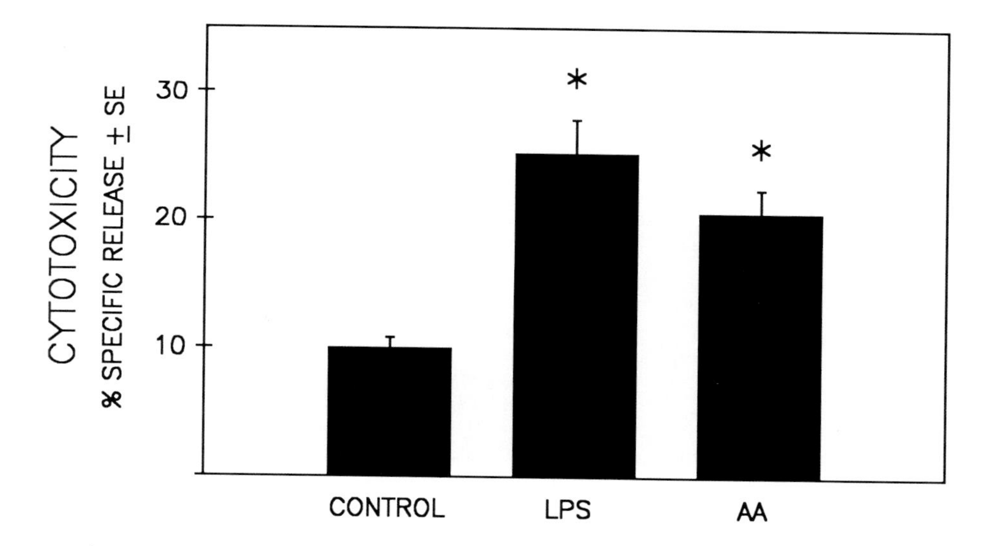

FIGURE 5. Comparison of the cytotoxic activity of macrophages from livers of control, lipopolysaccharide (LPS)-, or acetaminophen (AA)-treated rats. Macrophage cytotoxicity toward N1S1 hepatoma cells was quantified by ^{3}H-thymidine release as previously described.[114] Each bar represents the average of three samples ± S.E. Statistically significant differences ($p \leq 0.05$) between macrophages from control and treated rats are shown by asterisks.

TABLE 2
Effects of Lipopolysaccharide (LPS) Treatment of Rats on Release of Reactive Oxygen Intermediates by Hepatic Macrophages and Endothelial Cells

Treatment	Superoxide anion	Hydrogen peroxide
Macrophages		
Control	126	116
LPS	234	125
Endothelial cells		
Control	11	59
LPS	21	86

Note: Macrophages and endothelial cells were isolated from livers of rats treated with lipopolysaccharide (LPS) as described in the legend to Figure 1. Superoxide anion and hydrogen peroxide production by phorbol myristate acetate-stimulated cells was measured by flow cytometry using hydroethidine and dichlorofluorescin-diacetate, respectively, as previously described.[111,112] The results are presented as mean fluorescence channel number. On our flow cytometer, each histogram is presented on a 4-decade logarithmic scale which is divided into 1024 channels.

TABLE 3
Effects of Lipopolysaccharide (LPS) Treatment of Rats on IL-1 and IL-6 Production by Hepatic Macrophages and Endothelial Cells

Treatment	IL-1	IL-6
Macrophages		
Control	2,082 ± 160	3,194 ± 162
LPS	6,198 ± 1,245	3,549 ± 78
Endothelial cells		
Control	4,501 ± 432	3,927 ± 22
LPS	10,919 ± 122	4,967 ± 251

Note: Macrophages and endothelial cells were isolated from livers of rats treated with lipopolysaccharide (LPS) as described in the legend to Figure 1. Culture medium was collected from these cells after 4 h incubation and analyzed for IL-1 or IL-6 activity by proliferation of D10.G4.1 or B9 cells, respectively, as previously described.[110,113] Each value is the average counts per minute of three samples ± S.E. from one representative experiment.

cellular injury,[29,30] of injury to cultured endothelial cells,[31] and of injury during inflammatory joint diseases.[32] Following stimulation, activated macrophages release superoxide anion, which can react with water and other molecules to generate hydroperoxy and hydroxyl radicals.[33,34] These radicals are even more toxic than superoxide anion and have been linked to membrane and DNA damage, lipid peroxidation reactions, and to the induction of hepatocyte killing.[30,35-37] When macrophages are recruited to the liver following lipopolysaccharide or acetaminophen exposure, they are "activated" to release hydrogen peroxide and superoxide anion. Stimulation of these cells to produce additional reactive oxygen intermediates has been reported to augment hepatic injury induced by agents such as *Corynebacterium parvum* and galactosamine, whereas administration of antioxidants such as superoxide dismutase or allopurinol is hepatoprotective.[24,38,39] These studies support the hypothesis that oxygen-derived free radicals produced by macrophages in the liver contribute to the pathogenesis of hepatic injury.[12,24,38,39]

Recent reports have also implicated reactive nitrogen intermediates in cytotoxicity induced by macrophages.[10,40-43] Nitric oxide and its oxidation products, nitrate and nitrite, are produced by activated macrophages and endothelial cells. These products are hypothesized to be involved in Kupffer cell-mediated alterations in hepatic functions and in cytotoxicity following sepsis or trauma.[10,41] These macrophage and endothelial cell-derived mediators may also be involved in hepatotoxicity induced by acetaminophen or lipopolysaccharide. In this regard, N^G-monomethyl-L-arginine, an inhibitor of nitric oxide synthesis, has recently been reported to be effective in reversing endotoxin-mediated shock.[44]

Release of cytokines such as IL-1, IL-6, and TNF-α by activated Kupffer cells and inflammatory macrophages may also contribute to hepatotoxicity. These mediators can act directly on hepatocytes or may indirectly activate other nonparenchymal cells, as well as leukocytes that accumulate in the liver, thus amplifying the inflammatory response. IL-1 and IL-6 are low-molecular weight proteins that mediate a wide variety of biologic effects, including induction of proliferation and activation of B and T lymphocytes, endothelial cells, and epithelial cells as well as macrophages and natural killer cells.[45-49] IL-6 also acts on the liver to augment the production of acute-phase proteins, DNA synthesis, and lipid metabolism, but depresses albumin synthesis and cytochrome P-450 activity.[45,48,50-56] Similarly, IL-6, also known as hepatocyte stimulating factor-beta, induces acute-phase proteins and modifies fibronectin production.[52,56-59] Recently, there have been reports of the synthesis of both IL-1 and IL-6 by cells other than macrophages, including hepatic fat-storing cells and endothelial cells, as well as fibroblasts, epidermal cells, and glial cells.[45,46,48,49] The fact that these cytokines can affect so many different target tissues and that they are produced by a variety of cell types suggests that they are major mediators of inflammation and immune responses.

TNF-α is a secretory product of activated macrophages.[60,61] It has been implicated not only in the pathogenesis of shock and inflammation, but also in the regulation of acute-phase protein gene expression and in cellular proliferation.[56,61-65] TNF also stimulates the release of other immune mediators, including IL-1, IL-6, colony-stimulating factor, platelet-activating factor, and prostaglandins from inflammatory cells.[61,66-71] TNF may act in concert with these mediators to augment injury in the liver. For example, in endotoxemia associated with alcoholic cirrhosis and galactosamine-induced hepatitis, TNF is thought to be a major mediator of liver damage.[26,72,73] TNF is also known to have deleterious effects on endothelial cells.[74] In addition, this mediator sensitizes neutrophils and monocytes to produce reactive oxygen intermediates.[61,75] Taken together, these data indicate that reactive oxygen and nitrogen species as well as cytokines may act as primary mediators of tissue injury, and/or they may participate in the inflammatory response by initiating a cascade of additional immunologic reactions that result in tissue damage.

IV. ACTIVATION OF HEPATIC ENDOTHELIAL CELLS FOLLOWING TOXICANT EXPOSURE

Recent studies from our laboratory suggest that liver endothelial cells may also contribute to injury induced by hepatotoxicants. For example, following treatment of rats with lipopolysaccharide or acetaminophen, increased numbers of endothelial cells are recovered from rat livers (Table 1). As observed with liver macrophages, the endothelial cells appear to be "activated". They are larger and more granular than cells from untreated animals and produce increased levels of reactive oxygen intermediates, reactive nitrogen intermediates, IL-1, and IL-6 (Tables 2 and 3). The capacity of hepatic endothelial cells to produce these mediators may represent an important mechanism by which these cells participate in inflammatory and immune reactions associated with hepatotoxicity. Vascular endothelial cells are known to proliferate and to produce increased amounts of superoxide anion, as well as IL-1, IL-6, and IL-8 in response to macrophage-derived cytokines.[76-81] These cells also synthesize and secrete a monocyte chemoattractant.[82] Endothelial cells activated by inflammatory mediators are also known to release eicosanoids, thromboxane, platelet-activating factor, reactive nitrogen intermediates, plasminogen activator, and lysosomal enzymes.[66,83-89] These data suggest that inflammatory cells and their products may play a regulatory role in the functional and immunologic capacity of the microvasculature. Similarly, in the liver, it is possible that cytokines released by inflammatory macrophages and activated Kupffer cells promote growth and/or induce the release of reactive mediators from endothelial cells. These reactive mediators may directly injure the matrix of the vasculature and/or the surrounding hepatic tissue, as well as promote the accumulation of inflammatory cells. In support of this possibility, in preliminary studies we found that Kupffer cells, as well as hepatocytes and endothelial cells, release factors that stimulate proliferation of hepatic endothelial cells (Table 4). These data support the idea that leu-

TABLE 4
Release of Endothelial Cell Growth Factors by Parenchymal and Nonparenchymal Liver Cells

Conditioned medium from:	% Control
Hepatocytes[a]	132.6
Kupffer cells	140.3
Endothelial cells	126.6

Note: Endothelial cells from untreated rats were inoculated into 96-well dishes (5×10^4 cells per well). After 24 h incubation, the cells were washed and refed with conditioned medium from hepatocytes, Kupffer cells, or endothelial cells. Proliferation of endothelial cells was quantified 24 h later by ^{3}H-thymidine incorporation. The data are presented as the percentage proliferation of endothelial cells grown in cell-free control conditioned medium.

[a] Conditioned medium was collected from hepatocytes cultured for 48 h and from endothelial cells and Kupffer cells cultured for 8 h.

kocytes play a role in mediating structural changes in the endothelium during inflammation.

V. MECHANISM OF MACROPHAGE ACCUMULATION AND ACTIVATION IN THE LIVER

The mechanism(s) underlying the accumulation and activation of inflammatory cells in the liver following hepatotoxicant exposure is unknown. Damaged tissues and cells release chemotactic and activating factors for phagocytes which may be important for cell migration. These factors bind to specific receptors on phagocytic cells and induce directed cell movement. A number of factors have been characterized that are chemotactic for phagocytes, including complement fragments, products involved in the kinin and coagulation pathway, collagen and tissue breakdown products, arachidonic acid metabolites, in particular leukotriene B_4, as well as synthetic peptides related to bacterial-derived products.[90-94] Recent studies have also described leukocyte-derived chemoattractants for both neutrophils and macrophages.[95] Our laboratory has found that hepatocytes also release a factor that is chemotactic for Kupffer cells (Table 5). Similarly, Perez et al.[96] reported that hepatocytes treated with ethanol release a neutrophil chemotactic factor. These investi-

TABLE 5
Production of Kupffer Cell Chemotactic Factors by Parenchymal and Nonparenchymal Liver Cells

Conditioned medium from	Cells/10 oil fields ± SEM
Controls	18.5 ± 3.6
Untreated hepatocytes[a]	45.6 ± 4.4
Hepatocytes pretreated with acetaminophen	270.4 ± 14.5
Untreated Kupffer cells[b]	135.4 ± 8.9
Untreated endothelial cells	154.0 ± 10.6

Note: Conditioned medium from cultured hepatocytes, endothelial cells, Kupffer cells, or cell-free controls were analyzed for their ability to induce Kupffer cell chemotaxis using the modified Boyden chamber technique as previously described.[13,14] Each value represents the mean of four to six samples ± S.E. from one representative experiment.

[a] Hepatocytes cultured for 48 h were preincubated for 2 h with acetaminophen (50 μ*M*) or control, washed, and then refed with fresh culture medium. Hepatocyte-conditioned medium was collected 48 h later.

[b] Conditioned medium was collected from endothelial cells and Kupffer cell after 24 h incubation.

gators provided preliminary evidence that the chemotactic activity was leukotriene B_4. We found that the hepatocyte-derived monocyte and Kupffer cell chemotactic factor was stable at 4°C for up to 2 months and still active after freeze-thawing.[14] In addition, the chemotactic factor eluted at early times following size exclusion chromatography and was sensitive to trypsin treatment.[14] Taken together, these data suggest that the hepatocyte-derived macrophage chemotactic factor is a high-molecular weight protein.

In further studies, we found that unstimulated endothelial cells as well as Kupffer cells also produce factors that are chemotactic for hepatic macrophages and blood monocytes (Table 5). Production of these factors was time dependent, reaching a maximum after 24 h in culture. Interestingly, the nonparenchymal cell-derived factors appeared to be more biologically active and were released more rapidly than the factors derived from unstimulated hepatocytes. This suggests that nonparenchymal cells play a major role in the maintenance of immune homeostasis in the liver. Taken together, these data also suggest that the increase in the number of both macrophages and endothelial cells in the liver following exposure to hepatotoxicants is mediated by factors released from parenchymal as well as nonparenchymal cells.

Our data demonstrate that macrophages that accumulate in the livers of animals exposed to agents like acetaminophen or lipopolysaccharide are ''activated''. Since macrophage accumulation appeared to be mediated, at least in part, by hepatocyte-derived factors, it was of interest to determine if these factors also induced macrophage activation. To test this possibility, we analyzed the effects of factors derived from injured hepatocytes on macrophage functional responsiveness. We found that hepatocytes treated with acetaminophen release factors that "activate" Kupffer cells.[14] These factors induced morphologic changes in Kupffer cells that were characteristic of activated macrophages, including flattening and spreading of the cells on culture dishes and increased vacuolization. This was similar to the morphology of the activated macrophages isolated from acetaminophen (Figure 1) or lipopolysaccharide-treated rats.[12] Culture medium from hepatocytes treated with acetaminophen, but not acetaminophen by itself, also stimulated Kupffer cell phagocytosis, release of superoxide anion, and cytotoxicity towards normal as well as transformed hepatocytes.[14,97] These data suggest that activation of macrophages in the liver following exposure to chemicals such as acetaminophen is mediated, at least in part, by factors derived from injured hepatocytes. Studies are in progress to determine if hepatocyte-derived factors also mediate endothelial cell activation and if nonparenchymal cells release factors that activate hepatic macrophages.

VI. EFFECTS OF MODIFYING NONPARENCHYMAL CELL FUNCTION ON HEPATOTOXICITY

To further characterize the role of nonparenchymal cells in tissue injury, we also analyzed the effects of modifying macrophage and endothelial cell function on hepatotoxicity. In these experiments, rats were treated simultaneously with acetaminophen and agents that modify nonparenchymal cell activity. The results of these experiments demonstrate that the degree of hepatic injury induced by agents like acetaminophen is directly correlated with macrophage and endothelial cell function. For example, lipopolysaccharide and poly I:C are potent activators of liver macrophages.[11,98] Lipopolysaccharide also appears to activate hepatic endothelial cells.[99] Our laboratory has found that pretreatment of rats with lipopolysaccharide or poly I:C enhanced the toxicity of acetaminophen as evidenced by the appearance of severe hepatic necrosis within 24 h[100] and by increased levels of plasma L-alanine aminotransferase activity (not shown). None of these agents, by themselves, induced hepatotoxicity during this time. In similar experiments, pretreatment of rats with lipopolysaccharide has been reported to aggravate injury induced by hepatotoxicants like carbon tetrachloride, galactosamine, and *C. parvum*.[101,102] Furthermore, animals made tolerant to lipopolysaccharide or treated with the antibiotic polymyxin B, a positively charged detergent that binds to and neutralizes lipopolysaccharide, have been reported to be protected from hepatotoxicity induced by these agents.[102,103] In other experiments, Sipes et al.[104] have found that administration of large doses of vitamin A, which is known to activate Kupffer cells *in vivo*,[105,106] enhanced the hepatotoxicity of carbon tetrachloride. These investigators have proposed that this is due to reactive oxygen intermediates released from vitamin A-activated Kupffer cells. In this regard, methyl palmitate, which blocks Kupffer

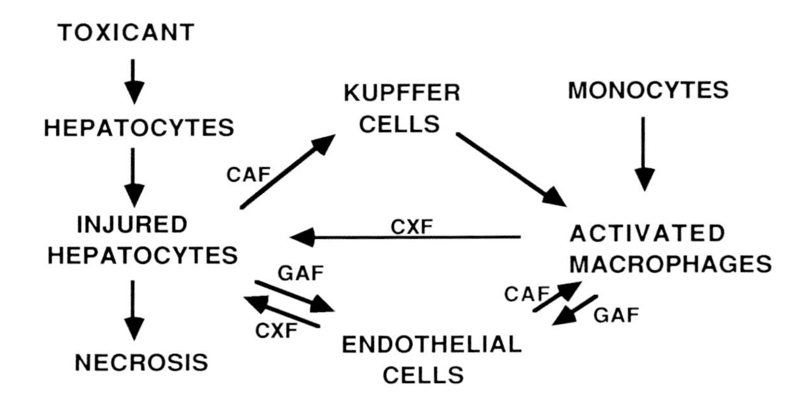

FIGURE 6. Model for the potential role of macrophages and endothelial cells in hepatotoxicity. Toxic doses of drugs or chemicals injure hepatocytes. These injured hepatocytes release factors that attract Kupffer cells to damaged areas of the liver. Additional mononuclear phagocytes are also recruited from blood and bone marrow precursors. Once localized in the liver, the macrophages become "activated" by parenchymal and nonparenchymal cell-derived factors and release mediators that induce proliferation and activation of endothelial cells. Activated macrophages and endothelial cells also release mediators that contribute to the damage initiated by toxicants. This eventually leads to cell death and necrosis. CAF, chemotactic/activating factors; CXF, cytotoxic factors; GAF, growth promoting/activating factors.

cell function, was found to abrogate the enhanced toxicity of carbon tetrachloride induced by vitamin A. Methyl palmitate has also been reported to exert a hepatoprotective effect during galactosamine-induced toxicity.[107] In agreement with these studies, our laboratory has found that accumulation of macrophages in the liver and subsequent toxicity of acetaminophen is blocked by pretreatment of rats with dextran sulfate or gadolinium chloride, agents also known to depress macrophage function.[108,109] Taken together, these data support the hypothesis that macrophages and endothelial cells play a role in hepatotoxicity.

VII. MODEL OF HEPATOTOXICITY

Based on our current experimental data, we have proposed a model of chemically induced hepatic injury (Figure 6). According to this model, hepatocytes injured by toxicants release factors that attract Kupffer cells to specific regions of the liver. Additional mononuclear phagocytes are also recruited from blood and bone marrow precursors. Once localized in the injured area, the macrophages become "activated" by parenchymal and nonparenchymal cell-derived factors and release mediators that induce proliferation and activation of endothelial cells. Activated macrophages and endothelial cells also release mediators that contribute to damage initiated by toxicants. This eventually leads to cell death and necrosis. The data from our laboratory and those of other investigators support this model of hepatotoxicity. Additional studies on the nature of the mediators released from nonparenchymal cells and their effects on hepatocytes will be particularly relevant for understanding mechanisms of liver injury.

VIII. CONCLUSIONS

Chemically induced hepatotoxicity involves participation of both parenchymal and nonparenchymal cells. Although exposure to chemicals and/or their metabolites can directly induce hepatocyte injury, they may also cause hepatotoxicity indirectly through their actions on nonparenchymal cells. A common feature of liver injury induced by a variety of agents is an increase in the content of nonparenchymal cell populations. These cells are ''activated'' to produce large quantities of inflammatory and cytotoxic mediators. Growth and differentiation promoting cytokines are also released from activated nonparenchymal cells, and this can further increase the proliferation of nonparenchymal cells as well as induce cellular ''activation''. This process can then contribute to injury induced by hepatotoxic chemicals. A precise understanding of the nature of these mediators and the factors that regulate their release will be particularly useful in designing treatment protocols to reduce liver damage following exposure to toxic chemicals.

ACKNOWLEDGMENT

This work was supported by USPHS National Institutes of Health grant GM34310.

REFERENCES

1. **Barker, J. D., DeCarle, D. J., and Anuras, S.,** Chronic excessive acetaminophen use and liver damage, *Ann. Intern. Med.*, 87, 299, 1977.
2. **Davidson, D. G. and Eastham, W. N.,** Acute liver necrosis following overdosage of paracetamol, *Br. Med. J.*, 2, 497, 1966.
3. **Johnson, G. K. and Tolman, K. G.,** Chronic liver disease and acetaminophen, *Ann. Intern. Med.*, 87, 302, 1977.
4. **Jollow, D. J., Mitchell, J. R., Potter, W. Z., Davis, D. C., Gillette, J. R., and Brodie, B. B.,** Acetaminophen-induced hepatic necrosis. II. Role of covalent binding *in vivo*, *J. Pharmacol. Exp. Ther.*, 187, 195, 1973.
5. **Mitchell, J. R., Jollow, D. J., Potter, W. Z., Davis, D. C., Gillette, J. R., and Brodie, B. B.,** Acetaminophen-induced hepatic necrosis. I. Role of drug metabolism, *J. Pharmacol. Exp. Ther.*, 187, 185, 1973.

6. **Mathison, J. C. and Ulevitch, R. J.,** The clearance capacity, tissue distribution and cellular location of intravenously injected lipopolysaccharide in rabbits, *J. Immunol.,* 123, 2133, 1979.
7. **Praaning-van Dalen, D. P., Brouwer, A., and Knook, D. L.,** Clearance capacity of rat liver Kupffer, endothelial and parenchymal cells, *Gastroenterology,* 81, 1036, 1981.
8. **Nolan, J. P.,** The role of endotoxin in liver injury, *Gastroenterology,* 69, 1346, 1975.
9. **Nolan, J. P.,** Endotoxin, reticuloendothelial function and liver injury, *Hepatology,* 1, 458, 1981.
10. **Billiar, T. R., Curran, R. D., West, M. A., Hoffman, K., and Simmons, R. L.,** Kupffer cell cytotoxicity to hepatocytes in coculture requires L-arginine, *Arch. Surg.,* 124, 1416, 1989.
11. **Pilaro, A. M. and Laskin, D. L.,** Accumulation of activated mononuclear phagocytes in the liver following lipopolysaccharide treatment of rats, *J. Leuk. Biol.,* 40, 29, 1986.
12. **Bautista, A. P., Meszaros, K., Bojta, J., and Spitzer, J. J.,** Superoxide anion generation in the liver during the early stage of endotoxemia in rats, *J. Leuk. Biol.,* 48, 123, 1990.
13. **Laskin, D. L. and Pilaro, A. M.,** Potential role of activated macrophages in acetaminophen hepatotoxicity. I. Isolation and characterization of activated macrophages from rat liver, *Toxicol. Appl. Pharmacol.,* 86, 204, 1986.
14. **Laskin, D. L., Pilaro, A. M., and Ji, S.,** Potential role of activated macrophages in acetaminophen hepatotoxicity. II. Mechanism of macrophage accumulation and activation, *Toxicol. Appl. Pharmacol.,* 86, 216, 1986.
15. **Nathan, C. F.,** Secretory products of macrophages, *J. Clin. Invest.,* 79, 319, 1987.
16. **Laskin, D. L.,** Nonparenchymal cells and hepatotoxicity, *Semin. Liver Dis.,* 10, 293, 1990.
17. **Laskin, D. L.,** Potential role of activated macrophages in chemical and drug induced liver injury, in *Cells of the Hepatic Sinusoid,* Vol. 2, E. Wisse, D. L. Knook, and K. Decker, Eds., Kupffer Cell Foundation, Amsterdam, 1989, 284.
18. **Gilette, J. R.,** A perspective on the role of chemically reactive metabolites of foreign compounds in toxicity. I. Correlation of changes in covalent binding of reactive metabolites with changes in the incidence and severity of toxicity, *Biochem. Pharmacol.,* 23, 2785, 1974.
19. **Hendriks, H. F. J., Horan, M. A., Durham, S. K., Earnst, D. L., Brouwer, A., Hollander, C. F., and Knook, D. L.,** Endotoxin-induced liver injury in aged and subacutely hypervitaminotic rats, *Mech. Aging Dev.,* 41, 241, 1987.
20. **Mochida, S., Ogata, I., Ohta, Y., Yamada, S., and Fujiwara, K.,** In situ evaluation of the stimulatory state of hepatic macrophages based on their ability to produce superoxide anions in rats, *J. Pathol.,* 158, 67, 1989.
21. **Johnson, W. J., Marino, P. A., Schreiber, R. D., and Adams, D. O.,** Sequential activation of murine mononuclear phagocytes for tumor cytolysis: differential expression of markers by macrophages in several stages of development, *J. Immunol.,* 131, 1038, 1983.

22. **Tanner, A., Keyhani, A., Reiner, R., Holdstock, G., and Wright, R.,** Proteolytic enzymes released by liver macrophages may promote hepatic injury in a rat model of hepatic damage, *Gastroenterology,* 80, 647, 1980.
23. **Abril, E. R., Simm, W. E., and Earnest, D. L.,** Kupffer cell secretion of cytotoxic cytokines is enhanced by hypervitaminosis A, in *Cells of the Hepatic Sinusoid,* Vol. 2, E. Wisse, D. L. Knook, and K. Decker, Eds., Kupffer Cell Foundation, Amsterdam, 1989, 73.
24. **Arthur, M. J. P., Bentley, I. S., Tanner, A. R., Kowalski, P., Millward-Sadler, G. H., and Wright, R.,** Oxygen-derived free radicals promote hepatic injury in the rat, *Gastroenterology,* 89, 1114, 1985.
25. **Ferluga, J. and Allison, A.,** Role of mononuclear infiltrating cells in the pathogenesis of hepatitis, *Lancet,* 2, 610, 1978.
26. **Lehman, V., Freudenberg, M. A., and Galanos, C.,** Lethal toxicity of lipopolysaccharide and tumor necrosis factor in normal and D-galactosamine-treated mice, *J. Exp. Med.,* 165, 657, 1987.
27. **Chojkier, M. and Fierer, J.,** D-Galactosamine hepatotoxicity is associated with endotoxin sensitivity and mediated by lymphoreticular cells in mice, *Gastroenterology,* 88, 115, 1985.
28. **Billiar, T. R., Curran, R. D., Harbrecht, B. J., Stuehr, D. J., Demetris, A. J., and Simmons, R. L.,** Modulation of nitrogen oxide synthesis in vivo: N^G-monomethyl-L-arginine inhibits endotoxin-induced nitrite/nitrate biosynthesis while promoting hepatic damage, *J. Leuk. Biol.,* 48, 565, 1990.
29. **Nathan, C. F., Silverstein, S. C., Brukner, L. H., and Cohn, Z. A.,** Extracellular cytolysis by activated macrophges and granulocytes. II. Hydrogen peroxide as a mediator of cytotoxicity, *J. Exp. Med.,* 149, 100, 1979.
30. **Fantone, J. C. and Ward, P. A.,** Role of oxygen-derived free radicals and metabolites in leukocyte-dependent inflammatory reaction, *Am. J. Pathol.,* 107, 397, 1982.
31. **Sacks, T., Moldow, C. F., Craddock, P. R., Bowers, T. K., and Jacob, H. S.,** Oxygen radicals mediate endothelial cell damage by complement-stimulated granulocytes. An in vitro model of immune vascular damage, *J. Clin. Invest.,* 61, 1161, 1977.
32. **McCord, J. M.,** Free radicals and inflammation: protection of synovial fluid by superoxide dismutase, *Science,* 185, 529, 1974.
33. **Babior, B. M.,** Oxidants from phagocytes: agents of defense and destruction, *Blood,* 64, 959, 1984.
34. **DelMaestro, R. F., Thaw, H., Bjork, J., Planker, M., and Arfors, K. E.,** Free radicals as mediators of tissue injury, *Acta Physiol. Scand. (Suppl.),* 492, 43, 1980.
35. **Halliwell, B. and Gutteridge, J. M. C.,** Oxygen toxicity, oxygen radicals, transition metals and disease, *Biochem. J.,* 219, 1, 1984.
36. **Black, H. S.,** Role of reactive oxygen species of inflammatory processes, in *Nonsteroidal Anti-Inflammatory Drugs. Pharmacology and the Skin,* Vol. 2, C. Hensby and N. J. Lowe, Eds., S. Karger, Basel, 1989, 1.
37. **Rubin, R. and Farber, J. L.,** Mechanisms of the killing of cultured hepatocytes by hydrogen peroxide, *Arch. Biochem. Biophys.,* 228, 450, 1984.
38. **Sugino, K., Dohi, K., Yamada, K., and Kawasaki, T.,** Changes in the levels of endogenous antioxidants in the liver of mice with experimental endotoxemia and the protective effects of the antioxidants, *Surgery,* 105, 200, 1989.

39. **Shiratori, Y., Kawase, T., Shiina, S., Okano, K., Sugimoto, T., Teraoka, H., Matano, S., Matsumoto, K., and Kamii, K.,** Modulation of hepatotoxicity by macrophages in the liver, *Hepatology,* 8, 815, 1988.
40. **Stuehr, D. J. and Nathan, C. F.,** Nitric oxide. A macrophage product responsible for cytostasis and respiratory inhibition in tumor target cells, *J. Exp. Med.,* 169, 1543, 1989.
41. **Billiar, T. R., Curran, R. D., Stuehr, D. J., West, M. A., Bentz, B. G., and Simmons, R. L.,** An L-arginine-dependent mechanism mediates Kupffer cell inhibition of hepatocyte protein synthesis *in vitro, J. Exp. Med.,* 169, 1467, 1989.
42. **Hibbs, J. B., Taintor, R. R., and Vavrin, Z.,** Macrophage cytotoxicity: role for L-arginine deaminase and imino nitrogen oxidation to nitrite, *Science,* 235, 473, 1987.
43. **Hibbs, J. B., Taintor, R. R., Vavrin, Z., and Rachlin, E. M.,** Nitric oxide: a cytotoxic activated macrophage effector molecule, *Biochem. Biophys. Res. Commun.,* 157, 87, 1988.
44. **Kilbourn, R. G., Jubran, A., Gross, S. S., Griffith, O. W., Levi, R., Adams, J., and Lodato, R. F.,** Reversal of endotoxin-mediated shock by N^G-methyl-L-arginine, an inhibitor of nitric oxide synthesis, *Biochem. Biophys. Res. Commun.,* 172, 1132, 1990.
45. **Dinarello, C. A.,** Interleukin-1 and its related cytokines, in *Macrophage-Derived Cell Regulatory Factors. Cytokines,* Vol. 1, C. Sorg, Ed., S. Karger, Basel, 1989, 105.
46. **Matsuda, T. and Hirano, T.,** Interleukin-6, *Biotherapy,* 2, 363, 1990.
47. **Onozaki, K., Matsushima, K., Kleinerman, E. S., Saito, T., and Oppenheim, J. J.,** The role of interleukin 1 in promoting human monocyte-mediated tumor cytotoxicity, *J. Immunol.,* 135, 314, 1985.
48. **Oppenheim, J. J., Kovacs, E. J., Matsushima, K., and Durum, S. K.,** There is more than one interleukin 1, *Immunol. Today,* 7, 45, 1986.
49. **Kishimoto, T.,** The biology of interleukin 6, *Blood,* 74, 1, 1989.
50. **Butterwith, S. C. and Briffin, H. D.,** The effects of macrophage-derived cytokines on lipid metabolism in chicken (*Gallus domesticus*): hepatocytes and adiopocytes, *Comp. Biochem. Physiol. (A),* 721, 1989.
51. **Subrahamanyan, L. and Kisilevsky, R.,** Effects of culture substrates and normal hepatic sinusoidal cells on *in vitro* hepatocyte synthesis of Apo-SAA, *Scand. J. Immunol.,* 27, 251, 1988.
52. **Hagiwara, T., Suzuki, H., Kono, I., Kashiwagi, H., Akiyama, Y., and Onozaki, K.,** Regulation of fibronectin synthesis by interleukin-1 and interleukin-6 in rat hepatocytes, *Am. J. Pathol.,* 136, 39, 1990.
53. **Rupp, R. G. and Fuller, G. M.,** The effects of leukocytic and serum factors on fibrinogen biosynthesis in cultured hepatocytes, *Exp. Cell Res.,* 118, 23, 1978.
54. **Hooper, D. C., Steer, C. J., Dinarello, C. A., and Peackock, A. C.,** Hepatoglobin and albumin synthesis in isolated rat hepatocytes, *Biochem. Biophys. Acta,* 653, 118, 1981.
55. **Katsumoto, F., Miyazaki, K., and Nakayama, F.,** Stimulation of DNA synthesis in hepatocytes by Kupffer cells after partial hepatectomy, *Hepatology,* 9, 405, 1989.

56. **Darlington, G. J., Wilson, D. R., and Lachman, L. B.,** Monocyte-derived conditioned medium, interleukin-1 and tumor necrosis factor stimulate the acute-phase response in human response in human hepatoma cells in vitro, *J. Cell Biol.,* 103, 787, 1986.
57. **Gauldie, J., Richards, C., Harnish, D., Landsdorp, P., and Bauman, H.,** Interferon $beta_2$/B cell stimulatory factor type 2 shares identity with monocyte-derived hepatocyte-stimulating factor, *Proc. Natl. Acad. Sci. U.S.A.,* 84, 7251, 1987.
58. **Geiger, T., Andus, T., Klapproth, J., Hirano, T., Kishimoto, T., and Heinrich, P. C.,** Induction of rat acute-phase proteins by interleukin-6 in vivo, *Eur. J. Immunol.,* 18, 717, 1988.
59. **Castell, J. V., Gomez-Lechon, M. J., David, M., Fabra, R., Trullenque, R., and Heinrich, P. C.,** Acute-phase response of human hepatocytes: regulation of acute-phase protein synthesis by interleukin-6, *Hepatology,* 12, 1179, 1990.
60. **Mannel, D. N., Moore, R. N., and Mergenhagen, S. E.,** Macrophages as a source of tumoricidal activity (tumor necrotizing factor), *Infect. Immun.,* 30, 523, 1980.
61. **Beutler, B. and Cerami, A.,** Cachectin: more than a tumor necrosis factor, *N. Engl. J. Med.,* 316, 379, 1987.
62. **Tracey, K. J., Beutler, B., Lowery, S. F., Merryweather, J., Wolpe, S., Milsark, I. W., Harri, R. J., Zentella, A., Albert, J. D., Sires, G. T., and Cerami, A.,** Shock and tissue injury induced by recombinant human cachectin, *Science,* 234, 470, 1986.
63. **Beutler, B. and Cerami, A.,** The biology of cachectin/TNF — a primary mediator of the host response, *Annu. Rev. Immunol.,* 7, 625, 1989.
64. **Perlmutter, D. H., Dinarello, C. A., Punsal, P. I., and Colten, H. R.,** Cachectin/tumor necrosis factor regulates hepatic acute-phase gene expression, *J. Clin. Invest.,* 78, 1349, 1986.
65. **Issekutz, A. C., Megyeri, P., and Issekutz, T. B.,** Role of macrophage products in endotoxin-induced polymorphonuclear leukocyte accumulation during inflammation, *Lab. Invest.,* 56, 49, 1987.
66. **Broudy, V. C., Kaushansky, K., Segal, G. M., Harlan, J. M., and Adamson, J. W.,** Tumor necrosis factor type alpha stimulates human endothelial cells to produce granulocyte/macrophage colony stimulating factor, *Proc. Natl. Acad. Sci. U.S.A.,* 83, 7467, 1986.
67. **Dinarello, C. A., Cannon, J. G., Wolff, S. M., Bernheim, H. A., Beutler, B., Cerami, A., Figari, I. S., Palladino, M. A., and O'Connor, J. V.,** Tumor necrosis factor (cachectin) is an endogenous pyrogen and induces production of interleukin 1, *J. Exp. Med.,* 163, 1433, 1986.
68. **Bachwich, P., Chensue, S., Larrick, J. W., and Kunkel, S. L.,** Tumor necrosis factor stimulates interleukin-1 and prostaglandin E_2 production in resting macrophages, *Biochem. Biophys. Res. Commun.,* 136, 94, 1986.
69. **Kohase, M., Henriksen-DeStefano, D., May, L. T., Vilcek, J., and Sehgal, P. B.,** Induction of $beta_2$-interferon by tumor necrosis factor: a homeostatic mechanism in the control of cell proliferation, *Cell,* 45, 659, 1986.
70. **Munkee, R., Gasson, J., Ogawa, M., and Koeffler, H. P.,** Recombinant TNF induces production of granulocyte-monocyte colony stimulating factor, *Nature,* 323, 79, 1986.

71. **Sheron, N., Lau, J. N., Hofman, J., Williams, R., and Alexander, G. J. M.,** Dose-dependent increase in plasma interleukin-6 after recombinant tumor necrosis factor infusion in humans, *Clin. Exp. Immunol.,* 82, 427, 1990.
72. **McClain, C. J. and Cohen, D. A.,** Increased tumor necrosis factor production by monocytes in alcoholic hepatitis, *Hepatology,* 9, 349, 1989.
73. **Hishinuma, I., Nagakawa, J., Hirota, K., Miyamoto, K., Tsukidate, K., Yamanaka, T., Katayama, K., and Yamatsu, I.,** Involvement of tumor necrosis factor-alpha in development of hepatic injury in galactosamine-sensitized mice, *Hepatology,* 12, 1187, 1990.
74. **Sato, N., Goto, T., Haranaka, K., Satomi, N., Nariuchi, H., Mano-Hirano, Y., and Sawasaki, Y.,** Actions of tumor necrosis factor on cultured vascular endothelial cells: morphologic modulation, growth inhibition, and cytotoxicity, *J. Natl. Cancer Inst.,* 76, 1113, 1986.
75. **Klebanoff, S. J., Vadas, M. A., Harlan, J. M., Sparks, L. H., Gamble, J. R., Agnosti, J. M., and Waltersdorph, A. M.,** Stimulation of neutrophils by tumor necrosis factor, *J. Immunol.,* 136, 4220, 1986.
76. **Ooi, B. S., MacCarthy, E. P., Hsu, A., and Ooi, Y.,** Human mononuclear cell modulation of endothelial cell proliferation, *J. Lab. Clin. Med.,* 102, 428, 1983.
77. **Martin, B. M., Gimbrone, M. A., Unanue, E. R., and Cotran, R. S.,** Stimulation of nonlymphoid mesenchymal cell proliferation by a macrophage-derived growth factor, *J. Immunol.,* 126, 1510, 1981.
78. **Matsubara, T. and Ziff, M.,** Increased superoxide anion release from human endothelial cells in response to cytokines, *J. Immunol.,* 137, 3295, 1986.
79. **Jirik, F. R., Podor, T. J., Hirano, T., Kishimoto, D. J., Loskutoff, D. A., and Lotz, M.,** Bacterial lipopolysaccharide and inflammatory mediators augment IL-6 secretion by human endothelial cells, *J. Immunol.,* 142, 144, 1989.
80. **Sironi, M., Breviario, F., Proserpio, P., Biondi, A., Vecchi, A., Van-Damme, J., Dejana, E., and Mantovani, A.,** IL-1 stimulates IL-6 production in endothelial cells, *J. Immunol.,* 142, 549, 1989.
81. **Strieter, R. M., Kunkel, S. L., Showell, H. J., and Marks, R. M.,** Monokine-induced gene expression of a human endothelial cell-derived neutrophil chemotactic factor, *Biochem. Biophys. Res. Commun.,* 156, 1340, 1988.
82. **Rollins, B. J., Yoshimura, T., Leonard, E. J., and Pober, J. S.,** Cytokine-activted human endothelial cells synthesize a monocyte chemoattractant, MCP-1/JE, *Am. J. Pathol.,* 136, 1229, 1990.
83. **Eyhorn, S., Schlayer, H. J., Henninger, H. P., Dieter, P., Hermann, R., Woort-Menker, M., Becker, H., Schaefer, H. E., and Decker, K.,** Rat hepatic sinusoidal endothelial cells in monolayer culture. Biochemical and ultrastructural characteristics, *J. Hepatol.,* 6, 23, 1988.
84. **Knook, D. L., Blansjaar, N., and Sleyster, E. C.,** Isolation and characterization of Kupffer and endothelial cells from the rat liver, *Exp. Cell Res.,* 99, 444, 1976.
85. **Cotran, R. S.,** New roles for the endothelium in inflammation and immunity, *Am. J. Pathol.,* 129, 407, 1987.
86. **Salvemini, D., Korbut, R., Anggard, E., and Vane, J. R.,** Lipopolysaccharide increases release of nitric oxide-like factor from endothelial cells, *Eur. J. Pharmacol.,* 171, 135, 1989.

87. **Tiku, M. L. and Tomasi, T. B.,** Enhancement of endothelial plasminogen activator synthesis by lymphokines, *Transplantation,* 40, 293, 1985.
88. **Anderson, N. D., Anderson, A. O., and Wylie, R. G.,** Microvascular changes in lymph nodes draining skin allografts, *Am. J. Pathol.,* 81, 131, 1975.
89. **Ramadan, F. M., Upchurch, G. R., Keagy, B. A., and Johnson, G.,** Endothelial cell thromboxane production and its inhibition by a calcium channel blocker, *Ann. Thorac. Surg.,* 49, 916, 1990.
90. **Ward, P. A. and Newman, L. J.,** A neutrophil chemotactic factor from C5a, *J. Immunol.,* 102, 93, 1969.
91. **Kaplan, A. P., Goetzl, E. J., and Austen, K. F.,** The fibrinolytic pathway of human plasma. II. The generation of chemotactic activity by activation of plasminogen activator, *J. Clin. Invest.,* 52, 2591, 1972.
92. **Laskin, D. L., Kimura, T., Sakakibara, S., Riley, D. J., and Berg, R. A.,** Chemotactic activity of collagen-like polypeptides for human peripheral blood neutrophils, *J. Leuk. Biol.,* 39, 255, 1986.
93. **Goetzl, E. and Pickett, W.,** Novel structural determinants of the human neutrophil chemotactic activity of leukotriene B, *J. Exp. Med.,* 153, 482, 1981.
94. **Schiffman, E., Corcoran, B., and Wahl, S.,** N-Formylmethionine peptides as chemoattractants for leukocytes, *Proc. Natl. Acad. Sci. U.S.A.,* 72, 1059, 1975.
95. **Matsushima, K. and Oppenheim, J. J.,** Interleukin 8 and MCAF: novel inflammatory cytokines inducible by IL-1 and TNF, *Cytokine,* 1, 2, 1989.
96. **Perez, H. D., Roll, F. J., Bissell, D. M., Shak, S., and Goldstein, I. M.,** Production of chemotactic activity for polymorphonuclear leukocytes by cultured rat hepatocytes exposed to ethanol, *J. Clin. Invest.,* 74, 1350, 1984.
97. **Laskin, D. L. and Pilaro, A. M.,** Activation of liver macrophages for killing of hepatocytes following acetaminophen treatment of rats, *Toxicologist,* 8, 32, 1988.
98. **Peterson, T. C. and Renton, K. W.,** The role of lymphocytes, macrophages and interferon in the depression of drug metabolism by dextran sulfate, *Immunopharmacology,* 11, 21, 1986.
99. **McCloskey, T. W., Tedaro, J. A., and Laskin, D. L.,** Lipopolysaccharide treatment of rats alters antigen expression and oxidative metabolism in hepatic macrophages and endothelial cells, *Hepatology,* in press.
100. **Laskin, D. L., Feder, L. S., McCloskey, T. W., and Gardner, C. R.,** Role of nonparenchymal cells in hepatotoxicity, in *Cells of Hepatic Sinusoid,* Vol. 3, E. Wisse, D. L. Knook, and R. McCuskey, Eds., Kupffer Cell Foundation, Leiden, 1991, 81.
101. **Galanos, C., Freudenberg, M. A., and Reuter, W.,** Galactosamine-induced sensitization of the lethal effects of endotoxin, *Proc. Natl. Acad. Sci. U.S.A.,* 76, 5939, 1979.
102. **Nolan, J. P. and Leibowitz, A. Z.,** Endotoxin and the liver. III. Modification of acute carbon tetrachloride injury by polymyxin B, an antiendotoxin, *Gastroenterology,* 75, 445, 1978.
103. **Nolan, J. P. and Ali, M. V.,** Endotoxin and the liver. II. Effect of tolerance on carbon tetrachloride induced injury, *J. Med.,* 4, 28, 1973.

104. **Sipes, I. G., El Sisi, A. E., Simm, W. W., Mobley, S. A., and Earnst, D. L.,** Role of reactive oxygen species by activated Kupffer cells in the potentiation of carbon tetrachloride hepatotoxicity by hypervitaminosis A, in *Cells of the Hepatic Sinusoid,* Vol. 2, E. Wisse, D. L. Knook, and K. Decker, Eds., Kupffer Cell Foundation, Amsterdam, 1989, 376.
105. **Sim, W. W., Abril, E. R., and Earnest, D. L.,** Mechanisms of Kupffer cell activation in hypervitaminosis A, in *Cells of the Hepatic Sinusoid,* Vol. 2, E. Wisse, D. L. Knook, and K. Decker, Eds., Kupffer Cell Foundation, Amsterdam, 1989, 91.
106. **Maier, R. V. and Ulevitch, R. J.,** The response of isolated rabbit hepatic macrophages to lipopolysaccharide, *Circ. Shock,* 8, 165, 1981.
107. **Al-Tuwaijri, A., Akdamar, K., and DiLuzio, N. R.,** Modification of galactosamine-induced liver injury in rats by reticuloendothelial system stimulation or depression, *Hepatology,* 1, 107, 1981.
108. **Souhami, R. L. and Bradfield, J. W.,** The recovery of hepatic phagocytosis after blockade of Kupffer cells, *J. Reticuloendothel. Soc.,* 16, 75, 1981.
109. **Husztik, E., Lazar, G., and Parducz, A.,** Electron microscopic study of Kupffer cell phagocytosis blockade induced by gadolinium chloride, *Br. J. Exp. Pathol.,* 61, 624, 1980.
110. **Feder, L. S., McCloskey, T. W., and Laskin, D. L.,** Characterization of interleukin-1 and interleukin-6 production by resident and lipopolysaccharide activated hepatic macrophages and endothelial cell, in *Cells of the Hepatic Sinusoid,* Vol. 3, E. Wisse, D. L. Knook, and R. McCuskey, Eds., Kupffer Cell Foundation, Leiden, 1991, 37.
111. **Kobzik, L., Godleski, J. J., and Brain, J. D.,** Oxidative metabolism in the alveolar macrophage: analysis by flow cytometry, *J. Leuk. Biol.,* 47, 295, 1990.
112. **Laskin, D. L., Robertson, F. J., Pilaro, A. M., and Laskin, J. D.,** Activation of liver macrophages following phenobarbital treatment of rats, *Hepatology,* 8, 1051, 1988.
113. **Helle, M., Boeiji, L., and Aarden, L. A.,** Functional discrimination between interleukin-6 and interleukin-1, *Eur. J. Immunol.,* 18, 1535, 1988.
114. **Gardner, C. R., Wasserman, A. J., and Laskin, D. L.,** Differential sensitivity of tumor targets to liver macrophage-mediated cytotoxicity, *Cancer Res.,* 47, 6686, 1987.

Chapter 8

LIVER ALLOIMMUNITY

G.L. Bumgardner and P.S. Almond

TABLE OF CONTENTS

ISBN 0-8493-6109-5

I. INTRODUCTION

The development of immunologic laboratory techniques in such areas as immunohistochemistry, fluorescence-activated cell sorting (FACS), tissue culture and cloning, monoclonal antibody (mAb) production, and molecular biology have allowed more in-depth investigation of the major histocompatibility complex (MHC) and the immunology of allograft rejection. One of the most striking findings from intensive research of cardiac, kidney, liver, skin, and lung transplantation is the existence of significant diversity in the host's immune response to different organ allografts from the same donor. Such organ-specific differences in graft immunogenicity and/or host and graft interaction are most pronounced in clinical and experimental liver transplantation. Similarly, alloimmune pathways operational for vascularized organ allografts may be quite distinct from parenchymal cellular transplants. Thus, it appears that studies which are directed to prolong the survival of a specific graft must study the specific organ or cell population of interest rather than infer similar mechanisms based on the results of studies from a different end-organ or cell population. In order to study liver alloimmunity, we will review the expression of MHC antigens of various cell populations in the liver of humans in the normal state and after transplantation during periods of quiescence as well as in the setting of acute liver allograft rejection. Furthermore, the histology of liver allograft rejection provides insight into potential alloimmune mechanisms involved in the rejection process. The results of clinical studies have the advantage of being directly applicable to transplantation of various allografts in humans, but at the same time carry the disadvantage of producing variable results due to variation in the human study population. Experimental systems employing various animal models to study liver alloimmunity can be more precisely controlled than clinical studies to answer specific questions, but the results may not be directly applicable to humans; nevertheless, results from experimental systems guide the design of clinical studies.

II. EXPRESSION OF MHC ANTIGENS ON LIVER CELLS (HEPATOCYTES, KUPFFER CELLS, ENDOTHELIAL CELLS, AND BILIARY EPITHELIAL CELLS)

A. HUMAN LIVER CELLS

In normal human liver tissue, MHC class I antigens have been detected on biliary epithelial cells, Kupffer cells, other interstitial cells, and endothelial cells; however, hepatocytes express little if any MHC class I antigens. In general, MHC class II antigens have been detected on Kupffer and interstitial cells and have not been detected on hepatocytes, while class II expression on biliary epithelial cells and endothelial cells has been variable. In diseased livers, such as with alcoholic hepatitis, chronic active hepatitis, or primary biliary cirrhosis, hepatocytes have been noted to express MHC class I antigens, and this correlated with the degree of interlobular inflammation.[1,2]

In contrast to the pretransplant normal liver, the transplanted liver, even during clinical quiescence, exhibits upregulation of MHC antigen expression, and during rejection even more intense MHC antigen expression is observed. Biopsies from fresh liver allografts have shown that hepatocytes express class I antigen weakly and do not express class II antigen. Biliary epithelium from fresh liver allografts have been reported to express class I strongly, but not class II antigens. However, during rejection hepatocytes have been shown to strongly express class I and beta-2-microglobulin, and there have been a few reports that some hepatocytes express class II weakly.[3-5] Biliary epithelial cells strongly express both class I and class II during rejection and also during viral infection.[3-5] This increase in MHC antigen expression declines with resolution of the rejection process.[4,5] Interferon-gamma and other lymphokines produced by activated T lymphocytes which accumulate in both rejection and viral infection have been postulated to result in upregulation of expression of MHC antigens.[5,6] Furthermore, liver biopsies taken during stable graft function have revealed decreased class II expression in the grafts which were transplanted under cyclosporine immunosuppression compared to those transplanted under immunosuppressive regimens without cyclosporine. This effect was attributed to the potent ability of cyclosporine to inhibit release of interferon-gamma.[2,6]

Human biliary epithelial cells have been harvested from donor liver allografts before transplantation and grown in tissue culture as a monolayer and stained for MHC antigens. Control cultures of biliary epithelial cells did not reveal staining for class II antigens. In contrast, cultures grown in the presence of exogenous interferon-gamma resulted in expression of class II antigens both in the cytoplasm and on the surface membrane.[7]

Despite the numerous reports describing increased MHC antigen expression at the time of clinical and histologic rejection episodes, it has also been established that allograft rejection does occur in the absence of upregulation of MHC antigen expression, and that upregulation of MHC antigen expression can occur in the absence of allograft rejection.[8] It is not clear whether upregulation of MHC antigens on liver cells observed during rejection reflects a role in initiating the alloimmune response or rather the consequences of alloimmune interactions. However, upregulation of MHC antigen expression results in cells which are more vulnerable targets of specifically sensitized effectors. Thus, clinical studies have established an association between rejection and increased MHC antigen expression. This finding is not absolute and the sequence of events which results in the association has not yet been clearly defined.

B. RAT AND MOUSE LIVER CELLS

Expression of MHC antigens on liver cells has also been extensively studied in animal models. There are rat strain combinations in which orthotopic liver allografts are either spontaneously accepted (DA to PVG) without immunosuppression, or rapidly rejected (DA to LEW). Weak expression of class I antigens was detected on hepatocytes and biliary epithelial cells in isograft controls. In both the spontaneously accepted and the rapidly rejected liver allografts, expression of class I antigens on hepatocytes and biliary epithelium was significantly increased over isograft controls. Class II expression for the most part was not detected on liver cells in isograft controls, but was detected on biliary epithelium, Kupffer cells, and a small number of hepatocytes (although expression was primarily intracellular rather than on the surface membrane) of both the spontaneously accepted and the rapidly rejected allografts without significant difference between the latter two. Therefore, differences in MHC antigen expression did not correlate with the markedly different alloimmune interactions in these two strain combinations.[9,10]

MHC antigen expression on the isolated cellular components of liver cell suspensions have been described in animals. Normal murine hepatocytes strongly express MHC class I antigens, but do not express MHC class II antigens.[11] After culture in the alloimmune environment of a mixed lymphocyte hepatocyte culture (MLHC), hepatocytes remain class I positive and class II negative.[11,12] On the other hand, the mixed population of nonparenchymal cells, which includes Kupffer, endothelial, and biliary epithelial cells, express both class I and class II antigens.[11]

III. HISTOLOGY OF HYPERACUTE, ACUTE, AND CHRONIC REJECTION

A. TARGETS OF LIVER ALLOGRAFT REJECTION

Despite the increasing numbers of liver transplants being performed, our understanding of the immune response to liver allografts is still quite limited. In man, the majority of our knowledge has come from the interpretation of serial tissue biopsies performed posttransplant. Although this approach provides a great deal of information, histologic analysis of tissue biopsies provides only static information. Therefore, we hypothesize about the dynamic interactions between host and graft from a collection of static pieces of information. The retrospective study of biopsies obtained during graft dysfunction has allowed pathologists to develop criteria which can assist the clinician in distinguishing between the different causes of graft dysfunction. More important clinically, these findings can be used to prospectively guide therapy and possibly predict outcome.

1. Hyperacute Rejection

Hyperacute rejection of organ allografts is thought to occur secondary to the presence of preformed antidonor cytotoxic antibodies within recipient sera.

In extrahepatic transplantation, these cytotoxic antibodies are detected pretransplant by the lymphocytotoxicity assay. Initially, in liver transplantation it was reported that the presence of preformed antibodies did not correlate with graft survival, and therefore, this assay has not been performed routinely.[13,14] Recently, however, cases have been reported in which the clinical course and liver pathology suggested that graft loss may have been due to preformed antibody.[15-18] The histology described in one case included the presence of submassive eosinophilic necrosis and the infiltration of polymorphonuclear leukocytes into the liver parenchyma and portal tracts.[15] In another case, early graft loss was associated with polymorphonuclear infiltrates, arteritis, ischemic parenchymal damage, and the deposition of antibody within the rejected graft.[16] Another three cases of emergency ABO incompatible human liver transplantation resulted in hyperacute rejection and were characterized histologically by edema, peliosis, and hemorrhage.[17]

Despite the relative infrequency of hyperacute liver allograft rejection, preformed lymphocytotoxic antibodies may have deleterious long-term effects. Recipients with preformed antidonor antibodies have a higher incidence of the vanishing bile duct syndrome (VBDS).[19] In addition, these recipients have a significantly higher rate of early graft failure ($<$3 weeks), 11 vs. 46% in the ABO-compatible vs. ABO-incompatible group. In a study of 14 patients with early graft loss, 4 were directly attributable to humoral rejection due to preformed isoagglutinins, while in 8 others humoral rejection was thought to have played a significant role in graft loss. Histological changes included focal fibrin masses attached to partially damaged vascular walls and widespread hemorrhagic necrosis; focal deposition of immunoglobulin M (IgM), C1q, and C3 were also detected on the arterial walls. Furthermore, the presence of donor-specific antibodies in a tissue eluate from the graft was demonstrated.[20]

2. Acute Rejection

The incidence and onset of acute liver allograft rejection has been reported to be high and to occur early after transplantation. Two studies, one retrospective and one prospective, suggested that no less than 50 to 77% of recipients experience at least one acute rejection episode.[16,21] The majority of rejection episodes occurred between posttransplant day 7 and day 10, rarely before day 4. The diagnosis of acute liver rejection is based on the correlation of clinical graft dysfunction with histology obtained by biopsy which meets the pathologic criteria of acute rejection. The pathologic criteria described by Snover et al.[22,23] includes the presence of a portal infiltrate, evidence of bile duct damage, and venous endothelialitis; arteritis is an additional finding in some cases. In the absence of endothelialitis, the diagnosis can still be made $>$50% of the bile ducts within the specimen are damaged. The portal infiltrate is typically a mixed inflammatory infiltrate consisting mainly of lymphocytes with occasional polymorphonuclear leukocytes and eosinophils. Bile duct damage is manifested by cellular necrosis, cytoplasmic vacuolation, and anisonucleosis of the biliary epithelium.[23,24] Endothelialitis has been observed in 63 to 100% of liver biopsies obtained from patients clinically undergoing acute rejection.[22-24] Endothelialitis is described as lymphocytic attachment to the endothelium and evidence of endothelial damage manifested by lifting of the endothelium off of the basement membrane.[23] Arteritis is defined by the presence of lymphocytes and/or polymorphonuclear leukocytes within the wall of the artery. Acute rejection has been classified into three grades: I (mild) includes mixed portal infiltrate, <50% bile duct damage, and endothelialitis; II (moderate) includes mixed portal infiltrate, >50% of the bile duct damaged, and with or without endothelialitis; III (severe) includes the findings in grades I or II with the addition of arteritis, decreased number of bile ducts, or central ballooning with confluent dropout of hepatocytes.[23]

A dominant feature of hepatic allograft rejection is progressive, nonsuppurative destructive cholangitis with eventual loss of interlobular bile duct structure. Furthermore, Ludwig et al.[25] defined an irreversible rejection process that affected hepatic allografts within 100 days after liver transplant, which was termed the acute VBDS. Biopsy showed destructive cholangitis, ductopenia, and eventually portal tracts without bile ducts. Subsequently, reports have emerged showing that this condition in some circumstances is reversible, with regeneration of bile ducts over time.[23,26] The occurrence of VBDS emphasizes once again that biliary epithelial cells appear to be a prime target of the rejection response. Donaldson et al.[27] correlated the frequency of VBDS with a complete mismatch for MHC class I antigens and a partial or complete match for class II antigens; furthermore, these patients developed a high titer of antibodies specific for donor MHC class I antigen. Both incompatibility and compatibility at MHC class II between donor and recipient have been described in association with acute liver rejection.[27-29]

3. Chronic Rejection

Chronic rejection is not as well characterized pathologically as acute rejection and may represent the result of inadequately treated or ongoing acute rejection, resulting in cell-mediated destruction of the liver and subsequent graft failure. Pathologic features associated with clinical chronic rejection include obliterative endarteritis, hepatocellular ballooning and dropout, fibrosis, and loss of bile ducts. Therefore, the pathology of hyperacute, acute, and chronic rejection in man suggests that endothelial cells (hyperacute and acute), biliary epithelial cells (acute, chronic), and hepatocytes (severe acute, chronic) are all potential targets of the alloimmune response to a liver allograft.[23]

B. HOST EFFECTORS IN LIVER ALLOGRAFT REJECTION

Another approach to studying alloimmune interactions after liver transplantation involves propagating and characterizing lymphocytes from sequential allograft biopsies. Lymphocyte cultures grown from liver allograft biopsies often contain alloreactive T cells specific for donor class I and/or class II MHC antigens. The degree of biliary duct damage has been correlated with the presence of class II-specific lymphocytes.[30] In the early posttransplant period, culture of liver allograft biopsies are more likely to propagate class I-specific cells, whereas later biopsies are more likely to yield class II-specific alloreactive T cells. Proliferation of these cells was inhibited by anticlass I or anticlass II mAb, respectively. Lymphocytes which have neither donor class I nor class II specificity and which did not lyse donor targets have been isolated from liver allograft, which raises the possibility of accumulation of lymphocytes with specificity for tissue-specific or endothelial antigens. Alternatively, these cells may represent a population of nonspecific inflammatory cells.[31]

Furthermore, immunohistology of serial liver allograft biopsies has revealed that stable graft function was associated with the lack of T lymphocytes or focal lobular accumulation of $CD8^+$ cytotoxic/suppressor cells (Leu-2a), whereas rejection was highly correlated with the presence of $CD4^+$ "helper" subset (Leu-3a) alone or in combination with $CD8^+$ cells in portal tracts. Furthermore, a recent study of 39 consecutive liver allograft recipients suggested that the presence of portal $CD4^+$ T cells with or without $CD8^+$ cells may predate the development of clinical and histologic rejection. The authors pointed out, however, that one must exclude cytomegalovirus hepatitis since it is also associated with accumulation of the $CD4^+$ "helper" T cell subset.[32] Some authors have reported a predominance of T cells in the inflammatory cell infiltrate during rejection, but with equal quantities of $CD4^+$ and $CD8^+$ cells; the majority of infiltrating T cells in these studies expressed class II antigen, but did not react with anti-IL-2r antibody.[33] Other reports suggest a predominance of $CD8^+$ T cells during rejection, and these cells were attached to or invading bile duct cells; 2:1 ratio of $CD8^+$ T lymphocytes to monocytes in acute rejection and an 8:1 ratio during chronic rejection were observed. All bile ducts in their specimens expressed MHC class II antigens. Although expression of MHC class I antigens was not examined in this study, the authors postulated that MHC class I antigens on biliary epithelial cells were the most likely target cell of liver allograft rejection based on their finding of predominance of $CD8^+$ T cells attached to bile duct cells in association with rejection and the known predilection of $CD8^+$ cells for recognition of MHC class I antigens.[34,35]

Demetris[36] studied 20 failed human liver allograft specimens obtained at the time of retransplantation and identified T lymphocytes as the predominant cells within the inflammatory cellular infiltrate, but B cells and macrophages were also present. Biliary epithelial cells, portal vein endothelium, central vein endothelium, and hepatic artery endothelium which were class II negative prior to transplantation, were presumably induced to express class II antigens following transplantation.

Compared to transplantation of other organ allografts, the liver appears relatively resistant to immune-mediated damage since hyperacute rejection is infrequent and many rejection episodes detected by protocol biopsies are mild and resolve easily. The immunogenicity of solid organ allografts has been attributed to the presence of class II positive "passenger leukocytes".[37] It has been noted by immunohistology (using donor- and recipient-specific class I and class II mAb) that in liver allografts, donor-specific "passenger leukocytes" such as Kupffer and interstitial dendritic cells are replaced by recipient cells of host bone marrow origin within the first month after transplant. Hepatocytes, biliary endothelial cells, and sinusoidal endothelial cells remained donor specific.[38,39] It has been postulated that the disappearance of highly immunogenic Kupffer cells from the graft and the replacement by recipient Kupffer cells may be an important factor in the weak immunogenicity of the human liver transplant in comparison to other organ allografts.[40]

IV. EXPERIMENTAL SYSTEMS TO STUDY HEPATOCYTE IMMUNOGENICITY

A. HEPATOCYTE VS. NONPARENCHYMAL CELL INDUCTION OF GRAFT REJECTION

The immunogenicity of liver allografts has also been examined in terms of its cellular components by isolating different cell populations (parenchymal vs. nonparenchymal cells, including biliary epithelial cells and Kupffer cells) from liver cell suspensions. Experiments designed to test the ability of parenchymal vs. a mixed population of nonparenchymal cells to sensitize a naive recipient and cause accelerated heart allograft rejection have been developed. Such studies demonstrated that control heart allografts were rejected on day 6.3 ± 1.2 d after transplant. Priming with liver nonparenchymal cells caused accelerated rejection of heart allografts (day 4.0 ± 0.0) and priming with hepatocytes did not affect graft survival in a rat model (day 5.5 ± 0.6). Furthermore, they found that class II antigen expression was primarily detected on Kupffer cells, endothelial cells, and passenger B and T lymphocytes, whereas hepatocytes expressed little or no class II antigens. These authors concluded that the immunogenicity of liver cells is primarily attributable to the nonparenchymal cells in their system.[41] It is important to realize that such results do not address the potential immunogenicity of parenchymal cell class I antigen or tissue-specific hepatocyte antigens. Furthermore, since the endpoint tested in this model is heart allograft rejection, the results may not be applicable to hepatocyte transplantation.

Unlike the aforementioned findings, Almond et al.[42] demonstrated that purified hepatocytes were capable of sensitizing murine recipients as demonstrated by accelerated rejection of a subsequent skin graft. C57BL/6 mice were immunized with DBA/2 hepatocytes or DBA/2 splenocytes either intraperitoneally or in sponge matrix allografts. Rejection times were the same in the DBA/2 hepatocytes and DBA/2 splenocytes group (mean 5.6 ± 1.0, 5.2 ± 1.1 d) and were both statistically significant in difference from control allogeneic skin graft rejection time (7.8 ± 0.5 d).

B. MIXED LYMPHOCYTE HEPATOCYTE CULTURE (MLHC)

In anticipation of clinical hepatocyte transplantation, experimental models to study hepatocyte (parenchymal cell) immunogenicity have been developed. These include the *in vitro* MLHC and the *in vivo* hepatocyte sponge matrix allograft (HC-SMA).[11,43,44] In both of these models, percoll-purified hepatocyte preparations were used. Purity of the cell populations is verified by light microscopy (99% hepatocytes), immunofluorescence of cytospins staining for class II (no class II$^+$ cells detected), and immunofluorescence for class I and class II using FACS analysis (no class II$^+$ cells detected). In MLHC, responder splenocytes and stimulator hepatocytes are cocultured. After 5 d, responder cytolytic T cells (CTL) specifically sensitized to MHC class I antigen of the stimulator hepatocytes have developed and specifically lyse tumor targets or Con A blasts bearing the allogeneic, but no syngeneic or third-party class I antigen. Similarly, these effectors injure allogeneic, but not syngeneic or third-party hepatocytes. The effectors in MLHC are CD8$^+$ CD4$^-$ T cells, and develop in a number of different strain combinations which differ in MHC disparity (class I or class I and class II, but not class II alone). Experiments using this system have shown that the allospecific cytolytic T cells (allo-CTL) that develop in MLHC in response to allogeneic hepatocytes are CD4 dependent, i.e., depletion of the CD4$^+$ T cell subset from the responder cell population abrogates the development of cytotoxic effector cells.[45] Furthermore, responder antigen-presenting cells (APC) are required for the development of allo-CTL in MLHC, suggesting that these APC may process hepatocyte class I antigen which is recognized by responder L3T4$^+$ T cells in the context of self class II molecules.[11] This pathway of alloantigen recognition in mixed lymphocyte culture (MLC) has been termed indirect antigen recognition in contrast to direct recognition of alloantigen or allogeneic APC.[46,47] This indirect pathway of alloantigen recognition may involve induction of cytokines which differ both qualitatively and temporally from the pattern of cytokine induction with direct allorecognition; interleukin 2 (IL-2) and IL-6, but not IL-4 nor TNF, were detected in the supernatants of allogeneic MLHC in a pattern distinct from syngeneic MLHC and allogeneic MLC.[48]

The potential for induction of MHC class II antigen on hepatocytes in culture or in HC-SMA in a cytokine-rich milieu has been addressed by testing for the development of class II-specific allo-CTL in class II-disparate MLHC and HC-SMA. Both direct cytotoxicity and limiting dilution analysis did not detect the development of allo-CTL, which suggests that purified hepatocytes do not express MHC class II antigen at the time of isolation and are not induced to express class II antigen in an alloimmune environment both *in vitro* and *in vivo*.[12]

C. HEPATOCYTE-SPONGE MATRIX ALLOGRAFT

The HC-SMA is a modification of the sponge matrix allograft model in which sponge matrix allografts bearing allogeneic hepatocytes are implanted into host animals. After 12 d, host cells infiltrating the sponge graft demonstrate allospecific cytotoxicity. These cytolytic effectors are $CD8^+$ $CD4^-$ cells.[44] The development of allo-CTL in HC-SMA appears to be dependent on host $CD4^+$ T cell subset as well as host APC, since depletion or inhibition of the $CD4^+$ subset with anti-CD4 mAb inhibits the development of allo-CTL as does inhibition of sponge macrophage function with local silica treatment.[49,50] Macrophages isolated from HC-SMA demonstrate ability to function as APC *in vitro*.[48] Thus, both *in vitro* studies in MLHC and *in vivo* studies in HC-SMA suggest that hepatocytes in the absence of class II positive "passenger leukocytes" are immunogenic and thus would be vulnerable to rejection.

The immunogenicity of MHC class I^+ hepatocytes has been compared to MHC class I^+ and class II^+ splenocytes in both MLHC and HC-SMA. In both models, fewer hepatocyte stimulator cells are sufficient to elicit a cytotoxic response compared to splenocyte stimulator cells. Subsequent studies attributed this finding to increased MHC class I antigen density on the surface of hepatocytes as compared to splenocytes using a ^{125}I assay. It should be noted that hepatocytes are approximately three times the size of splenocytes, and increased expression of class I on hepatocytes may simply reflect their greater surface area.[51]

D. HEPATOCYTE TRANSPLANTATION

Most research involving hepatocyte transplantation has focused on the functional result or histology after transplantation, and the question of immunogenicity of the cellular transplant has not been completely addressed. The majority of hepatocyte transplant studies have used unpurified populations of liver cells which probably included hepatocytes, biliary epithelial, endothelial, and Kupffer cells. Under these circumstances, immunosuppression was required to maintain survival of the transplanted hepatocytes.[52-54]

In a recent study, purified hepatocytes were transplanted in a hypercholesterolemic rabbit model. A short course of cyclosporine immunosuppression was used to maintain survival of the transplanted hepatocytes (measured by the decrease in serum LDL cholesterol). The requirement for immunosuppression after transplantation of purified hepatocytes may be indirect evidence of *in vivo* immunogenicity of purified hepatocytes.[55]

We have also transplanted percoll-purified syngeneic and allogeneic hepatocytes into the neck pads of murine recipients. Brown nodules were present 1 and 2 months posttransplantation at the transplantation site in the hosts which received syngeneic, but not allogeneic hepatocytes (unpublished observation). Histologic examination of the nodules revealed hepatocytes arranged in cords and clusters as well as bile ductules. This is similar to the report of Jirtle et al.,[56] who described detailed morphologic and histochemical analysis of syngeneic hepatocytes transplanted in rats. Further immunohistological studies in both groups which received syngeneic or allogeneic hepatocytes may reveal a pattern of host inflammatory vs. alloimmune response.

UV irradiation has been shown to be effective in reducing the immunogenicity of unpurified hepatocytes in a rat hepatocyte-transplantation model.[57] The effect of UV irradiation on enhanced graft survival has been attributed to its effect on allogeneic class II antigen. However, UV radiation also results in changes in calcium flux, which may mediate its immunosuppressive effects.[58,59] UV irradiation has been reported to affect the function of APC,[60] but its effect on expression of MHC class I antigen has not been reported.

V. ENDOGENOUS IMMUNOSUPPRESSIVE PROPERTIES OF THE LIVER

A. CELLULAR FACTORS

The liver has long been recognized as an immunologically privileged organ. One of the earliest demonstrations of this phenomenon was the long-term survival of orthotopic liver allografts in nonimmunosuppressed porcine recipients.[61] Since then, this and other animal models have been used to investigate the potential immunosuppressive effects of liver allografts.[62-65]

With the development of microsurgical techniques, the initial observations made in large animals were extended to rat models.[62] These small animal models, although perhaps more technically demanding, offered several advantages over large animal models. At present, there are very few immunologically defined inbred porcine or canine colonies. Therefore, it is difficult to interpret and reproduce results obtained in one animal with those obtained in another. Likewise, there are very few porcine or canine immunologic reagents available.

Mechanisms by which liver allografts could exert immunosuppressive effects include: (a) the elaboration of an immunosuppressive factor by the allograft; (b) the production of an immunosuppressive factor by the recipient in response to antigens on the graft/donor lymphocytes; or (c) the tolerance/deletion of donor reactive T cells. This last mechanism may be the result of either differences in antigen processing between the liver and other lymphoid organs or sequestration of donor reactive cells by the graft.

Kamada[66] reproduced the results observed in large animal models of liver transplantation by demonstrating that in the DA to PVG rat strain combinations, orthotopic, but not heterotopic liver allografts can survive indefinitely. This effect is a characteristic of the liver allograft and not the strain combination since DA heart and kidney allografts are promptly rejected at 8 and 12 d posttransplant, respectively. Tolerance of the DA liver then leads to a state of donor-specific, systemic tolerance in which subsequent nonliver DA allografts are not rejected.[67] There is evidence that both the B and T lymphocyte compartments contribute to the observed tolerance. In the B cell compartment, PVG recipients of DA livers develop antibodies to donor class I and class II antigens. The antibodies to class I antigens are transient, whereas those to class II antigens persist. Adoptive transfer of whole, isolated IgG fraction, or class I absorbed serum from PVG allograft recipients to naive PVG rats confers tolerance to future DA allografts.[66,68,69] This suggest that the anticlass II antibodies are responsible for the induction of donor-specific tolerance in this model. At the T cell level, tolerance may be due to either deletion or unresponsiveness of DA reactive $CD8^+$ cytotoxic effector cells. Adoptive transfer of T cells form PVG recipients of a DA liver transfers tolerance to the naive recipient. In these tolerant animals, the ability to proliferate, but not to lyse DA targets is retained, suggesting that the tolerance/deletion is at the effector (CD8) and not the inducer (CD4) level.

The immunohistology of liver allograft acceptance in long-surviving strains and liver allograft rejection in rapidly rejecting stains has revealed a relative predominance of $CD4^+$ T cells in the former case and a relative predominance of $CD8^+$ T cells in the latter case.[70,71] The potential immunosuppressive effects of liver transplantation have also been suggested by studies of liver allografting into the sensitized recipient in rats. For example, liver allografts into sensitized recipients were not rejected, and in fact induced tolerance as they had in naive recipients (DA to PVG). However, this phenomenon was not uniform and occurred in approximately 50% of the animals.[72,73] Liver allografts in the same strain combination apparently reversed ongoing rejection of donor-specific heart allografts.[74]

Other investigators have compared the effects of intravenous (i.v.) vs. portal venous (p.v.) injection of donor lymphocytes on subsequent allograft survival. Early studies demonstrated that p.v., but not i.v. injection of donor lymphocytes resulted in a donor-specific decrease in delayed type hypersensitivity and antibody production *in vivo.* This effect was critically dependent on the antigen dose.

More recently, others have assessed the effect of recipient pretreatment with donor antigen given via the portal vein. In these models, the donor antigen was supplied by injection,[64,65] portal venous drainage of the graft,[63,75] or a combination of the two.[76] Recipient pretreatment with donor, but not third-party lymphocytes, given p.v., but not i.v., significantly prolonged survival of subsequently transplanted heart[65] and renal[64] allografts. Serum, but not spleen cells from these animals inhibited proliferation to donor, but not third-party stimulators in one-way mixed lymphocyte culture.[64] Similarly, transfer of serum from p.v.-treated animals into naive recipients resulted in the prolonged survival of donor, but not third-party grafts.[64]

To determine the mechanism of this tolerance, graft-infiltrating cells (GIC) were harvested, phenotyped, and tested for cytotoxicity and the ability to produce cytokines.[76] In p.v.-treated animals, the percentage of $OX8^+$ (cytotoxic) T cells and $OX6^+$ (class II^+) cells was significantly lower than in untreated controls. In addition, the ability of GIC to lyse donor targets was decreased. Finally, IL-2, IL-3, and B cell stimulating factor production by Con A-activated GIC was decreased compared to untreated controls. These data would suggest that pretransplant injection of donor lymphocytes into the recipient's portal venous system leads to functional inactivation, not clonal deletion of donor reactive cells. In separate, but similar experiments, portal venous drainage of heart and renal allografts led to prolonged survival.[74,77] The effect of combined portal vein injection and portal venous drainage on allograft survival is additive.[75]

Some evidence exists to implicate the development of both nonspecific and specific suppressor cells for the observed tolerance of liver allografts in some rat strains.[78] In other cases, the observed tolerance was hypothesized to result from the release of humoral blocking factors, release of soluble donor MHC class I antigen which, alone or in antigen antibody complexes, may maintain or induce tolerance, and from IgG antibodies against donor class II antigens.[79-82]

Despite the apparent low immunogenic potential of liver allografts in various species and strain combinations in other studies, the outcome of liver allotransplantation in pigs and dogs was quite dependent on swine lymphocyte antigen and dog lymphocyte antigen histocompatibility.[84-86]

These data support several conclusions. First, the immune response to liver allografts is largely dependent on the immune response genes of the recipient and therefore varies with each donor/recipient combination. In certain strain combinations, donor reactive T cell clones may be either eliminated or rendered functionally inactive. Serum from liver allograft recipients contains a factor(s) that can prolong donor, but not third-party allografts when transferred to naive animals. This factor may be anticlass II antibodies.

B. LIVER AND HEPATOCYTE CYTOSOL

In addition to the potential endogenous immunosuppressive properties of the liver suggested by animal studies, it is also of note that patients with liver failure and presumably ongoing hepatocellular death are immunocompromised hosts and demonstrate defects in both humoral and cell-mediated immunity.[87-89] Furthermore, histologic analysis of acute rejection after human orthotopic liver transplantation reveals a portal inflammatory infiltrate, endothelialitis, and bile duct damage, but relatively intact hepatocytes until late stages.[23] Such observations have suggested that perhaps hepatocytes release substances which are immunocytoprotective in small amounts, but if released in larger quantities by massive hepatocellular death, result in an immunocompromised host. This has led to the study of hepatocyte cytosolic constituents. Crude preparations of the cytosol of purified hepatocytes, liver nonparenchymal cells, and whole liver have been tested for immunosuppressive properties in *in vitro* and *in vivo* assays. We found that hepatocyte and whole liver, but not nonparenchymal cell cytosol, inhibited proliferation in MLC and the development of allo-CTL in MLC in a titratable manner. This effect was observed only if the cytosol was added early to the cultures (day 0 to day 2), which suggested that its effect was exerted upon the sensitization phase of alloimmune responses *in vitro*. Furthermore, the inhibitory effect of cytosol appeared to target responder lymphocytes rather than stimulator cells, since pretreatment of the stimulator cells with cytosol and later removal of the cytosol prior to the addition of responder lymphocytes abrogated the inhibitory effects.[90]

Liver cytosol has also been shown to inhibit alloimmunity *in vivo*. That is, host animals treated with liver cytosol develop a significantly decreased humoral response to sheep red blood cell compared to saline controls as measured by the Jerne plaque assay. In addition, liver cytosol injected locally into SMA resulted in the abrogation of the development of allo-CTL in the sponge grafts. Systemic administration of liver cytosol did not inhibit the local development of allo-CTL in sponge grafts.[91]

Attempts to purify the active component in liver and hepatocyte cytosol have implicated arginase as the inhibitory factor. First, molecular weight studies indicate that the active factor is in the molecular weight range of 100 K, which approximates the molecular weight of arginase. Pure arginase reproduces the *in vitro* inhibitory effects, and arginine, the substrate for arginase, reverses the inhibitory effect of liver and hepatocyte cytosol as well as that of pure arginase.[92] There is also evidence that increased serum arginase occurs in hepatic failure both in rats[93] and in pigs (unpublished observation). Therefore, local and/or systemic release of hepatocyte cytosolic factor(s), including arginase, may mediate some of the observed endogenous immunosuppressive properties of the liver and the relative resistance of hepatocytes in liver allografts to immune damage. This mechanism, however, would not explain the observed donor-specific tolerance observed in the rat liver transplant model. The liver is a complex organ and many as-of-yet undefined secreted factors and cellular interactions may be responsible for its apparent immunosuppressive effects.

VI. POTENTIAL HEPATOCYTE-KUPFFER CELL INTERACTIONS IN ALLOGRAFT REJECTION AND ENDOGENOUS IMMUNOSUPPRESSION

A number of studies have demonstrated the ability of Kupffer cells to perform accessory cell functions in various *in vitro* immunologic systems. For example, murine Kupffer cells function as accessory cells in antigen-specific T cell proliferation.[94] In addition, Kupffer cells have been demonstrated to function as accessory cells *in vitro* in the induction of primary antibody responses and in T cell proliferation to Con A.[95] Furthermore, Kupffer cells are as stimulatory as splenic adherent cells in one-way mixed lymphocyte reactions.[95] Other studies have isolated murine hepatic accessory cells from the bulk population of hepatic nonparenchymal cells and characterized such cells as Ia^+, FcR^+, $33D1^+$, firmly adherent to plastic culture dishes, sensitive to radiation *in vivo,* but insensitive to radiation *in vitro,* phagocytic, nonspecific esterase positive, and downregulated by activated natural killer cells. These cells were capable of performing accessory cell function in primary antibody responses *in vitro* as well as primary proliferative responses in one-way mixed lymphocyte reactions.[96] The cells with accessory cell function were isolated within a population of cells highly enriched for Kupffer cells, but it was not clear in these studies whether or not the hepatic accessory cell and the Kupffer cell were one and the same. Guinea pig liver sinusoidal lining cells (LSLC) comprised of Kupffer and endothelial cells have been demonstrated to function as APC. Specifically, LSLC which were exposed to a specific antigen *in vivo* by p.v. injection were capable of processing the antigen *in vivo* and presenting the specific antigen to primed T cells *in vitro* in a genetically restricted fashion.[97] It is known that the liver sequesters antigen administrated by the intravenous or oral route, but a number of studies have suggested that antigens entering the liver result in tolerance rather than immune stimulation.[98] Therefore, although LSLC appear capable of processing and presenting antigen to T lymphocytes, the fact that this does not occur *in vivo* under normal circumstances suggests that hepatic microenvironmental factors (perhaps elaborated by hepatic parenchymal cells) may suppress the accessory function of LSLC or LSLC interaction with circulating lymphocytes. After transplantation, alteration of the normal hepatic microenvironment may result in the predominance of immunostimulatory rather than immunosuppressive pathways and subsequent processing of alloantigen and rejection.

Given the demonstrated accessory cell function of Kupffer cells and the fact that donor Kupffer cells are later replaced by recipient Kupffer cells, it is possible that in the clinical transplant setting, donor Kupffer cells (present early after transplantation) participate in direct presentation of allogeneic class II antigens, and recipient Kupffer cells (present later after transplantation) participate in the potential pathway of indirect antigen presentation described in the experimental systems of MLHC and HC-SMA. In the latter situation, one could postulate that recipient Kupffer cells expressing syngeneic class II antigens process and present donor hepatocyte (or biliary or endothelial cells) class I antigens to recipient helper T lymphocytes which become activated and produce lymphokines. The lymphokines stimulate the maturation of precytolytic effector T cells which directly recognize donor hepatocyte (or biliary or endothelial cell) class I antigens and injure these target cell populations.

VII. SUMMARY

In summary, both human and animal studies have shown that a number of cell populations within the liver (including hepatocytes, Kupffer cells, endothelial cells, and biliary epithelial cells) express MHC antigens which may be induced or amplified during rejection. These cells which express MHC antigens appear to be the targets of rejection. The effectors of allograft rejection appear to involve both helper and cytotoxic T cells. Both direct and indirect mechanisms of alloantigen recognition may occur during allograft rejection. Endogenous immunosuppressive properties of the liver may be attributable to secreted factors as well as cellular interactions. Finally, hepatocyte and Kupffer cell interactions are implicated in both allograft rejection as well as endogenous immunosuppression mediated by the liver.

REFERENCES

1. **Thung, S. N., Schaffner, F., and Gerber, M. A.,** Expression of HLA antigen on hepatocytes in liver disease, *Transplant. Proc.,* 20, 722, 1988.
2. **Fuggle, S. V.,** MHC antigen expression in vascularize organ allografts: clinical correlations and significance, in *Transplantation Reviews,* Vol. 3, P. J. Morris and N. J. Tilney, Eds., W. B. Saunders, Philadelphia, 1988, 81.
3. **Steinhoff, G., Wonigeit, K., Ringe, B., Lauchart, W., Kemnitz, J., and Pichlmayr, R.,** Modified patterns of major histocompatibility complex antigen expression in human liver grafts during rejection, *Transplant. Proc.,* 19, 2466, 1987.
4. **So, S. K. S., Platt, J. L., Ascher, N. L., and Snover, D. C.,** Increased expression of class I major histocompatibility complex antigens on hepatocytes in rejecting human liver allografts, *Transplantation,* 43, 79, 1987.
5. **Nagafuchi, Y., Thomas, H. C., Hobbs, K. E. F., and Scheuer, P. J.,** Expression of beta-2-microglobulin on hepatocytes after liver transplantation, *Lancet,* 1, 551, 1985.
6. **Gridelli, B., Colledan, M., Grendele, M., Rossi, G., Fassati, L. R., Galmarini, D., Lazzro, A., Malavasi, F., and Ippoliti, G.,** Induction of class I and class II major histocompatibility complex (MHC) in rejecting human liver grafts, *Transplant. Proc.,* 21, 2215, 1989.
7. **Demetris, A. J., Markus, B., Saidman, S., Fung, J., Nalesnik, M., Makowka, L., Duquesnoy, R., and Starzl, T.,** Establishment of primary cultures of human biliary epithelium and induction of class II major histocompatibility complex antigens by interferon gamma, *Transplant. Proc.,* 20, 728, 1988.
8. **Wood, K. J., Hopley, A., Dallman, M. J., and Morris, P. J.,** Lack of correlation between the induction of donor class I and class II major histocompatibility complex antigens and graft rejection, *Transplantation,* 45, 759, 1988.
9. **Roser, B. J., Kamada, N., Zimmermann, F., and Davies, H. F. S.,** Immunosuppressive effect of experimental liver allografts, in *Liver Transplantation,* R. Z. Y. Calne, Ed., Grune & Stratton, London, 1987, 35.
10. **Houssin, D., Gigou, M., Franco, D., Szekely, A. M., and Gismuth, H.,** Spontaneous long-term survival of liver allograft in inbred rats, *Transplant. Proc.,* 11, 567, 1979.

11. **Bumgardner, G. L., Chen, S., Hoffman, R. A., Cahill, D., So, S. K. S., Platt, J., Bach, F. H., and Ascher, N. L.,** Afferent and efferent pathways in T cell responses to MHC class I^+ II^- hepatocytes, *Transplantation,* 47, 163, 1989.
12. **Clemmings, S., Alan, T., Bumgardner, G. L., and Ascher, N. L.,** Lack of class II antigen expression on hepatocytes profoundly affects CTL development *in vitro* and *in vivo, Transplant. Proc.,* 23, 817, 1991.
13. **Iwatsuki, S., Iwatsaki, Y., Kano, T., Klintmalm, G., Koep, L. J., Weil, R., and Starzl, T. T.,** Successful liver transplantation from cross-match positive donors, *Transplant. Proc.,* 13, 286, 1981.
14. **Gordon, R. D., Fung, J. J., Markus, B., Fox, I., Iwatsuki, S., Esquivel, C. O., Tzakis, A., Todo, S., and Starzl, T. E.,** The antibody crossmatch in liver transplantation, *Surgery,* 100, 705, 1986.
15. **Bird, G., Friend, P., Donaldson, P., O'Grady, J., Portmann, B., Calne, R., and Williams, R.,** Hyperacute rejection in liver transplantation: a case report, *Transplant. Proc.,* 21, 3742, 1989.
16. **Snover, D. C., Freese, D. K., Sharp, H. L., Bloomer, J. R., Najarian, J. S., and Ascher, N. L.,** Liver allograft rejection: an analysis of the use of biopsy in determining outcome of rejection, *Am. J. Surg. Pathol.,* 11, 1, 1987.
17. **Gugenheim, J., Samuel, D., Fabiani, B., Saliba, F., Castaing, D., Reynes, M., and Bismuth, H.,** Rejection of ABO incompatible liver allograts in man, *Transplant. Proc.* , 21, 2223, 1989.
18. **Hanto, D. W., Snover, D. C., Norreen, H. J., Sibley, R. K., Gajo-Peczalska, K. J., Najarian, J. S., and Ascher, N. L.,** Hyperacute rejection of a human orthotopic liver allograft in a presensitized recipient, *Clin. Transplant.,* 1, 304, 1987.
19. **Batts, K. P., Moore, S. B., Perkins, J. D., Weisner, R. H., Grambesch, P. M., and Krom, R. A. F.,** Influence of positive lymphocyte crossmatch and HLA mismatching on vanishing bile duct syndrome in human liver allografts, *Transplantation,* 45, 376, 1988.
20. **Demetris, A. J., Jaffe, R., Tzakis, A., Ramsey, G., Todo, S., Belle, S., Esquivel, C., Shapiro, R., Zjako, A., Markus, B., Morozec, E., Van Thiel, D. H., Sysyn, G., Gordon, R., Makowka, L., and Starzl, T. E.,** Antibody mediated rejection of human liver allografts: transplantation across ABO blood group barriers, *Transplant. Proc.,* 21, 2217, 1989.
21. **Eggink, H. F., Hofstee, N., Gips, C. H., Krom, R. A. F., and Houth, H. J.,** Histopathology of serial graft biopsies from liver transplant recipients, *Am. J. Pathol.,* 114, 18, 1984.
22. **Snover, D. C., Sibley, R. K., Freese, D. K., Sharp, H. L., Bloomer, J. R., Najarian, J. S., and Ascher, N. L.,** Orthotopic liver transplantation: a pathological study of 63 serial liver biopsies from 17 patients with special reference to the diagnostic features and natural history of rejection, *Hepatology,* 4, 1212, 1984.
23. **Snover, D.,** Liver transplantation, in *The Pathology of Organ Transplantation,* G. E. Sale, Ed., Butterworths, Boston, 1990, 103.
24. **Wight, D. G. D. and Portmann, B.,** Pathology of liver transplantation, in *Liver Transplantation,* 2nd ed., R. Calne, Ed., Grune & Stratton, New York, 1987, 385.

25. **Ludwig, J., Wiesner, R. H., Batts, K. P., Perkins, J. D., and Krom, R. A. F.,** The acute vanishing bile duct syndrome (acute irreversible rejection) after orthotopic liver transplantation, *Hepatology,* 7, 476, 1987.
26. **Ruebner, B. H., Paul, M. P., Pimstone, N., and Ward, R. E.,** Histopathology of early and late human allograft rejection, *Hepatology,* 7, 203, 1987.
27. **Donaldson, P. T., O'Grady, J., and Portmann, B.,** Evidence for an immune response to HLA class I antigens in the vanishing bile duct syndrome after liver transplantation, *Lancet,* 1, 945, 1987.
28. **Markus, B., Demetris, A. J., Fung, J. J., et al.,** Allospecificity of liver allograft-derived lymphocytes and correlation with clinicopathologic findings, *Transplant. Proc.,* 20(2), 219, 1988.
29. **Grond, J., Gouw, A. S. H., and Poppema, S.,** *Transplant. Proc.,* 18, 128, 1986.
30. **Saidman, S., Markus, B., Demetris, J., Fung, J., Zeevi, A., Murase, N., Starzl, T., and Duquesnoy, R.,** HLA specificity of lymphocytes propagated from liver transplant biopsies: correlation with histological and clinical findings, in *Abstracts of the International Organ Transplant Forum,* Pittsburgh, 1987.
31. **Duquesnoy, R. J., Saidman, S., Markus, B. H., Demetris, A. M., and Zeevi, A.,** Role of HLA in intragraft cellular immunity in human liver transplantation, *Transplant. Proc.,* 20, 724, 1988.
32. **Perkins, J. D., Wiesner, R. H., Banks, P. M., LaRusso, N. F., Ludwig, J., and Krom, R. A. F.,** Immunohistologic labeling as an indicator of liver allograft rejection, *Transplantation,* 43, 105, 1987.
33. **Gouw, A. S. H., Guitema, S., Grond, J., Slooff, M. J. H., Klompmaker, I. J., Gips, C. H., and Poppema, S.,** Early induction of MHC antigens in human liver grafts, *Am. J. Pathol.,* 133, 82, 1988.
34. **McCaughan, G. W., Davies, S., Waugh, J., Painter, D., Gallagher, N. D., Thompson, J., and Sheil, A. G. R.,** Cell surface phenotype of mononuclear cells infiltrating bile ducts during acute and chronic liver allograft rejection, *Transplant. Proc.,* 21, 2201, 1989.
35. **Si, L., Whiteside, T. L., VanThiel, D. H., and Rabin, B. S.,** Lymphocyte subpopulations at the site of "piecemeal" necrosis in end stage chronic liver diseases and rejecting liver allografts in cyclosporine-treated patients, *Lab. Invest.,* 50, 341, 1984.
36. **Demetris, A. J., Lasky, S., VanThiel, D. H., Starzl, T., and Whiteside, T.,** Induction of DR/IA antigens in human liver allografts, *Transplantation,* 40, 504, 1985.
37. **Lafferty, K. J., Prowse, S. J., Simeonovic, C. J., and Warren, H. S.,** Immunobiology of tissue transplantation: a return to the passenger leukocyte theory, *Annu. Rev. Immunol.,* 1, 143, 1983.
38. **Gouw, A. S. H., Houthoff, H. J., Hyitema, S., Beelen, J. M., Gips, C. H., and Poppema, S.,** Expression of major histocompatibility complex antigens and replacement of donor cells by recipient ones in human liver grafts, *Transplantation,* 43, 291, 1987.
39. **Gale, R. P., Sparkes, R. S., and Golde, D. W.,** Bone marrow origin of hepatic macrophage (Kupffer cells) in humans, *Science,* 201, 937, 1978.
40. **Bollinger, R.,** Immunological aspects of liver transplantation, in *Transplantation Reviews,* Vol. 2, P. J. Morris and N. J. Tilney, Eds., W.B. Saunders, Philadelphia, 1988, 109.

41. **Lautenschlager, I., Nyman, N., Vaanagen, H., Lehto, V. P., Virtanen, I., and Hayry, P.,** Antigenic and immunogenic components in rat liver, *Scand. J. Immunol.,* 17, 61, 1983.
42. **Almond, P. S., Bumgardner, G. L., Chen, S., and Matas, A. J.,** Major histocompatibility complex I^+ class II^- hepatocytes stimulate the generation of "memory" cells in vivo, *J. Surg. Res.,* submitted.
43. **So, S. K., Wilkes, L. M., Platt, J. L., Ascher, N. L., and Simmons, R. L.,** Purified hepatocytes can stimulate allospecific cytotoxic T lymphocytes in a mixed lymphocyte-hepatocyte culture, *Transplant. Proc.,* 19, 251, 1987.
44. **Bumgardner, G. L., Matas, A. J., Chen, S., Cahill, K., Cunningham, T. R., Payne, W. D., Bach, F. H., and Ascher, N. L.,** Analysis of the in vivo and in vitro immune response to purified hepatocytes, *Transplantation,* 49, 429, 1990.
45. **Bumgardner, G. L., Cahill, D., Chen, S., Bach, F. H., and Ascher, N. L.,** $L3T4^+$,$Lyt2^+$ and $Lyt2^+$,$L3T4^-$ T cells participate in the generation of allospecific cytotoxicity in response to MHC class I^+ hepatocytes, *Transplant. Proc.,* 21, 421, 1989.
46. **Singer, A., Kruisbeek, A. M., and Andrysiak, P. M.,** T Cell-accessory cell interactions that initiate allospecific cytotoxic T lymphocyte responses: existence of both Ia-restricted and Ia-unrestricted cellular interaction pathways, *J. Immunol.,* 132, 2199, 1984.
47. **Golding, H. and Singer, A.,** Role of accessory cell processing and presentation of shed H-2 alloantigens in allospecific cytotoxic T lymphocyte responses, *J. Immunol.,* 133, 597, 1984.
48. **Martinez, O., Burke, E., Alan, T, and Ashby, T.,** Allogeneic hepatocytes and allogeneic lymphocytes stimulate different immunoregulatory pathways, *Transplant. Proc.,* 23, 805, 1991.
49. **Bumgardner, G. L., Chen, S., Almond, P. S., Ascher, N. L., Payne, W. D., and Matas, A. J.,** Role of macrophages in the immune response to hepatocytes, *J. Surg. Res.,* 48, 568, 1990.
50. **Bumgardner, G. L., Almond, P. S., Chen, S., Ascher, N. L., and Matas, A. J.,** Cell subsets responding to purified hepatocytes and evidence of indirect recognition of hepatocyte MHC class I antigen: role of L3T4 T cell in the development of allospecific cytotoxicity in hepatocyte-sponge matrix allografts, *Transplantation,* in press.
51. **Almond, S. P., Bumgardner, G. L., Chen, S., Cunningham, T. R., Payne, W. D., and Matas, A. J.,** Correlation of class I antigen expression with immunogenicity, *Curr. Surg.,* 47, 262, 1990.
52. **Demetriou, A. A., Whiting, J. F., Feldman, D., Levenson, S. M., Chowdhury, N. R., Moscioni, A. D., Kram, M., and Chowdhury, J. R.,** Replacement of liver function in rats by transplantation of microcarrier attached hepatocytes, *Science,* 233, 1190, 1986.
53. **Makowka, L., Lee, G., Cobourn, C. S., Farber, E., Falk, J. A., and Falk, R. E.,** Allogeneic hepatocyte transplantation in the rat spleen under cyclosporine immunosuppression, *Transplantation,* 42, 537, 1986.
54. **Bumgardner, G. L., Fasola, C., and Sutherland, D. E. R.,** Prospects for hepatocyte transplantation, *Heptology,* 8, 1158, 1988.

55. **Wang, J., Pollack, R., and Bartholemew, A.,** Sustained reduction of serum cholesterol levels following allotransplantation of parenchymal hepatocytes in WHHL rabbits, *Transplant. Proc.,* 23, 894, 1991.
56. **Jirtle, R. L., Biles, C., and Michalopoulous, G.,** Morphologic and histochemical analysis of hepatocytes transplanted into syngeneic hosts, *Am. J. Pathol.,* 101, 115, 1980.
57. **Patel, A., Chowdhury, N. R., Thompson, E., Wilson, J., Hardy, M., and Chowdhury, J. R.,** Long-term acceptance of hepatocyte allografts by ultraviolet irradiation and culture of donor rat liver cells, *Hepatology,* 8, 1219, 1988.
58. **Aprile, J. A., Deeg, H. J., Castner, T., Miller, L., and Storb, R.,** Impaired expression of class II antigens on canine dendritic cells following ultraviolet irradiation, *Exp. Hematol.,* 15, 452, 1987.
59. **Cereb, N., June, C., and Deeg, H. J.,** Effect of gamm and ultraviolet irradiation on mitogen induced intracellular calcium mobilization in human peripheral blood leukocytes, *Clin. Res.,* 356, 801A, 1987.
60. **Greene, M. I., Sy, M. S., Kripke, M., and Benacerraf, B.,** Impairment of antigen-presenting cell function by ultraviolet radiation, *Proc. Natl. Acad. Sci. U.S.A.,* 76, 6591, 1979.
61. **Calne, R. Y., White, H. J. O., Yoffa, D. E., Binns, R. M., Maginn, R. R., Herbertson, B. M., Millare, P. R., Molina, V. P., and Davis, D. R.,** Observations of orthotopic liver transplantation in the pig, *Br. Med. J.,* 2, 478, 1967.
62. **Kamada, N. and Calne, R. Y.,** A surgical experience with 530 liver transplants in the rat, *Surgery,* 93, 64, 1983.
63. **Boeck, W., Sobis, H., Lacquet, A., Gruwez, J., and Vandeputte, M.,** Prolongation of allogeneic heart graft survival in the rat after implantation in portal vein, *Transplantation,* 19, 145, 1975.
64. **Yoshimura, N., Matsui, S., Hamahima, T., Lee, C., Ohsaka, Y., and Oka, T.,** The effects of perioperative portal venous inoculation with donor lymphocytes on renal allograft survival in the rat: specific prolongation of donor grafts and suppressor factor in the serum, *Transplantation,* 49, 167, 1990.
65. **Lowry, R. P., Kenick, S., and Lisbona, R.,** Speculation on the pathogenesis of prolonged cardiac allograft survial following portal venous inoculation of allogeneic cells, *Transplant. Proc.,* 19, 3451, 1987.
66. **Kamada, N.,** The immunology of experimental liver transplantation, *Immunology,* 55, 369, 1985.
67. **Kamada, N., Brons, G., and Davies, H. S.,** Fully allogeneic liver grafting in rats induces a state of systemic nonreactivity to donor transplantation antigens, *Tranplantation,* 29, 429, 1980.
68. **Kamada, N., Shinomiya, T., and Tamaki, T,.** Specific inhibiton of heart allograft rejection by serum from liver grafted rats, *Transplant. Proc.,* 19, 570, 1987.
69. **Kamada, N., Shinomiya, T., Tamaki, T., and Ishiguro, K.,** Immunosuppressive activity of serum from liver grafted rats, *Transplantation,* 42, 581, 1986.
70. **Wolfe, J. A., Knechtle, S. J., Burchette, J., Bollinger, R. R., and Sanfilippo, F.,** Phenotype and patterns of inflammatory cell infiltration association with rejection of acceptance of rat liver allografts, *Transplant. Proc.,* 19, 364, 1987.

71. **Knechtle, S. J., Wolfe, J. A., Burchette, J., Sanfilippo, F., and Bollinger, R. R.,** Infiltrating cell phenotypes and patterns associated with hepatic allograft rejection or acceptance, *Transplantation,* 43, 169, 1987.
72. **Kamada, N. and Shinomiya, T.,** Clonal deletion as the mechanism of abrogation of immunological memory following liver grafting in rats, *Immunology,* 55, 85, 1985.
73. **Kamada, N., Davies, H. S., and Roser, B.,** Reversal of transplantation immunity by liver grafting, *Nature,* 292, 840, 1981.
74. **Kamada, N. and Wight, D. G. D.,** Antigen-specific immunosuppression induced by liver transplantation in the rat, *Transplantation,* 38, 217, 1984.
75. **Holman, J.,** Enhanced survival of heterotopic rat heart allografts with portal venous drainage, *Transplantation,* 49, 229, 1990.
76. **Kamei, T., Callery, M. P., and Flye, M. W.,** Pretransplant portal vein administration of donor antigen and portal venous allograft drainage synergistically prolong rat cardiac allograft survival, *Surgery,* 108, 415, 1990.
77. **Hamashima, T., Yoshimura, N., Matsui, S., Lee, C., Ohsaka, Y., and Oka, T.,** The effects of perioperative portal venous inoculation with donor lymphocytes on renal allograft survival in the rat: phenotypic and functional analysis of graft-infiltrating cells, *Transplantation,* 49, 171, 1990.
78. **Sakai, A.,** Role of the liver in kidney allograft rejection in the rat, *Transplantation,* 9, 333, 1970.
79. **Tsuchimoto, S., Kakita, A., Uchino, J., Mizuno, K., Niyamna, T., Fuji, H., Matsuno, Y., Natori, T., and Aizawa, M.,** Mechanism of tolerance in rat liver transplantation: evidence for the existence of suppressor cells, *Transplant. Proc.,* 19, 514, 1987.
80. **Houssin, D., Charpentier, B., Lang, P. H., Tamisier, D., Gugenheim, J., Gigou, M., and Bijsmuth, H.,** In vivo and in vitro correlates of the specific transplantation tolerance induced by spontaneously tolerated liver allografts in inbred strains of rats, *Transplant. Proc.,* 13, 619, 1981.
81. **Kamada, N., Davies, H. S., and Roser, B. J.,** Fully allogeneic liver grafting and the induction of donor-specific unreactivity, *Transplant. Proc.,* 13, 837, 1981.
82. **Kamada, N.,** Transfer of specific immunosuppression of graft rejection using lymph from tolerant liver grafted rats, *Immunology,* 55, 241, 1985.
83. **Kamada, N., Shinomiya, T., Tamaki, T., and Ishiguro, K.,** Immunosuppressive activity of serum from liver grafted rats: passive enhancement of fully allogeneic heart grafts and induction of systemic tolerance, *Transplantation,* 42, 581, 1986.
84. **Vaiman, M., Bacourt, F., Villiers, P. A., and Garnier, H.,** The influence of the major histocompatibility complex on liver allograft survival in the pig, *Transplantation,* 22, 402, 1976.
85. **Noorloos, A. A. B., Visser, J. J., Drexhage, H. A., Meijer, S., and Hoitsma, H. F. W.,** Liver allograft rejection in pigs: histology of the graft and role of swine leukocyte antigen-D, *J. Surg. Res.,* 37, 269, 1984.
86. **Beuvers, C. B., Terpstra, O. T., Ten Kate, F. W. J., Kooy, P. P. M., Molenaar, J. C., and Jeekel, J.,** Long term survival of auxiliary partial liver grafts in DLA-identical littermate beagles, *Transplantation,* 39, 113, 1985.

87. **Young, G. P., Dudley, F. J., and Van Der Weyden, M. B.,** Suppressive effect of alcoholic liver disease sera on lymphocyte transformation, *Gut,* 20, 833, 1979.
88. **Brattig, N. W. and Berg, P. A.,** Immunosuppressive serum factors in viral hepatitis. I. Characterization of serum inhibition factor(s) as lymphocyte activators, *Hepatology,* 3, 638, 1983.
89. **Brattig, N. W., Schrempf-Decker, G. E., Brockl, C. W., and Berg, P. A.,** Immunosuppressive serum factors in viral hepatitis. II. Further characterization of serum inhibition factor as an albumin-associated molecule, *Hepatology,* 3, 647, 1983.
90. **Bumgardner, G. L., Billiar, T. R., So, S. K., Chen, S., Dunn, G., Payne, W., and Ascher, N. L.,** In vitro immunosuppressive effects of murine hepatocyte cytosol, *Transplant. Proc.,* 21, 1154, 1989.
91. **Bumgardner, G. L., Dunn, G., Chen, S., Cahill, D., So, S. K., Payne, W., and Ascher, N. L.,** Immunosuppressive effects of liver cytosol in vivo, *Surg. Forum,* 39, 370, 1988.
92. **Bumgardner, G. L., Ascher, N. L., Chen, S., Cunningham, T. R., Payne, W. D., and Matas, A. J.,** Immunosuppressive effects of hepatocyte cytosol is mediated by arginase, *Surg. Forum,* 40, 391, 1989.
93. **Emery, G. N. and Beveridge, J. M. R.,** The cause of the disappearance of arginine from the blood of rats with acute hepatic necrosis induced by dietary means, *Can. J. Biochem. Physiol.,* 39, 977, 1961.
94. **Richman, L. K., Klingenstein, R. J., Richman, J. A., Strober, W., and Berzofsky, J.,** The murine Kupffer cell. I. Characterization of the cell serving accessory function in antigen-specific T cell proliferation, *J. Immunol.,* 123, 2602, 1979.
95. **Nadler, P. I., Klingenstein, R. J., Richman, L. K., and Ahmann, G. B.,** The murine Kupffer cell. II. Accessory cell function in in vitro primary antibody responses, mitogen-induced proliferation, and stimulator of mixed lymphocyte responses, *J. Immunol.,* 125, 2521, 1980.
96. **Shah, P. D., Rowley, D. A., Latta, S. L., and Magilavy, D. B.,** A comparison of murine hepatic accessory cells and splenic dendritic cells, *Cell. Immunol.,* 118, 394, 1989.
97. **Rubinstein, D., Roska, A. K., and Lipsky, P. E.,** Antigen presentation by liver sinusoidal lining cells after antigen exposure in vivo, *J. Immunol.,* 138, 1377, 1987.

Chapter 9

NONPARENCHYMAL CELL AND HEPATOCYTE INTERACTION IN ANTIMALARIAL ACTIVITY: A CENTRAL ROLE FOR NITRIC OXIDE

A.K. Nussler

TABLE OF CONTENTS

ISBN 0-8493-6109-5

I. INTRODUCTION

One of the major parasitic diseases of man and animals is malaria. The name "malaria" (or "bad air") derives from the original association of the disease with the bad air of marshlands. Malaria is a mosquito-borne infection caused by protozoa of the genus *Plasmodium*. Presently, about half the population of the world lives in areas where malaria is endemic. It is estimated that there are around 100 million clinical cases annually, and malaria is assumed to be responsible for at least 1 million deaths per year.[1] The clinical pattern of the disease is related to its multiplication of blood stage parasites, and is manifested by repeated episodes of chills, fever, anemia, and splenomegaly. The most severe complication is the cerebral syndrome due to *P. falciparum,*[2] shown in some reports to be responsible for up to 80% of fatal falciparum cases.[1] The infectious agent was discovered in 1880 by Laveran.[3] Grassi[4] and Ross[5] identified that malaria is transmitted via a bite of an infected female *Anopheles* mosquito. The preerythrocytic stages of malaria were discovered in 1948.[6,7] Today, four *Plasmodium* species (*P. falciparum, P. malariae, P. ovalae, P. vivax*) are known to infect humans.

Various attempts have been made to disrupt the life cycle (Figure 1) of the parasite in its mosquito and human host. The two main interventional methods are vector control and the use of antimalaria drugs. However, in many parts of the world, a combination of these methods for malaria control has proven unsuccessful.[8] The development of malaria vaccines target the preerythrocytic, erythrocytic, and sexual stages of the parasite (Figure 1). Research on the preerythrocytic stages of malaria became more widespread after *in vitro* establishment of the liver stage in cultured hepatocytes.[9,10] Investigation of this stage became even more interesting when it was shown that protection during the liver stage could prevent the clinical pattern of the disease. It is now well established that both humoral and cell-mediated mechanisms are involved in malaria immunity. After presented with malaria antigen, sensitized B and T cells may function in several ways. Among these is the production of soluble substances referred to as cytokines which regulate the proliferation, differentiation, and maturation of lymphocytes and hemopoietic cells. These cytokines induce a whole cascade of immunological functions and mediate malaria immunity directly or indirectly, as has been recently documented.[11]

II. INTERACTIONS BETWEEN SPOROZOITE AND HEPATIC CELLS

The details of how sporozoites invade liver parenchymal cells (Figure 1) remain controversial and poorly understood. Currently, there are two predominant hypotheses. One supports the idea that the sporozoite invades hepatocytes by passing through the space of Disse, the lumen between endo-

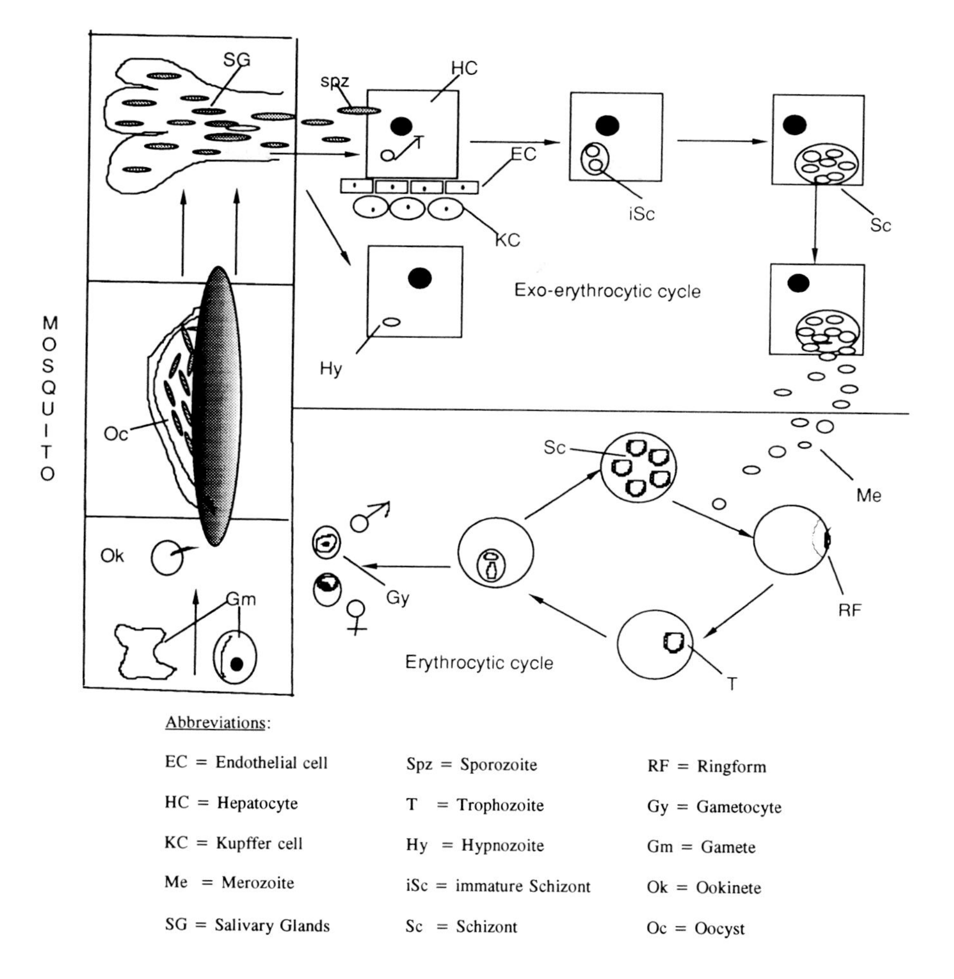

FIGURE 1. Life cycle of *Plasmodium*. The life cycle of *Plasmodium* has two distinct phases, sexual and asexual. In humans and other mammals, the asexual phase starts when sporozoites are released from the salivary glands of infected mosquitoes. Within minutes of being injected into the bloodstream, the sporozoites enter parenchymal liver cells. Here they are transformed into uninuclear trophozoites and then into exoerythrocytic schizonts over several days for the human malaria species *P. falciparum,* and in less than 48 h for rodent malaria species such as *P. yoelii.* Mature liver schizonts then release thousands of merozoites. On release, merozoites invade red blood cells, where they develop into ring-stage forms, trophozoites, and schizonts. The daughter merozoites produced by schizont-infected cells are released from erythrocytes and invade other red blood cells in a number of successive cycles. The sexual phase of the parasite starts by differentiation of some erythrocytic parasites into female and male gametocytes. When these gametocytes are ingested by a feeding mosquito, the sexual phase is initiated within the insect. Gametocytes undergo gametogenesis in the mosquito midgut. The male gamete exflagellates and fertilizes the female counterpart to produce the zygote 10 to 20 min. later. Within 24 h after fertilization, zygotes become ookinetes. Mature ookinetes penetrate the midgut wall and develop through the oocyst stage into sporozoites, thereby completing the cycle.

thelial cells and hepatocytes.[12] The second hypothesis argues that the sporozoite travels via sinusoidal Kupffer cells (KC) to reach the hepatocytes. Meis and co-workers[13] described KC as the macrophages which are responsible for the uptake of sporozoites from the bloodstream, and following uptake the KC then present the parasite to hepatocytes. Another group found a minimal role for KC in the clearance of parasites from the bloodstream in nonimmune hosts.[14] However, the same group found that KC harvested from animals immunized with irradiated sporozoites are almost resistant to invasion of *Leishmania* amastigotes, whereas normal KC were easily invaded. This finding suggests that immunization with irradiated sporozoites against the hepatic stages of malaria induces nonspecific, as well as specific, immunomechanisms and underlines the crucial role of nonparenchymal cells in the acquisition of this immunity.

In KC from immunized animals, this nonspecific resistance to invading *Leishmania* parasites was not as effective against malaria sporozoites. However, the small difference might have limited importance under natural conditions where generally small numbers of sporozoite actually reach the liver and invade liver parenchymal cells. KC have been described as playing an important role in the elimination of other microorganisms[15] and viruses.[16] KC are hepatic sinusoidal macrophages which derive from circulating blood monocytes. They are considered to be antigen-presenting cells[17,18] and seem to be a major source of cytokine production during experimental endotoxemia.[19] The combined findings to date clearly show the range of problems involved in understanding malaria immunity in the hepatic stages, since cells other than the hepatocyte itself might play an important role. These other cells, termed nonparenchymal cells (NPC), include endothelial cells, KC, Ito cells, and pit cells. Invading cells such as T cells, natural killer (NK) cells, and fibroblasts must also be considered. KC, Ito cells, and possibly endothelial cells secrete several different cytokines[20,21] when activated and thereby may promote antimalarial activity in the hepatocyte.[22-25]

III. NON-SPECIFIC EFFECTOR MECHANISMS AGAINST THE HEPATIC STAGES OF MALARIA

A. ROLE OF NITRIC OXIDE

Macrophage activation during infection occurs through a T cell-mediated process involving the synthesis of several growth factors.[26] During this activation, macrophages produce an array of biologically active products, e.g., reactive oxygen intermediates (ROI),[27] tumor necrosis factor alpha (TNF-α),[28] and interleukins (IL), and become capable of killing several pathogens and lysing tumor cells, as has recently been reviewed.[29] In 1985, it was shown that lipopolysaccharide (LPS)-stimulated murine peritoneal macrophages released nitrite (NO_2^-) and nitrate (NO_3^-).[30] The observed NO_2^-/NO_3^- synthesis carried out by activated macrophages was related to its cytocidal effect on bacteria.[31] In 1987, Hibbs et al.[32] found that L-arginine served as substrate for this nitrogen oxide synthesis and later it was shown that inorganic nitric oxide (·N=O) is synthesized from the terminal guanido nitrogen atom of L-arginine and transformed by oxidation into NO_2^- + NO_3^-. ·N=O can bind to iron at the catalytic site of enzymes controlling mitochondrial respiration,[33-35] which results in intracellular iron loss, causes inhibition of aconitase, nicotinamide-adenine dinucleotide, ubiquinone oxidoreductase (complex I), and succinate-ubiquinone oxidoreductase (complex II) in the citric acid cycle, and destruction of iron sulfur centers.[36] Furthermore, it has been reported that this action, which decreases total protein synthesis in primary hepatocyte cultures or cocultivated KC,[37,38] is caused by ·N=O. The observed cytostasis in certain tumor cell lines[9,19] and inhibition of several infectious pathogens might thus be due to ·N=O-mediated mitochondrial respiration inhibition in tumor target cells,[35,40] hepatocytes,[41] or in the infectious agent itself.[42]

Several cell types, including macrophages,[43,44] endothelial cells,[45,46] cerebellar neurons,[46] neutrophils,[48] KC, and hepatocytes[37,38] have been reported to possess the biochemical pathway for formation of ·N=O. In addition to the cytostatic/cytotoxic action of ·N=O, this mediator is known to act as a messenger molecule through the stimulation of soluble guanylate cyclase. It has been shown that increased ·N=O levels leads to a substantial increase of cyclic guanylate monophosphate (cGMP) levels and was first associated with smooth-muscle relaxation.[45,49] Moreover, it is related to inhibition of platelet aggregation[50] and adherence,[51] and was also found in high quantities in hepatocyte supernatant upon cytokine exposure.[52,53] In 1987, a factor which was previously referred to as endothelium-derived relaxing factor was identified as ·N=O.[49] It was known that this substance lead to a substantial increase of cGMP in endothelial cells, promoted vascular relaxation, and was responsible for hypotension in endotoxin and shock.[54] Furthermore, it was reported that the induction of cGMP might play a crucial role in the regulation of neutrophil chemotaxis.[55]

Within the liver, it was first shown that KC, in response to LPS, were able to block protein synthesis in cocultivated hepatocytes through conversion of L-arginine to nitrogen oxides.[37] Subsequently, hepatocytes themselves have been demonstrated to synthesize ·N=O in response to various immunostimuli, including IL-1, interferon-gamma (IFN-γ), TNF-α, and LPS.[56] Several recent reports have since shown that such an L-arginine-dependent effector mechanism (LADEM) can be induced *in vitro* by different immunostimuli to control the development of intra- and extracellular pathogens.[57]

TNF-α, a known potent inducer of the ·N=O pathway in macrophages and tumor cells,[33] did not show antimalaria activity when added to freshly isolated murine hepatocytes through the cytotoxic ·N=O pathway[58,59] despite clear TNF-induced hepatic malaria inhibition when administered *in vivo*.[25,60] However, in *in vitro* cocultures (murine hepatocytes plus NPC) TNF exerted an obvious inhibitory effect. Analysis of TNF-stimulated coculture supernatants revealed elevated IL-6 concentrations, whereas IL-1 and IFN-γ remained undetectable. The IL-6 associated killing was later found to be mediated in part through LADEM.[58]

The mechanism by which oxygenated L-arginine-derived intermediates inhibit intrahepatic malaria parasite development remains unclear. It has been proposed that L-arginine-derived ·N=O binds to intracellular iron at the catalytic sites of enzymes controlling mitochondrial respiration in the liver,[41] thus leading to intracellular energy loss, which could be one possible mechanism. Mitochondria or mitochondria equivalents have been the target in malaria parasites for several growth-inhibiting drugs.[61] It has been shown that multiplication of *P. falciparum*-infected erythrocytes is decreased after treatment with the mitochondrial uncoupler dinitrophenol.[62] Whether or not L-arginine-dependent depletion of intracellular adenyltriphosphate could explain the inhibition of intrahepatic malaria parasites requires further study. Furthermore, it is unclear how this mechanism is controlled and/or influenced by NPC.

IFN-γ has been shown *in vitro*[22,23] and *in vivo*[22,63] to reduce parasite development in the liver. Its antimicrobicidal role has been established for several intra- and extracellular pathogens.[27] Besides its specific effects in the induction of specific immunity, its exact role in cytotoxicity is only partially understood. IFN-γ-enhanced cytotoxic effects observed in monocytes and macrophages[64,65] might be explained by the induction of superoxide anion production[66] and/or by the induction of LADEM.[32]

Indeed, Curran et al.[56] clearly showed the participation of IFN-γ in the induction of the ·N=O pathway in hepatocytes. A recent report shows that IFN-γ reduces intrahepatic malaria parasite growth through this effector mechanism.[59] Most interesting is the finding that NO_2^-/NO_3^- production could not be demonstrated after cytokine stimulation in primary hepatocyte culture supernatants, despite a clear demonstration of the ·N=O pathway induction and the reduction of intrahepatic malaria forms. This may suggest that the antiparasitic effect of ·N=O requires only small amounts of ·N=O or that ·N=O is consumed by the parasite.

Initial studies into *in vivo* relevance of this antimicrobial effector mechanism are just underway. Increased NO_3^- urine levels were found in animals treated with Bacillus Calmette-Guérin (BCG).[31] The origin of the increased NO_3^- levels were later identified as being induced through conversion of L-arginine into citrulline and ·N=O.[32] This effect was completely abrogated by giving animals N^G monomethyl-L-arginine (NMA), the inhibitor of ·N=O synthesis, in their drinking water.[67] A further study from Liew and co-workers[68] underlined for the first time the *in vivo* importance of NMA in parasitic diseases. Treatment of footpath-infected CBA mice with NMA increased the spread of the *L. major* parasite development. Enhancement of pathogens was first described in fungi infections *in vitro* after NMA treatment[42] and later in sporozoite-infected hepatocytes.[58] Injection of BCG and *Corynebacterium parvum* has been shown to protect mice against a challenge with *P. berghei* sporozoites.[69] Today, this protection can be associated with increased circulating NO_2^-/NO_3^- levels as shown after *C. parvum* injection in rats[70] or mice.[71] However, the interrelationship of the induction of LADEM *in vivo* and its role in protection in animals challenged with malaria sporozoites has yet to be proven. Indeed, injection of 5 mg LPS per mouse intraperitoneal resulted in protection against a challenge with 3500 sporozoites of *P. yoelii*. This inhibitor effect was reversed by injection of 5 mg NMA at the time of LPS injection, supporting a role for the LADEM *in vivo*. Also, intravenous injection of NMA (5 mg per mouse) alone increased blood stage parasite development by approximately 15%.[72] *In vitro* studies on the liver stage, using freshly isolated hepatocytes, revealed that *P. yoelii* intrahepatic parasite development was inhibited by LPS through the LADEM, suggesting that LPS administration *in vivo* induces ·N=O production to a degree adequate to inhibit parasite infection. These data demonstrate the *in vivo* importance of ·N=O in plasmodia infections at the hepatic level. This is at least true in an animal model where the intrahepatic life cycle of malaria takes place within 48 h. This liver stage mechanism might be of greater importance in humans where the hepatic life cycle takes 7 to 10 d, allowing even more time for induction of ·N=O within the liver. The ·N=O effect on malarial growth may not be limited to the liver. A recent research report showed that *P. falciparum*-infected erythrocytes can also be killed through LADEM *in vitro*.[73]

B. ACUTE PHASE PROTEINS AND SPOROZOITE INTERACTION

Two other cytokines have been reported to have potent antimalarial activities against intrahepatic parasites. Incubation of IL-1 strongly inhibits *P. falciparum* sporozoite development in primary hepatocyte cultures.[22] Despite its multifunctional properties,[29] IL-1 is associated with the induction of a subset of proteins called acute-phase proteins (APP).[74] APP secretion had been reported in bacterial[75] and in murine malaria infection.[76] In mice, serum amylum P-component, complement factor 3, and fibrinogen are the major APP,[77] whereas in humans it is the C-reactive protein (CRP) and alpha-2-macroglobulin.[78]

CRP produced by hepatocytes in response to IL-1[79] has been shown to interfere with bacterial and parasitic infections, and it has been shown to be elevated in malaria patients.[80] CRP were found to play a protective role in *Streptococcus* infection in mice[81] and in schistosomiasis in rats.[82] In addition to these results, CRP were found to bind to various bacterial membranes[83] at a phosphorylcholin binding site.[84] Furthermore, it has been shown that bound CRP activates the complement system[85] which could possibly be involved in the lysis of malarial sporozoites or influence its penetration into the hepatocyte.[24] CRP seem to prevent parasite division inside the hepatocyte, impeding the evolution towards multinucleated parasites.[86] This observation may be compared to an intracellular effect of antibodies directed against the circumsporozoite protein (CS). Like anti-CS antibodies, CRP can penetrate with the sporozoite and block translocation of the CS protein onto the parasitophorous vacuole membrane. This effect could lead to an acidification of the vacuole and to the lysosomal digestion of the uninuclear forms.[87] Besides these *in vitro* effects, *in vivo* CRP might enhance monocyte respiratory burst activity[88] and increase macrophage activity.[89]

C. OXIDATIVE BURST AND HEPATIC PARASITE DEVELOPMENT

Another important killing mechanism in malarial exoerythrocytic infections is the oxidative burst.[90] Increased activity of the oxidative burst by IL-6 occurs at the level of superoxide anion-generating system which in turn generates free radical and reactive oxygen intermediates to produce highly toxic molecules. One of them, hydrogen peroxide, appears, to be critical in the process of growth inhibition on the erythrocytic stages of malaria.[91] The mechanism of ROI-induced changes has not been established, but it has been proposed that DNA damage and lipid peroxidation in membranes may contribute to the inhibition of parasite multiplication or induced lysis in IL-6-stimulated cells.[92] It is interesting that IL-6 may induce both the respiratory burst and LADEM in the hepatocyte at the same time. In both cases, the observed inhibitory activity on the development of exoerythrocytic parasite forms was reversed by using either scavengers of the hydrogen peroxide and superoxide anion[90] or a competitive inhibitor of ·N=O formation.[58]

D. ANTIBODY-DEPENDENT CELL-MEDIATED CYTOTOXICITY AND THE HEPATIC STAGES OF MALARIA

Cytotoxic T cells ($CD8^+$) have been shown to kill cultured liver stage parasites in an antigen-specific and MHC-restricted manner.[93] The discovery of a malaria heat shocklike determinant expressed on infected hepatocyte surface provided a new view on how intrahepatic parasites could be eliminated.[94] The presence of neoantigens expressed on the cell surface of infected hepatocytes can induce the antibody-dependent cell-mediated cytotoxicity which has been shown to destroy tumor cells[95] or virus-infected cell.[96] It was shown that a mixed cell population of NPC added to the infected hepatic culture lysed about 70% of the infected hepatocytes in the presence of a monoclonal antibody.[94] This lysis was highly specific for the presence of NPC, since under the same conditions spleen cells showed only 25% lysis. It is likely that KC were the effector cells in this lysis. KC are highly concentrated in the periportal region[97] where most of the liver schizont develop *in vivo*.[98] Whether the oxidative burst or LADEM interferes at this level is unclear, but as already mentioned, KC secrete an array of cytokines, including IL-6, which can induce effector mechanisms on the hepatic stage. Another cell population is the NK cells which, once triggered, might release cytotoxins and proteases[99] to exert a schizonticide effect. However, NK cells release IFN-γ which has been shown to eliminate intrahepatic malarial forms.

In malaria, various cytokines are shown to be elevated during blood stage infection,[100,101] and thus those cytokines might modulate the infection by the liver stages. However, *in vitro* data presented here support a crucial role for NPC in malaria immunity in the hepatic stage.

REFERENCES

1. **World Health Organization (WHO),** Tropical diseases in media spotlight, *Trop. Dis. Res. News,* 31, 3, 1990.
2. **Maegraith, B.,** *Pathological process in malaria and blackwater fever,* Oxford: Blackwell, London, 1948.
3. **Laveran, A.,** Nouveau parasite du sang, *Bull. Acad. Natl. Med. Paris,* 9, 1235, 1880.
4. **Grassi, B.,** Studi di un zoologo sulla malaria, *Mem. R. Accad. Lincei,* 3, 299, 1900.
5. **Ross, R.,** On some peculiar pigmented cells found in two mosquitoes fed on malaria blood, *Br. Med. J.,* 2, 1786, 1897.
6. **Shortt, H. E. and Garnham, P. C. C.,** The pre-erythrocytic development of *Plasmodium cynomolgi* and *P. vivax, Trans. R. Soc. Trop. Med. Hygiene,* 41, 785, 1948.
7. **Hawking, F., Perry, W. L. M., and Thurston, J. P.,** Tissue forms of a malaria parasite, *Lancet,* 1, 783, 1948.
8. **World Health Organization (WHO),** Resistance of vectors and reservoirs of disease to pesticides: tenth report of the WHO expert committee on vector biology and control, *WHO Tech. Rep. Ser.,* 737, 9, 1986.
9. **Lambiotte, M., Landau, I., Thierry, N., and Miltgen, F.,** Developpement de schizontes dans des hepatocytes de rat adulte en culture apres infestation *in vitro* par des sporozoites de *Plasmodium yoelii, C.R. Acad. Sci. Paris,* 293, 431, 1981.
10. **Mazier, D., Landau, I., Miltgen, F., Druihle, P., Lambiotte, M., Baccam, D., and Gentilini, M.,** Infestation *in vitro* d'hepatocytes de Thamnomys adulte par des sporozoites de *Plasmodium yoelii:* schizogonie et liberation de merozoites infestants, *C.R. Acad. Sci. Paris,* 294, 963, 1982.
11. **Good, M. F.,** The implications for malaria vaccine programs if memory T cells from non-exposed humans can respond to malaria antigens, *Curr. Opin. Immunol.,* 3, 496, 1991.
12. **Shin, S. C., Vanderberg, J., and Terzakis, J. A.,** Direct infection of hepatocytes by sporeozoites of *Plasmodium berghei, J. Protozool.,* 29, 448, 1982.
13. **Meis, J. F. G. M., Jap, P. H. K., Hollingdale, M. R., and Verhave, J. P.,** An ultrastructural study on the role of Kupffer cells in the process of infection by *Plasmodium berghei* sporozoites in rats, *Parasitology,* 86, 231, 1983.
14. **Seguin, M. C., Ballou, W. R., and Nacy, C. A.,** Interaction of *Plasmodium berghei* sporozoites and murine Kupffer cells *in vitro, J. Immunol.,* 143, 1716, 1989.

15. **Crofton, R. W., Diesselhoff-Den Dulk, M. C., and Van Furth, R.,** The origin, kinetics, and characteristics of the Kupffer cells in normal steady state, *J. Exp. Med.,* 148, 1, 1978.
16. **Keller, F., Wild, M.-T., and Kirn, A.,** *In vitro* antiviral properties of endotoxin-activated rat Kupffer cells, *J. Leuk. Biol.,* 38, 293, 1985.
17. **Poulfort, K. and Souhami, R. L.,** The surface properties and antigen-presenting function of hepatic non-parenchymal cells, *Clin. Exp. Immunol.,* 46, 581, 1981.
18. **Richman, L. K., Klingenstein, R. J., Richman, J. A., Strober, W., and Bersofsky, J. A.,** The murine Kupffer cell. I. Characterization of the cell serving accessory in antigen-specific T cell proliferation, *J. Immunol.,* 123, 2602, 1979.
19. **Chensue, S. W., Terebuh, P. D., Remick, D. G., Scales, W. E., and Kunkel, S. L.,** *In vivo* biologic and immunohistochemical analysis of interleukin-1, alpha, beta and tumor necrosis factor during experimental endotoxemia, *Am. J. Pathol.,* 138, 395, 1991.
20. **Kohase, M. D., Hendriksen-DeStefano, D., May, L., Vilcek, J., and Segal, P. B.,** Induction of beta-2-interferon by tumor necrosis factor: a homeostatic mechanism in the control of cell proliferation, *Cell,* 45, 659, 1986.
21. **Schindler, R., Macilla, J., Endres, S., Ghorbani, R., Clark, S. C., and Dinarello, C. A.,** Correlation and interactions in the production of interleukin-6 (IL-6), IL-1, tumor necrosis factor alpha (TNF) in human blood mononuclear cells: IL-6 suppress IL-1 and TNF, *Blood,* 75, 40, 1990.
22. **Ferreira, A., Schofield, L., Enea, V., Schellekens, H., Van Der Meide, P., Collins, W. E., Nussenzweig, R. S., and Nussenzweig, V.,** Inhibition of development of exo-erythrocytic forms of malaria parasites by gamma interferon, *Science,* 232, 881, 1986.
23. **Mellouk, S., Maheshwari, R. K., Rhodes-Feuillette, A., Beaudoin, R. L., Berbiguier, N., Matile, H., Miltgen, F., Landau, I., Pied, S., Chigot, J. P., Friedman, R. M., and Mazier, D.,** Inhibitory activity of interferons and interleukin-1 on the development of Plasmodium falciparum in human hepatocyte cultures, *J. Immunol.,* 139, 4192, 1987.
24. **Pied, S., Nussler, A., Pontet, M., Miltgen, F., Renia, L., Gentilini, M., and Mazier, D.,** C-Reactive protein protects against pre-erythrocytic stages of malaria, *Infect. Immun.,* 57, 278, 1989.
25. **Nussler, A., Pied, S., Goma, J., Renia, L., Miltgen, F., Grau, G., and Mazier, D.,** TNF inhibits malaria hepatic stages *in vitro* via synthesis of IL-6, *Int. Immunol.,* 3, 317, 1991.
26. **Mossman, T. and Coffman, R. F.,** Heterogenicity of cytokine secretion pattern and functions of helper T cell, *Adv. Immunol.,* 46, 111, 1988.
27. **Nathan, C. F., Murray, H. W., Wiebe, E., and Rubin, B. Y.,** Identification of interferon-gamma as the lymphokine that activates human macrophages oxidative metabolism and antimicrobial activity, *J. Exp. Med.,* 158, 670, 1983.
28. **Urban, J. L., Shepard, H. M., Rothstein, J. L., Sugarman, B. J., and Schreiber, H.,** Tumor necrosis factor: a potent effector molecule for tumor cell killing by activated macrophages, *Proc. Natl. Acad. Sci. U.S.A.,* 83, 5233, 1986.

29. **Akira, S., Hirano, T., Taga, T., and Kishimoto, T.,** Biology of multifunctional cytokines: IL-6 and related molecules (IL-1 and TNF), *FASEB J.*, 4, 2867, 1990.
30. **Stuehr, D. J. and Marletta, M. A.,** Mammalian nitrate biosynthesis: mouse macrophages produce nitrite and nitrate in response to *Escherichia coli* lipopolysaccharide, *Proc. Natl. Acad. Sci. U.S.A.*, 82, 7783, 1985.
31. **Stuehr, D. J. and Marletta, M. A.,** Induction of nitrite/nitrate synthesis in murine macrophages by BCG infection, lymphokines, and interferon-gamma, *J. Immunol.*, 139, 518, 1987.
32. **Hibbs, J. B., Jr., Taintor, R. R., and Vavrin, Z.,** Macrophages cytotoxicity: role for L-arginine deiminase and imino nitrogen oxidation to nitrite, *Science*, 235, 473, 1987.
33. **Drapier, J. C. and Hibbs, J. B., Jr.,** Differentiation of murine macrophages to express nonspecific cytotoxicity for tumor cells results in L-arginine dependent inhibition of mitochondrial iron-sulfur enzymes in the macrophages effector cells, *J. Immunol.*, 140, 2407, 1988.
34. **Pellat, C., Henry, Y., and Drapier, J. C.,** IFN-gamma activated macrophages: detection by electron paramagnetic resonance of complexes between L-arginine-derived nitric oxide and non-heme iron proteins, *Biochem. Biophys. Res. Commun.*, 166, 119, 1990.
35. **Lancaster, J. R., Jr. and Hibbs, J. B., Jr.,** EPR demonstration of iron-nitrosyl complex formation by cytotoxic activated macrophages, *Proc. Natl. Acad. Sci. U.S.A.*, 87, 1223, 1990.
36. **Drapier, J. C. and Hibbs, J. B., Jr.,** Murine cytotoxic activated macrophages inhibit aconitase in tumor cells, *J. Clin. Invest.*, 78, 790, 1986.
37. **Billiar, T. R., Curran, R. D., Stuehr, D. J., West, M. A., Bentz, B. G., and Simmons, R. L.,** An L-arginine dependent mechanism mediates Kupffer cell influences on hepatocyte protein synthesis *in vitro*, *J. Exp. Med.*, 169, 1467, 1989.
38. **Curran, R. D., Billiar, T. R., Stuehr, D. J., Hofmann, K., and Simmons, R. L.,** Heptocytes produce nitrogen oxides from L-arginine in response to inflammatory stimuli, *J. Exp. Med.*, 170, 1769, 1989.
39. **Hibbs, J. B., Jr., Taintor, R. R., and Vavrin, Z.,** L-Arginine is required for expression of the activated macrophage effector mechanism causing selective metabolic inhibition in target cells, *J. Immunol.*, 138, 550, 1987.
40. **Stuehr, D. J. and Nathan, C. F.,** Nitric oxide: a macrophage product responsible for cytostasis and respiratory inhibition in tumor target cells, *J. Exp. Med.*, 169, 1543, 1989.
41. **Stadler, J., Billiar, T. R., Curran, R. D., Stuehr, D. J., Ochoa, J. B., and Simmons, R. L.,** Effect of authentic and cell-generated nitric oxide on mitochondrial respiration of rat hepatocytes, *Am. J. Physiol.*, 260, C910, 1991.
42. **Granger, D. L., Hibbs, J. B., Jr., Perfect, J. R., and Durack, D. T.,** Specific amino acid (L-arginine) requirement for the microbiostatic activity of murine macrophages, *J. Clin. Invest.*, 81, 1129, 1987.
43. **Marletta, M. A., Yoon, P. S., Iyengar, R., Leaf, C. D., and Wishnok, J. S.,** Macrophage oxidation of L-arginine to nitrite and nitrate: nitric oxide is an intermediate, *Biochemistry*, 27, 8706, 1988.

44. **Hibbs, J. B., Jr., Taintor, R. R., Vavrin, Z., and Rachlin, E. M.,** Nitric oxide: a cytotoxic activated macrophage effector molecule, *Biochem. Biophys. Res. Commun.*, 157, 87, 1988.
45. **Ignarro, L. J., Buga, G. M., Wood, K. S., Byrns, R. E., and Chaudhouri, G.,** Endothelium-derived relaxing factor produced and released from artery and vein is nitric oxide, *Proc. Natl. Acad. Sci. U.S.A.*, 84, 9265, 1987.
46. **Palmer, R. M. J., Ashton, D. S., and Moncada, S.,** Vascular endothelial cells synthesize nitric oxide from L-arginine, *Nature,* 333, 664, 1988.
47. **Bredt, D. S., Hwang, P. M., and Snyder, S.,** Localization of nitric oxide synthase indicating a neural role for nitric oxide, *Nature,* 347, 786, 1990.
48. **McCall, T. B., Boughton-Smith, N. K., Palmer, R. M. J., Whittle, B. J. R., and Moncada, S.,** Synthesis of nitric oxide from L-arginine by neutrophils: release and interaction with superoxide anion, *Biochem. J.*, 261, 293, 1989.
49. **Palmer, R. M. J., Ferrige, A. G., and Moncada, S.,** Nitric oxide release accounts for the biologic activity of endothelium derived relaxing factor, *Nature,* 357, 524, 1987.
50. **Mellion, B. T., Ignarro, L. J., Ohlstein, E. H., Pontecorvo, E. G., Hyman, A. L., and Kadowitz, P. J.,** Evidence for the inhibitory role of guanosine 3′,5′ monophosphate in ADP-induced human platelet aggregation in the presence of nitric oxide and related vasodilators, *Blood,* 57, 946, 1981.
51. **Radomski, M. W., Palmer, R. M. J., and Moncada, S.,** Endogenous nitric oxide inhibits platelet adhesion to vascular endothelium, *Lancet,* 2, 1057, 1987.
52. **Billiar, T. R., Curran, R. D., Harbrecht, B. G., Stadler, J., Williams, D. L., Ochoa, J. B., Di Silvio, M., Simmons, R. L., and Murray, S. A.,** The association between the synthesis and release of cGMP and nitric oxide biosynthesis by hepatocytes, *Am. J. Physiol.*, in press.
53. **Nussler, A. K., Di Silvio, M., Billiar, T. R., Hoffman, R. A., Selby, R., Madariaga, J., and Simmons, R. L.,** Stimulation of the nitric oxide pathway in human hepatocytes by cytokines and endotoxin, submitted.
54. **Rees, D. D., Palmer, R. M. J., and Moncada, S.,** Role of endothelium derived nitric oxide in the regulation of blood pressure, *Proc. Natl. Acad. Sci. U.S.A.*, 86, 3375, 1989.
55. **Kaplan, S. S., Billiar, T. R., Curran, R. D., Zdziarski, U. E., Simmons, R. L., and Basford, R. E.,** Inhibition of neutrophil chemotaxis with N^G monomethyl-L-arginine: a role for cyclic GMP, *Blood,* 74, 1885, 1989.
56. **Curran, R. D., Billiar, T. R., West, M. A., Bentz, B. G., and Simmons, R. L.,** Effect of interleukin 2 on Kupffer cell activation: interleukin 2 primes and activates Kupffer cells to suppress hepatocyte protein synthesis *in vitro, Arch. Surg.*, 123, 1373, 1988.
57. **Nathan, C. F. and Hibbs, J. B., Jr.,** Role of nitric oxide synthesis in macrophage antimicrobial activity, *Curr. Opin. Immunol.*, 3, 65, 1991.
58. **Nussler, A., Drapier, J. C., Renia, L., Pied, S., Miltgen, F., Gentilini, M., and Mazier, D.,** L-Arginine-dependent destruction of intrahepatic malaria parasites in response to tumor necrosis factor and/or interleukin-6 stimulation, *Eur. J. Immunol.*, 21, 227, 1991.
59. **Mellouk, S., Green, S. J., Nacy, C. A., and Hoffman, S. L.,** IFN-gamma inhibits development of *Plasmodium berghei* exoerythrocytic stages in hepatocytes by an L-arginine-dependent effector mechanism, *J. Immunol.*, 146, 3971, 1991.

60. **Schofield, L., Ferreira, A., Nussenzweig, R. H., and Nussenzweig, V.,** Antimalarial activity of alpha tumor necrosis factor and gamma interferon, *Fed. Proc., Fed. Am. Soc. Exp. Biol.,* 46, 760, 1987.
61. **Peters, W., Ellis, R., Boulard, Y., and Landau, I.,** The activity of a new 8-aminoquinoline, WR 225, 448 against exo-erythrocytic schizonts of *Plasmodium yoelii yoelii, Ann. Trop. Med. Parasitol.,* 78, 467, 1984.
62. **Kiatfuengfoo, R., Suthiphongchai, T., Prapunwattana, P., and Yuthavong, Y.,** Mitochondria as the site of action of tetracycline on *Plasmodium falciparum, Mol. Biochem. Parasitol.,* 34, 109, 1989.
63. **Schofield, L., Villaquiran, J., Ferreira, A., Schellekens, H., Nussenzweig, R. H., and Nussenzweig, V.,** Gamma-interferon, CD8 + T-cells and antibodies required for immunity to malaria sporozoites, *Nature,* 330, 664, 1987.
64. **Meltzer, M. S.,** Macrophage activation for tumor cytotoxicity: characterization of primering and trigger signals during lymphokine activation, *J. Immunol.,* 127, 179, 1981.
65. **Cottrell, B., Pye, C., and Butterworth, A.,** Cytotoxic effects *in vitro* of human monocytes and macrophages on schistosomula of *Schistosoma mansoni, Parasite Immunol.,* 11, 91, 1989.
66. **Davila, D. R., Edwards, C. K., III, Arkins, S., Simon, J., and Kelley, K. W.,** Interferon-gamma induced primering for secretion of superoxide anion and tumor necrosis factor-alpha declines in macrophages from aged rats, *FASEB J.,* 4, 2906, 1990.
67. **Granger, D. L., Hibbs, J. B., Jr., and Broadnax, L. M.,** Urinary nitrate excretion in relation to murine macrophage activation: influence of dietary L-arginine and oral NG-monomethyl-L-arginine, *J. Immunol.,* 146, 1294, 1991.
68. **Liew, F. Y., Millot, S., Parkinson, C., Palmer, R. M. J., and Moncada, S.,** Macrophage killing of *Leishmania* parasite *in vivo* is mediated by nitric oxide, *J. Immunol.,* 144, 4794, 1990.
69. **Nussenzweig, R. S.,** Increased non-specific resistance to malaria produced by administration of killed *Corynebacterium parvum, Exp. Parasitol.,* 21, 224, 1967.
70. **Billiar, T. R., Curran, R. D., Stuehr, D. J., Stadler, J., Simmons, R. L., and Murray, S. A.,** Inducible cytosolic enzyme activity for the production of nitrogen oxides from L-arginine in hepatocytes, *Biochem. Biophys. Res. Commun.,* 168, 1034, 1990.
71. **Billiar, T. R., Lysz, T. W., Curran, R. D., Bentz, B. G., Machiedo, G. W., and Simmons, R. L.,** Hepatocyte modulation of prostaglandin E_2 production by Kupffer cells *in vitro, J. Leuk. Biol.,* 47, 304, 1990.
72. **Nussler, A. K., Pasquetto, V., Renia, L., Miltgen, F., Billiar, T. R., and Mazier, D.,** Endotoxin administration inhibits the hepatic stages of malaria via the induction of the nitric oxide pathway, submitted.
73. **Rockett, K. A., Awburn, M. M., and Clark, I. A.,** Killing of *Plasmodium falciparum in vitro* by nitric oxide derivatives, *Infect. Immun.,* 59, 3280, 1991.
74. **Koj, A.,** Cytokines regulating acute inflammation and synthesis of acute phase proteins, *Blut,* 51, 267, 1985.
75. **Czuprynski, C. J., Brown, J. F., Young, K. M., Cooley, A. J., and Kurtz, R. S.,** Effects of murine recombinant interleukin 1 alpha on the host response to bacterial infection, *J. Immunol.,* 140, 962, 1988.

76. **Adler, J. D., Brooks-Adler, B., and Kreier, J. P.,** *Plasmodium berghei* malaria: effect of acute phase serum on immunity generated in rats by infection and vaccination, *Parasitol. Res.,* 74, 116, 1987.
77. **Mortensen, R. F., Shapiro, J., Lin, B. F., Douches, S., and Neta, R.,** Interaction of recombinant IL-1 and recombinant tumor necrosis factor alpha in the induction of mouse acute phase proteins, *J. Immunol.,* 140, 2260, 1988.
78. **McIntyre, S., Kushner, I., and Salmos, D.,** Secretion of C-reactive protein becomes more efficient during the course of the acute phase response, *J. Biol. Chem.,* 260, 4169, 1983.
79. **Darlington, G. J., Wilson, D. R., and Lachtan, L. B.,** Monocyte conditioned medium, interleukin-1, and tumor necrosis factor stimulate the acute phase response in human hepatoma cells *in vitro, J. Cell. Biol.,* 103, 187, 1986.
80. **Naik, P. and Voller, A.,** Serum C-reactive protein levels and *falciparum* malaria, *Transact. R. Soc. Trop. Med. Hygiene,* 78, 812, 1984.
81. **Mold, C., Nakajama, S., Holzer, T. J., Gewurz, H., and Du Clos, T. W.,** C-Reactive protein is protective against *Streptococcus pneumoniae* infection in mice, *J. Exp. Med.,* 24, 154, 1703, 1981.
82. **Bout, D., Joseph, M., Pontet, M., Vorang, H., Deslee, D., and Capron, A.,** Rat resistance to schistosomiasis: platelet-mediated cytotoxicity induced by C-reactive protein, *Science,* 231, 153, 1986.
83. **Mold, C., Rodgers, P. C., Kaplan, R., and Gewurz, H.,** Binding of human C-reactive protein to bacteria, *Infect. Immun.,* 38, 392, 1982.
84. **Volankis, J. E. and Kaplan, M. H.,** Specificity of C-reactive protein for choline phosphate residues of pneumococcal C-polysaccharide, *Proc. Natl. Acad. Sci. U.S.A.,* 85, 4350, 1971.
85. **Claus, D. R., Siegle, J., Petras, K., Osmand, A. P., and Gewurz, H.,** Interaction of the C-reactive protein with the first component of the human complement, *J. Immunol.,* 119, 187, 1977.
86. **Nussler, A., Pied, S., Pontet, M., Miltgen, F., Gentilini, M., and Mazier, D.,** Inflammatory status and pre-erythrocytic stages of malaria: role of the C-reactive protein, *Exp. Parasitol.,* 72, 1, 1991.
87. **Nudelman, S., Renia, L., Charoenvit, Y., Yuan, L., Miltgen, F., Beaudoin, R. L., and Mazier, D.,** Dual action of anti-sporozoite antibodies *in vitro, J. Immunol.,* 143, 996, 1989.
88. **Zeller, J., Landay, A. L., Lint, T. F., and Gewurz, H.,** Enhancement of human peripheral blood monocyte respiratory burst activity by aggregated C-reactive protein, *J. Leuk. Biol.,* 40, 796, 1986.
89. **Zahedi, K. and Mortensen, R. F.,** Macrophage tumoricidal activity induced by human C-reactive protein, *Cancer Res.,* 46, 5077, 1986.
90. **Pied, S., Renia, L., Nussler, A., Miltgen, F., and Mazier, D.,** Inhibitory activity of IL-6 on malaria hepatic stages, *Parasite Immunol.,* 13, 211, 1991.
91. **Ockenhouse, C. F. and Shear, A. L.,** Oxidative killing of the intraerythrocytic malaria parasite *Plasmodium yoelii* by activated macrophages, *J. Immunol.,* 132, 424, 1984.
92. **Halliwell, B. and Gutteridge, J. M. C.,** Oxygen toxicity, oxygen radicals, transition metals and disease, *Biochem. J.,* 219, 1, 1984.
93. **Weiss, W. R.,** Host-parasite interactions and immunity to irradiated sporozoites, *Immunol. Ltrs.,* 25, 39, 1990.

94. **Renia, L., Mattei, D., Goma, J., Pied, S., Dubois, P., Miltgen, F., Nussler, A., Matile, H., Menegaux, F., Gentilini, M., and Mazier, D.,** A malaria heat-shock-like determinant expressed on the infected hepatocyte surface is the target of antibody-dependent cell-mediated cytotoxic mechanisms by non-parenchymal cells, *Eur. J. Immunol.,* 20, 1445, 1990.
95. **Ralph, P. and Nakoinz, I.,** Cell mediated lysis of tumor targets directed by murine monoclonal antibodies of IgM and all IgG isotypes, *J. Immunol.,* 131, 1028, 1983.
96. **Rager-Zinsman, B. and Bloom, B. R.,** Immunological destruction of *Herpes simplex* virus I infected cells, *Nature,* 251, 542, 1974.
97. **Sleyster, E. C. and Knook, D. L.,** Relation between localization and functions of rat liver Kupffer cells, *Lab. Invest.,* 47, 484, 1982.
98. **Meis, J. F. G. M. and Verhave, J. P.,** *Adv. Parasitol.,* 27, 1, 1988.
99. **Herberman, R. B. and Ortaldo, J. R.,** Natural killer cells: their role in defenses against diseases, *Science,* 214, 24, 1981.
100. **Kwiatkowski, D., Hill, A. V. S., Sambou, I., Twumasi, P. M., Castracane, J., Manogue, K., Cerami, A., Brewster, D. R., and Greenwood, B.,** TNF concentration in fatal cerebral, non-fatal, cerebral, and uncomplicated *Plasmodium falciparum* malaria, *Lancet,* 1201, 1990.
101. **Ringwald, P., Peyron, F., Vuillez, J. P., Touze, J. E., Le Bras, J., and Deleron, P.,** Levels of cytokines during *Plasmodium falciparum* malaria attacks, *J. Clin. Microbiol.,* 29, 2076, 1991.

Section III

The Liver in Sepsis

Chapter 10

REGULATION OF KUPFFER CELL ACTIVATION

M.A. West and M.L. Heaney

TABLE OF CONTENTS

ISBN 0-8493-6109-5

I. CONCEPT OF ACTIVATION

The liver occupies a critical role in the host response to sepsis. When significant liver dysfunction occurs the mortality is 40 to 50% with isolated hepatic failure[1] and exceeds 90% if other organs have also failed.[2,3] Kupffer cells are the largest reservoir of fixed-tissue macrophages and are quantitatively the most important for removal of circulating bacteria, endotoxin, and other circulating microbial debris.[4,5] Quantitatively, Kupffer cell phagocytosis is the most important route of bacterial clearance in sepsis.[5] Kupffer cells, like other macrophages throughout the body, are derived from a common bone marrow precursor cell.[6,7] Despite their common origin, macrophages are adapted to function in different microenvironments. We hypothesize that many of the differences in the functional attributes of different macrophagelike cells may be due to differences in their ''activation state''.[8]

Activation refers to how ''stimulated'' a macrophagelike cell is. This concept has been delineated and conceptualized to understand *in vitro* stimulation of macrophages and peripheral blood mononuclear cells.[9-11] A spectrum of stimulation is seen from unstimulated states (baseline) to highly activated states of cellular function. Macrophagelike cells that have not been specifically stimulated are described as ''resident'' cells.[9] Between resident and fully activated states many other functions are gained or lost depending on the timing and type of stimulation. Unstimulated or resident macrophages are difficult to study, as most methods of isolation cause some degree of stimulation.[9] *In vivo* and *in vitro* treatment of macrophages with thioglycolate, zymosan, or muramyl dipeptide (MDP) stimulate macrophages to a higher level, referred to as inflammatory[9,12] or elicited.[10] As macrophage stimulation increases from a resting state, other biological characteristics of the cell are altered; there is increased expression of cell surface Ia antigen,[13] enhanced respiratory burst activity,[14] increased production of reactive oxygen intermediates,[15] and the release of lysosomal enzymes.[16]

In vitro treatment with lipopolysaccharide (LPS),[11,17] phorbol myristate acetate (PMA),[18] or calcium ionophore A23187[19] under proper conditions stimulate macrophages to a fully activated level. Very high levels of macrophage activation are required to lyse tumor cells. This tumor cytolytic capacity of macrophages has been a *sine quo non* of complete macrophage activation. A two-stage mechanism for macrophage activation has been described involving a ''priming'' stimulus followed by a second ''triggering'' stimulus through *in vitro* studies of tumor cytolysis.[11,17] Very potent stimuli are needed to activate macrophages in the absence of a priming agent.[17] The most well-studied priming agent, or macrophage activating factor, is interferon-gamma (IFN-γ). Synergism has been demonstrated between LPS and cytokines to activate macrophagelike cells for tumor cytolysis. Cytotoxic mechanisms, per se, probably do not play a role in sepsis, but changes in tumor cytolytic activity reflect changes in activation state of these cells. Fully "activated" macrophages[11] display a variety of functional changes.[12,16,20] They have an increased capacity for phagocytosis, decreased migration,[9] increased myeloperoxidase activity,[10] increased production of reactive oxygen intermediates,[15] and increased microbicidal activity.[21] Activated macrophages also synthesize and secrete a variety of biochemical messengers, including prostaglandins,[22,23] leukotrienes,[22] interleukin-1 (IL-1),[24] tumor necrosis factor (TNF),[25] and other protein cytokines.[26] High levels of macrophage activation also result in macrophage-mediated tumor cytolysis or cytostasis.[11,17,27,28]

Kupffer cells are derived from and share many properties with other macrophagelike cells.[29-31] Therefore, it is advantageous and appropriate to utilize the conceptual framework of "activation" to understand the alterations in Kupffer cell function, which occur after they are exposed to stimuli that "activate" other macrophagelike cells. The concept of Kupffer cell activation may be particularly relevant to sepsis because the stimuli that activate macrophagelike cells (LPS, bacteria, and cytokines) circulate during surgical infections.[3,8] Indeed, levels of these inflammatory stimuli may be especially high in the portal venous blood.[32] When macrophagelike cells are stimulated *in vitro* with LPS, there is a coordinated secretory protein response.[33] Some secretory proteins are increased, whereas others appear to be simultaneously depressed. Synthesis of dozens of macrophage proteins are coordinately altered following LPS treatment, but the functions of only a small number of these are known at the present time.[33] Many of the secretory proteins stimulated by LPS have been shown to have regulatory effects on other cells; such mediators are now referred to as cytokines.[34,35] Soluble cytokines from lymphocytes, such as IFN-γ,[36] IL-4,[13] IL-2,[35] and others[35] substantially alter the macrophage activation state and have been shown to act synergistically with LPS or other inflammatory stimuli.[17,35] Both inflammatory stimuli and high levels of splenic lymphocyte-derived cytokines may be present in portal venous blood.

II. KUPFFER CELL ACTIVATION

A. INTRODUCTION

For the purpose of this discussion, Kupffer cell activation will be defined as the stimulated functional state which occurs after treatment with stimuli that have been associated with activation of macrophagelike cells. In the context of sepsis, such *in vitro* stimulation would most often be with LPS. Until recently, it was difficult and cumbersome to study Kupffer cells *in vitro*.[37,38] In addition, the important role which Kupffer cells are hypothesized to play in mediating host septic responses was not appreciated. Factors such as these may be responsible for the paucity of work specifically studying activation of Kupffer cells. Thus, while few reports have specifically addressed the concept of Kupffer cell activation, a large number of studies have examined the effect of LPS treatment on various Kupffer cell functions. In

TABLE 1
Kupffer Cell Functions Associated With Activation

Inhibition of hepatocyte protein synthesis
Enhanced production of nitric oxide (arginine-dependent mechanism)
Increased production of PGE_2, PGD_2
Increased IL-1 release
Increased TNF release
Increased IL-6 release
Increased tumor cell cytotoxicity
Enhanced expression of Ia antigen
Increased oxidative metabolism

this section, the alterations in Kupffer cell function seen after stimulation with "activating agents" will be summarized. These alterations, summarized in Table 1, include: inhibition of hepatocyte protein synthesis,[39,40] production of reactive nitrogen intermediates,[41,42] increased synthesis of cytokines,[43] release of arachidonic acid metabolites,[44] enhanced oxidative metabolism,[45] increased phagocytosis,[46] and augmented tumor cytotoxicity.[47]

B. INHIBITION OF HEPATOCYTE PROTEIN SYNTHESIS

One of the most well-documented effects of activated Kupffer cells is inhibition of cocultured hepatocyte protein synthesis.[39,40,48,49] In these experiments, LPS was found to have no measurable direct effect on hepatocyte protein synthesis. However, when hepatocytes were cocultivated with Kupffer cells, there was a Kupffer cell-dependent stimulation of hepatocyte protein synthesis in the absence of inflammatory stimuli.[40] When LPS was added to cocultures, a significant Kupffer cell-mediated inhibition of hepatocyte protein synthesis was seen.[39] The dose-response relationship for inhibition of hepatocyte protein synthesis was a highly reproducible and characteristic "square wave pattern". No inhibition was seen until LPS concentration reached approximately 100 ng/ml, after which an abrupt shift from stimulation to inhibition of hepatocyte protein synthesis was seen.[40,50] As LPS concentrations were further increased, no additional inhibition was seen. In general, this inhibition of hepatocyte protein synthesis was not associated with alterations in hepatocellular viability.[48,51] More recently, Billiar et al.[41] has shown that high effector to target cell ratios and high concentrations of LPS can result in Kupffer cell-mediated death of some hepatocytes.

Although total protein synthesis by cocultured hepatocytes was decreased, the response was not uniform when protein synthesis was examined with SDS-polyacrylamide gel electrophoresis. Hepatocytes cocultured with LPS-triggered Kupffer cells produced significantly decreased quantities of proteins with molecular weights of 64kDa[30-34,64] (albumin) and 73 kDa compared to control cocultured hepatocytes.[49] Furthermore, despite the overall decrease in cocultured hepatocyte protein synthesis, there was a significant enhancement of several protein bands, particularly proteins of 23, 44, 58, and 68 kDa.[49] Fibronectin, an acute-phase reactant with a molecular weight of 23 kDa, has been shown to have increased production after Kupffer cell cultures were exposed to LPS.[52]

Preliminary investigations into the mechanism of inhibition suggested that it was due to a 15- to 30-kDa factor which could be transferred in the supernatant of LPS-triggered Kupffer cells.[48] This inhibition could not be reproduced by addition of recombinant human IL-1 or TNF to hepatocytes alone. In addition, antibody to either TNF or IL-1 did not prevent or ameliorate the inhibition of hepatocyte function seen in coculture with LPS-triggered Kupffer cells. These data strongly suggest that IL-1 or TNF were not primary mediators of the inhibitory response.[48] Currant et al.[53] have now shown that this inhibition of hepatocyte protein synthesis can be reproduced in the absence of Kupffer cells using a mixture of cytokines and LPS. When the combination of LPS and three recombinant cytokines (IFN-γ, IL-1, and TNF) was added to hepatocytes cultured alone, there was significant inhibition of hepatocyte protein synthesis. The kinetics of this inhibition were identical to those reported previously, specifically, a 6- to 8-h lag after addition of inhibitory supernatant prior to significant inhibition of hepatocyte protein synthesis.[48]

C. ARGININE-DEPENDENT EFFECTOR MECHANISM

Further investigations into alterations of hepatocyte protein synthesis by LPS-activated Kupffer cells, carried out by Billiar et al.,[41,42] revealed a critical role for the arginine-dependent effector cell mechanism. This mechanism involves Kupffer cell metabolism of arginine into reactive nitrogen intermediates, particularly nitric oxide.[42] Treatment of Kupffer cell:hepatocyte cocultures with N^G-monomethyl-L-arginine (NMA), a competitive inhibitor of the arginine-dependent effector cell mechanism, blocked LPS-triggered Kupffer cell-mediated inhibition of hepatocyte protein synthesis.[42] Inhibition of hepatocyte protein synthesis using a cytokine cocktail was also blocked when NMA was added to hepatocytes.[53]

D. CYTOKINE SYNTHESIS

Kupffer cells stimulated with LPS have been shown to produce significant quantities of several cytokines, including IL-1,[54] TNF,[43] and IL-6.[55] Significant levels of IL-1 production were seen even in the absence of *in vitro* triggering with LPS.[54] In addition, Kupffer cells exposed to LPS produce a procoagulant factor.[56] The kinetics of cytokine production by Kupffer cells, and all macrophagelike cells, depends upon the particular cytokine being produced. It has been noted that TNF synthesis appears to occur prior to IL-1 or IL-6.[57,58]

E. PROSTAGLANDIN PRODUCTION

When Kupffer cells are exposed to activating stimuli such as LPS, there is a significant increase in production of arachidonic acid metabolites.[23,44]

Prostaglandin E_2 (PGE_2) production has been most extensively studied. LPS concentrations in the range of 10 ng/ml significantly increase PGE_2 release by Kupffer cells, with maximal increases seen 24 to 48 h after the triggering stimulus.[23] Increased production of PGE_2 can also be seen in the absence of complete activation, when Kupffer cells are stimulated with compounds such as calcium ionophore A23187.[44,59] Other investigators have suggested that PGD_2, whose production is also stimulated by inflammatory stimuli, may be a quantitatively more important arachidonic acid metabolite, accounting for up to 55% of total arachidonic acid metabolites.[60]

F. OXIDATIVE METABOLISM

In general, most macrophagelike cells produce various toxic oxygen metabolites when they are exposed to inflammatory stimuli. In particular, production of hydrogen peroxide and superoxide anion are increased. Kupffer cells exposed to zymosan particles[45] had markedly increased oxygen uptake and superoxide release. Release of oxygen metabolites by mature Kupffer cells occurred immediately after triggering with inflammatory stimuli. Lepay et al.[61] noted that Kupffer cells which had been "primed" by exposure to *Listeria monocytogenes* produced significantly less hydrogen peroxide than control Kupffer cells when triggered with LPS.

G. CELL-MEDIATED CYTOTOXICITY

Macrophage-mediated cytotoxicity is the *sina quo non* of complete macrophage activation.[9,11] Considerably less work has been performed examining Kupffer cell cytotoxicity, however.[62] Nonetheless, Kupffer cells do cause lysis of TNF-resistant tumor cells, and this activity is enhanced by activating stimuli like LPS or IFN-γ.[47,63] Cytotoxicity and cytostatic activities of Kupffer cells and other macrophages have not been directly compared, but it appears that Kupffer cell-mediated tumor lysis is greater than that of peritoneal macrophages.[64]

III. REGULATION OF KUPFFER CELL ACTIVATION

At the present time, relatively little detailed information is available concerning Kupffer cell activation, and considerably less is known about factors that regulate their activation state. In this section, an attempt will be made to present data concerning regulation of Kupffer cell activation in the Kupffer cell: hepatocyte coculture model system, summarize the available literature on Kupffer cell activation, and make inferences, where appropriate, from the literature on regulation of similar functions in other macrophagelike cells.

Using the hepatocyte:Kupffer cell coculture system, preliminary investigations into the mechanisms by which Kupffer cell-mediated alterations in hepatocellular function are regulated were performed.[49,50,65-68] In order to detect subtle variations in the regulatory response, qualitative and quantitative

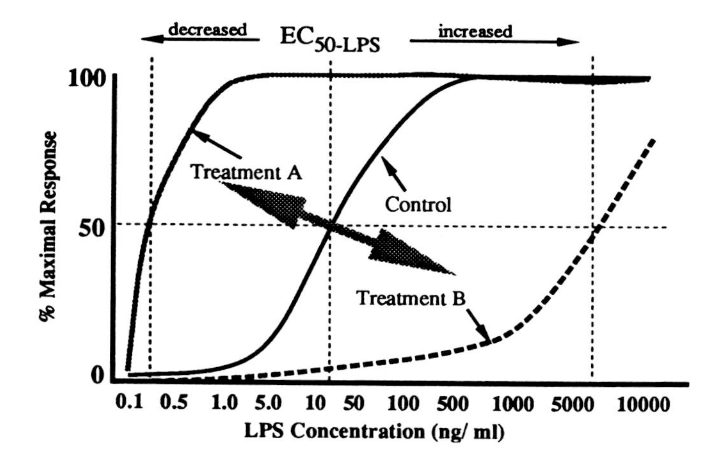

FIGURE 1. Depiction of how the LPS dose-response relationship can be utilized to examine regulation of Kupffer cell function. The control LPS dose response for this function has a 50% maximal response at an LPS concentration of ≈15 ng/ml. Treatment A changes the Kupffer cell response so that lower doses of LPS achieve the same effect. Thus, the LPS concentration for a 50% maximal response after treatment A is now ≈0.2 ng/ml. In contrast, treatment B decreases the Kupffer cell LPS responsiveness because higher levels of LPS, ≈6000 ng/ml, are required for 50% maximal inhibition. In this way, subtle differences in Kupffer cell responsiveness to LPS can be appreciated and compared. To simplify interpretation of the results, inhibition of ^{3}H-leucine incorporation into hepatocyte protein by Kupffer cells has been expressed on a scale of percent maximal inhibition (protein synthesis = 100% maximal inhibition).

alterations in the highly reproducible LPS dose-response curve were noted. In this characteristic curve, no inhibition of hepatocyte protein synthesis was observed until a reproducible threshold concentration of LPS was achieved. After exceeding the threshold concentration, there was a relatively abrupt switch from increased to decreased hepatocyte protein synthesis. Figure 1 shows how treatment which resulted in a rightward shift of the LPS dose-response curve could be interpreted as representing a decrease in the responsiveness of Kupffer cells to endotoxin. That is, higher concentrations of LPS were required to achieve the same activation state. In a similar manner, a leftward shift in the dose response to LPS would represent increased sensitivity of Kupffer cells to LPS.

A. *IN VIVO* PRETREATMENT WITH LPS ALTERS KUPFFER CELL ACTIVATION STATE

The effect of repetitive LPS stimulation had not been addressed previously with Kupffer cells or other types of macrophages. Clinically it is probable that patients have multiple episodes of endotoxemia or bacteremia.[3] We wondered whether or not repeated stimulation with LPS would enhance the LPS-stimulated effects we had observed in Kupffer cell:hepatocyte coculture. In an effort to study the effect of LPS stimulation on subsequent LPS-respon-

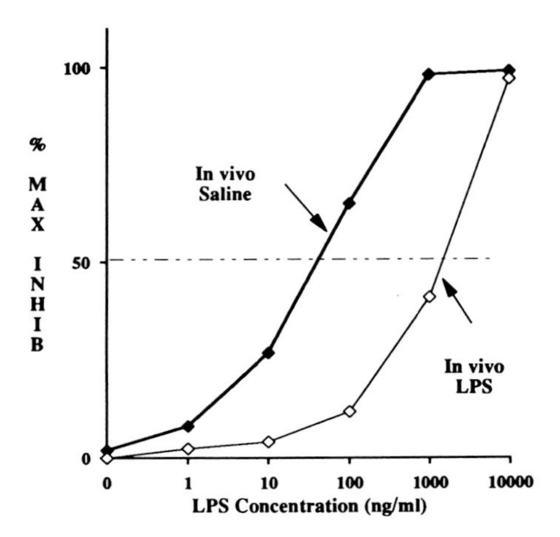

FIGURE 2. Comparison of dose-response relationship of Kupffer cells obtained from animals that received i.p. LPS (0.1 mg/100 g body weight of *Escherichia coli* 011B4 LPS) vs. control rats receiving i.p. saline. *In vivo* exposure to LPS significantly raised the threshold for subsequent LPS activation of Kupffer cells during *in vitro* coculture with hepatocytes. Inhibition of hepatocyte protein synthesis was measured and is expressed at percent maximal inhibition. Cocultures were performed as described previously.[39]

siveness *in vitro,* animals were inoculated with 0.1 mg/100 g body weight of LPS (0111B4 from *Escherichia coli*) or saline i.p., and 24 h later the Kupffer cells were harvested and added to hepatocytes to establish coculture in the usual manner. This degree of LPS treatment was relatively well tolerated by the animals with no mortality. Kupffer cells from animals that received i.p. LPS were compared to Kupffer cells from rats which received i.p. saline in coculture with hepatocytes and ^{3}H-leucine incorporation into protein was assessed following stimulation with a wide range of LPS concentrations. Figure 2 shows these results, with the data expressed as percent maximal inhibition, as a function of LPS concentration. It is readily apparent that markedly different dose responses were observed in cocultures with Kupffer cells that had been exposed to LPS *in vivo* compared to cocultures of control Kupffer cells. Kupffer cells obtained from rats which received i.p. saline had a normal LPS dose-response curve, with the characteristic threshold of activation seen between 10 to 100 ng/ml. *In vivo* LPS treatment significantly increased the concentration of LPS required to inhibit hepatocyte protein synthesis. It is somewhat more difficult to assess whether or not the upward shift of the curve represents a significant effect. These results were surprising because it was thought that low-dose LPS pretreatment might "prime" the Kupffer cells, with the result that subsequent LPS responses would be exaggerated. Many factors could have been stimulated *in vivo* that might account for decreased LPS responsiveness, including factors from other cells, such as lymphocytes. Therefore, to more thoroughly investigate the factors responsible, additional experiments were performed *in vitro*.

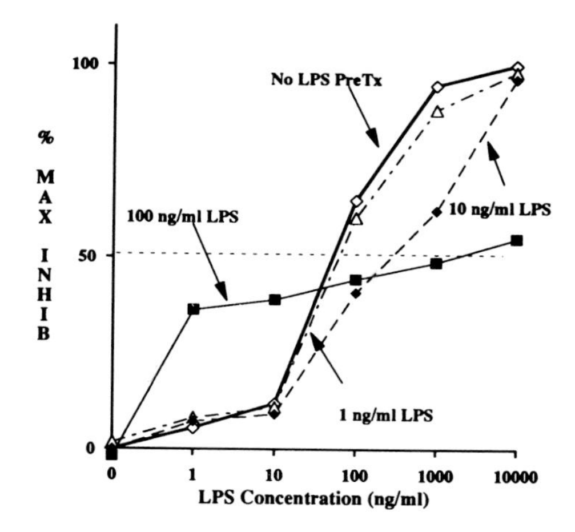

FIGURE 3. The effect of *in vitro* pretreatments of Kupffer cells with 1, 10, or 100 ng/ml of *E. coli* 011B4 LPS during the 24-h interval prior to performing the LPS dose-response curve is shown. A dose-dependent loss of Kupffer cell LPS responsiveness can be appreciated.

It is not surprising that *in vivo* exposure to LPS alters Kupffer cell function. McCuskey et al.[69] have noted a significant alteration in Kupffer cell morphology following *in vivo* LPS exposure. In the process of investigating induction of LPS tolerance, it was noted that animals exposed to a 10% lethal dose (LD_{10}) of LPS had a significant decrease in sinusoidal blood flow. When subsequently exposed to a LD_{70} dose of LPS, the survival of these animals was enhanced. This improvement in survival was thought to be associated with an increased activation state of the remaining Kupffer cells. In a study comparing species sensitivity to endotoxin, a significant correlation between the LD_{50} for all species and the number of Kupffer cells in periportal and centrilobular regions was noted.[70]

B. *IN VITRO* PRETREATMENT WITH LPS ALTERS KUPFFER CELL ACTIVATION STATE

The *in vitro* system utilized for culturing Kupffer cells included a 24 h *in vitro* incubation period prior to stimulation with LPS, which was intended as an interval to allow these cells to recover from the isolation procedure.[50] In our efforts to further examine LPS pretreatment, this interval was utilized for *in vitro* exposure to potential regulatory substances. Figure 3 shows the results obtained when Kupffer cells were pretreated with various concentrations of LPS for 24 h *in vitro* before they were stimulated with the range of LPS concentrations used to construct the LPS dose-response curve. The specific LPS concentrations used for these pretreatment experiments, 1, 10, and 100 ng/ml, were concentrations that were below the threshold concentrations that triggered significant Kupffer cell activation. A dose-dependent loss of

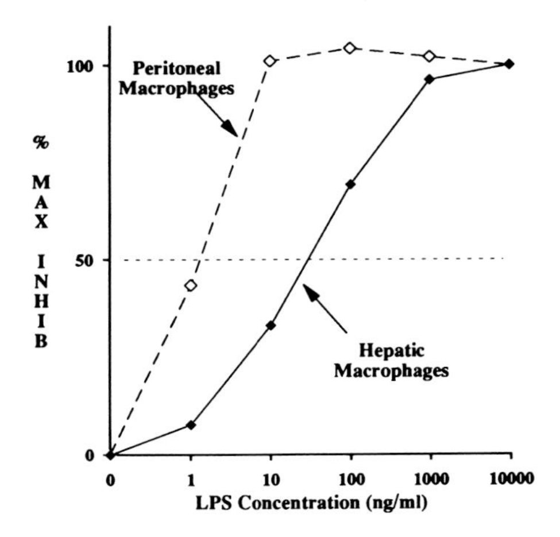

FIGURE 4. Comparison of LPS dose responsiveness of Kupffer cells and peritoneal macrophages. Although maximal response of the two cell types were nearly identical, significant differences in the threshold for LPS triggering were seen.

LPS responsiveness was seen after 24 h of LPS pretreatment, with statistically significant differences seen following exposure to 10 ng/ml ($p < 0.01$) and 100 ng/ml ($p < 0.001$) of LPS. Indeed, cocultures with Kupffer cells which had been pretreated with 100 ng/ml of LPS did not respond to subsequent LPS treatment.

C. PERITONEAL MACROPHAGES AND KUPFFER CELLS DIFFER IN ACTIVATION STATE

The anatomic position of Kupffer cells, within the portal venous circulation, exposes them to several factors, such as low levels of translocating bacteria and circulating LPS,[5,32,71] that may have important regulatory influences. We speculated that Kupffer cells, which are exposed to higher levels of inflammatory stimuli than peritoneal macrophages, might have decreased LPS responsiveness, similar to that seen after *in vivo* or *in vitro* LPS pretreatment. Figure 4 shows a comparison of LPS dose response of peritoneal macrophages and Kupffer cells. In this experiment, both types of macrophages were cultured for 24 h *in vitro* without stimulation to allow recovery before performing the LPS dose response. The subsequent responsiveness to LPS was strikingly different for these two types of macrophages. Peritoneal macrophages required a far lower concentration of LPS for complete activation, assessed as maximal inhibition of hepatocyte protein synthesis, than Kupffer cells ($p < 0.001$). This finding is also consistent with the concept that Kupffer cell preexposure to LPS *decreased* subsequent LPS responsiveness.

Other functional properties of Kupffer cells and peritoneal macrophages have also been directly compared.[46] Kupffer cells were found to be two to three times more active in phagocytosis of sheep red blood cells than peritoneal macrophages. No specific differences in chemotaxis to C5a were observed, but Kupffer cells were chemotactically responsive to PMA, whereas peritoneal macrophages did not respond to this chemoattractant. These investigators also found that stimulated release of superoxide anion by peritoneal macrophages was greater than Kupffer cell release.[46] Production of H_2O_2, in the absence of exogenous stimulation, has also been shown to be greater by peritoneal macrophages when compared to Kupffer cells. Several authors have speculated that decreased oxidative response by Kupffer cells may be advantageous to prevent parenchymal cell damage.[61] The lysosomal enzyme activity of Kupffer cells has also been compared to that of peritoneal macrophages and found to be increased, suggesting *in vivo* exposure to inflammatory stimuli.[72]

D. INDOMETHACIN PREVENTS DECREASE OF KUPFFER CELL ACTIVATION AFTER LPS PRETREATMENT

Other investigators had also noted decreased macrophage function 24 h after treatment with LPS. Taffet et al.,[73] and Pace and Russel[17] examined the time course of macrophage-mediated tumor cytolysis after LPS stimulation. Their studies showed significant tumor cytotoxicity 6 h after treatment with LPS, but very little cytolytic activity 24 h later. They demonstrated that the LPS-stimulated macrophages synthesized large quantities of PGE_2, which has been shown to antagonize LPS stimulation of macrophage functions.[73] When these investigators treated the macrophages with both LPS and indomethacin to prevent prostaglandin synthesis, they showed no loss of tumoricidal activity after 24 h. We speculated that a similar mechanism may be operative in the loss of Kupffer cell activation after LPS stimulation. To examine whether or not the loss of cocultivated Kupffer cell LPS responsiveness was mediated via synthesis of arachidonic acid metabolites, indomethacin was added immediately prior to pretreatment with 100 ng/ml of LPS. Figure 5 shows that treatment with 1 μM/l indomethacin did largely restore the LPS responsiveness of Kupffer cells pretreated with 100 ng/ml of LPS. In particular, there was restoration of LPS responsiveness at both the upper and lower regions of the LPS dose-response curve. It should be noted that pretreatment with low-dose LPS had little effect when Kupffer cells were subsequently triggered with very high doses of LPS (10 μg/ml). This concentration is routinely utilized by many individuals examining Kupffer cell activation. Insights into factors important for regulation of Kupffer cell activation were only forthcoming when the midportion of the LPS dose-response curve was studied.

E. PRETREATMENT WITH PGE_2 DECREASES KUPFFER CELL ACTIVATION STATE

Significant quantities of PGE_2 are produced by LPS-activated Kupffer cells.[23,44] To examine whether or not PGE_2 was responsible for loss of Kupffer cell LPS responsiveness, various concentrations of PGE_2 were added 30 min

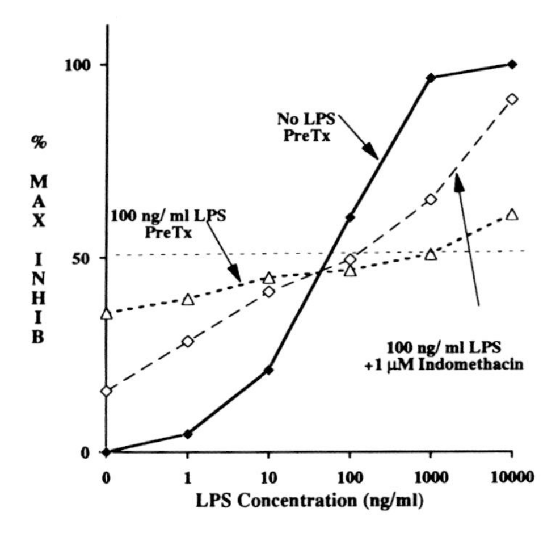

FIGURE 5. Effect of indomethacin treatment on loss of Kupffer cell LPS responsiveness which occurred following preexposure to 100 ng/ml of LPS *in vitro*. Pretreatment with 100 ng/ml significantly impaired the capacity of Kupffer cells to respond to LPS; however, the presence of 1 μM indomethacin treatment prevented this effect.

prior to performing the LPS dose-response curve. PGE_2 was added immediately prior to LPS triggering because it is not stable for prolonged periods in hepatocyte:Kupffer cell coculture.[74] Figure 6 shows that PGE_2 caused a significant, dose-dependent shift of the LPS dose-response relationship. If PGE_2 antagonized LPS stimulation of Kupffer cells, a rightward shift in the LPS dose responsiveness would be expected. Indeed, an increased threshold

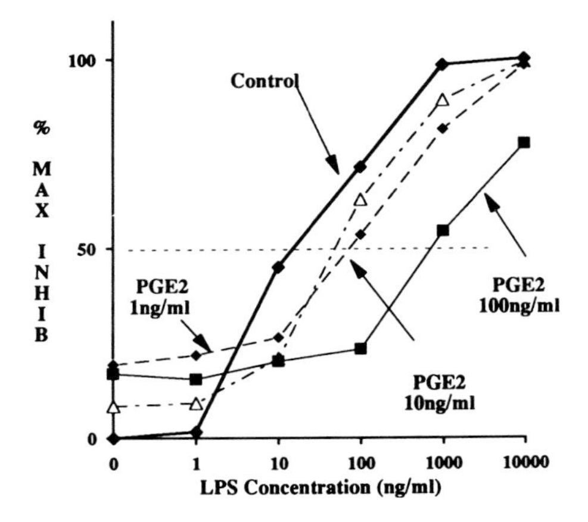

FIGURE 6. Effect of exogenous PGE_2 in Kupffer cell LPS dose-response relationships. PGE_2 resulted in a dose-dependent rightward shift of the Kupffer cell LPS dose-response curve. Thus, PGE_2 significantly raised the threshold for LPS-triggered Kupffer cell activation.

for Kupffer cell activation was observed, suggesting that *in vivo* or *in vitro* LPS pretreatment stimulates Kupffer cell PGE_2 release which mediates the loss of Kupffer cell LPS responsiveness. It may be that *in vivo* Kupffer cell PGE_2 production, in response to low portal venous bacteremia or endotoxemia, is responsible for the decreased LPS responsiveness of Kupffer cells compared to peritoneal macrophages. Peritoneal macrophages would not normally be exposed to bacteria or endotoxin *in vivo*.[4]

The production of PGE_2 by macrophages can be inhibited by leukotriene B_4 (LTB_4)[59] and IFN-γ.[75] There is no evidence for feedback inhibition of prostaglandin synthesis, inasmuch as exogenously added PGE_1 or PGE_2 had no effect on prostaglandin synthesis.[59] Inhibition of PGE_2 synthesis by IFN-γ appears to involve inhibition of the enzyme phosphorylase, which slows the release of arachidonic acid from cell membranes with a decline in production of all eicosanoids, including PGE_2.[76] A similar degree of inhibition of PGE_2 release is seen with IFN-β, via a mechanism involving Ca^{++} and increased intracellular levels of cyclic adenosine monophosphate (cAMP).[77] Reider et al.[78] noted that PGE_2 has many significant effects on a variety of Kupffer cell functions. Treatment of Kupffer cells with PGE_2 caused a decrease in the Ia receptor expression while increasing Fc receptor-mediated phagocytosis and secretion of collagenase. These studies clearly showed that PGE_2 could significantly affect subsequent Kupffer cell activation by LPS and suggested that it may be an important *in vivo* regulator of Kupffer cell activation state.

F. CYTOKINES REGULATE KUPFFER CELL ACTIVATION STATE

There is considerable evidence that cytokines affect the activation state of macrophagelike cells. Chen et al.[79] showed that LPS-stimulated macrophage-mediated tumor cytolysis requires the presence of LPS, TNF, and IL-1. In these studies, LPS-stimulated macrophages were found to cause a significant degree of tumor cytolysis using a standard assay. However, if the LPS-triggered cultures were treated with a neutralizing antibody to either IL-1 or TNF, there was no cytolytic activity. Addition of recombinant cytokine could restore tumoricidal activity. In the absence of LPS, TNF and IL-1 were not sufficient, in and of themselves, to mediate tumor cytolysis.[79] Kupffer cell functions stimulated by LPS are probably independently regulated, despite the cooperative interactions implied in the results presented above. Billiar et al.[55] recently examined the effect of the specific inhibitor (NMA) of the arginine-dependent effector mechanism which inhibits hepatocyte protein synthesis, on Kupffer cell cytokine production. Figure 7 shows that despite complete blockade of Kupffer cell-mediated inhibition of hepatocyte protein synthesis, NMA had no effect on LPS-stimulated Kupffer cell synthesis of IL-1 or TNF. These studies support the hypothesis that LPS-stimulated Kupffer cell functions are independently regulated. Furthermore, these results

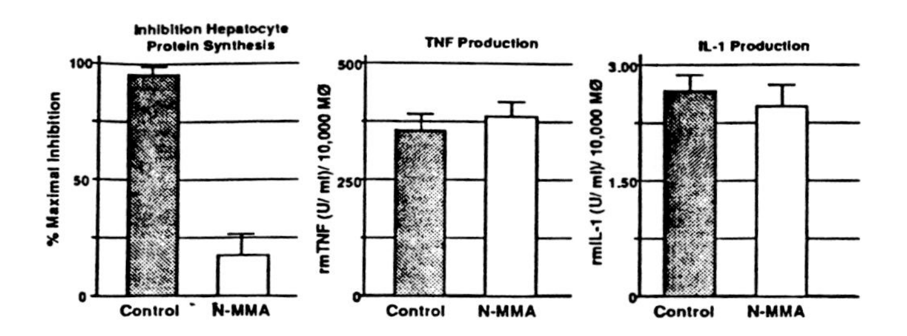

FIGURE 7. Effect of NMA on Kupffer cell inhibition of hepatocyte protein synthesis, TNF production, and IL-1 production in response to LPS. NMA blocked LPS-triggered inhibition of hepatocyte protein synthesis, but had no significant effect on TNF or IL-1 production.

suggest that regulation of functions associated with Kupffer cell activation are likely far more complex than initially appreciated.

Lymphokines and IFN-γ have both been found to enhance macrophage nitrite and nitrate production, suggesting that they stimulate the arginine-dependent mechanism.[80] A large number of other cytokines, including IFN-α, IFN-β, TNF-α, TNF-β, granulocyte-macrophage colony stimulating factor (GM-CSF), macrophage colony stimulating factor (M-CSF), IL-1-β, IL-2, IL-4, and transforming growth factor-beta (TGF-β) had no effect on murine peritoneal macrophage arginine metabolism.[81] Furthermore, IFN-γ was found to be more potent than LPS in macrophages,[80] although the presence of low concentrations of LPS was obligatory for induction of the arginine-dependent mechanism.[82] At the present time, IFN-γ is the only cytokine which has been clearly shown to enhance the L-arginine-dependent effector cell mechanism. This mechanism has been shown to be responsible for Kupffer cell inhibition of hepatocyte synthesis.[41,42]

There is a large body of literature indicating that various cytokines lower the threshold for LPS or other stimuli to activate macrophagelike cells. Furthermore, a considerable amount of direct and indirect evidence suggests that a wide variety of cytokines have significant regulatory effects on Kupffer cell activation. Kupffer cells may normally be exposed to higher concentrations of lymphocytic cytokines *in vivo,* from both the splenic venous effluent and from mesenteric lymph nodes. We examined the effect of *in vitro* pretreatment of Kupffer cells with a lymphokine-rich supernatant from Con-A-stimulated splenic lymphocytes on subsequent LPS dose responsiveness. Figure 8 shows that preexposure to lymphokines significantly lowered the threshold for complete Kupffer cell activation ($p < 0.001$). it is interesting to note that no such effect was seen in earlier experiments in which splenic lymphocytes were added to cocultures,[39] suggesting that the presence of lymphocytes in coculture may result in more complex regulatory interactions. Subsequent experiments by Curran et al.[83] showed that a similar dramatic lowering of the Kupffer cell activation threshold could be demonstrated, utilizing defined recombinant cytokines. Both IL-2 and IFN-γ resulted in a 3- to 4-log decrease in the LPS

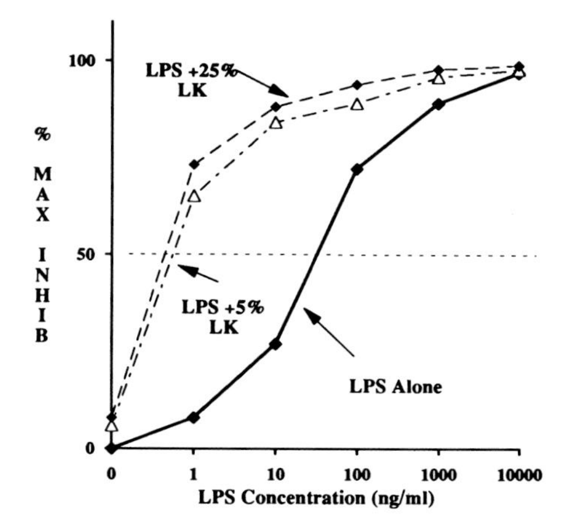

FIGURE 8. Effect of *in vitro* pretreatment of Kupffer cells with lymphokine-rich supernatant. A significant leftward shift of the LPS dose-response relationship can be seen vs. control.

concentration required to trigger maximal Kupffer cell-mediated inhibition of hepatocyte protein synthesis. In the Kupffer cell:hepatocyte coculture system, the addition of IL-2 had no effect on hepatocyte protein synthesis unless the coculture was exposed to LPS.[83] Pretreatment with IL-2 for less than 6 h failed to induce protein synthesis alterations.

It has been noted that IL-2 receptors appear only on primed macrophages and these receptors take 12 to 24 h, after exposure to a priming agent, to appear.[84,85] This may explain the failure of short-course IL-2 pretreatment to affect Kupffer cell function. IFN-γ, LPS, and IL-2 are all capable of inducing IL-2 receptor expression.[84] Macrophagelike cells which were exposed to IL-2 *in vitro* had significantly increased production of IL-1-α and IL-1-β.[86] LPS also increased production of IL-1-β; however, the combination of IL-2 and LPS produced a larger increase in IL-1-β than either stimulant alone. Treatment of macrophages with IL-2 also augmented LPS-triggered PGE_2 production.[87] This effect was only seen in mixed monocyte populations; fractionation of monocytes into plastic adherent and nonadherent cell groups revealed no response if the groups were studied separately. An interaction between two subpopulations, an IL-2-responsive and an IL-1-β-producing cell group, was postulated to be necessary for this response.[87] IL-2 has also been found to synergize with IFN-γ for induction of macrophage TNF production.[88]

Experiments with crude lymphokine,[40] recombinant IL-2,[83] and recombinant IFN-γ in the Kupffer cell coculture system confirmed that lymphocytic cytokines could upregulate Kupffer cell LPS responsiveness. Investigations by Billiar et al.[67] suggested that the absence of the same factors might downregulate Kupffer cell response to LPS. The effect of splenectomy on subsequent *in vitro* Kupffer cell LPS responsiveness was examined and it was found that splenectomy exerted a biphasic influence on Kupffer cell response to LPS. Initially, Kupffer cell response to LPS was unchanged or very slightly increased postsplenectomy, as measured by inhibition of hepatocyte protein synthesis. However, after 3 d and until the completion of the study period (60 d), splenectomy resulted in nearly complete less of Kupffer cell LPS responsiveness. Splenectomy had no effect on the number of Kupffer cells obtained.

Other cytokines from lymphocytic cells, such as IL-3, IL-4, and TGF-β, profoundly alter macrophage function *in vitro*. IL-3 has been shown to be a potent activator of macrophagelike cells. IL-3 alone increased macrophage production of IL-1-α and IL-1-β and acted synergistically with other inflammatory stimuli, such as LPS, IFN-γ, and GM-CSF.[89] Induction of IL-1 by IL-3 occurred via a mechanism different than that usually associated with LPS activation. No evidence of a synergistic interaction between IL-3 and TNF for macrophage activation was observed. IL-4 seemed to exert both stimulatory and inhibitory influences on macrophage functions. IL-4 has been shown to augment macrophage-mediated tumor cytotoxicity, Ia expression, and microbicidal activity. On the other hand, production of TNF, IL-1,[90] and PGE_2[91] have been shown to be downregulated by exposure of mononuclear cells to IL-4. This downregulation appears to be mediated via an IL-4 effect on gene transcription for these proteins.[90] The stimulatory effects of IFN-γ for macrophage activation are also antagonized by IL-4 in several model systems.[90-92] IL-4 has also been shown to significantly inhibit superoxide anion production by human macrophages in response to a variety of inflammatory triggering agents.[92] TGF-β[93] is another cytokine which has been shown to be a potent inhibitor of macrophage hydrogen peroxide production.[94] However, this inhibition could be overcome with IFN-γ, TNF-α, or TNF-β. Although IFN-γ has predominantly stimulatory effects, pretreatment of macrophages with IFN-γ was found to decrease the production of IL-1 in response to LPS.[95]

Cytokines from LPS-stimulated macrophages, or Kupffer cells themselves, might influence Kupffer cell LPS responsiveness. To investigate this possibility, the effect of pretreatment with recombinant cytokines on subsequent activation of rat Kupffer cells by LPS was examined. We had previously shown that addition of recombinant TNF or IL-1 to hepatocytes cultured alone did not recreate the inhibition of hepatocyte protein synthesis observed in LPS-triggered Kupffer cell coculture.[49] Furthermore, addition of IL-1 or TNF to coculture did not influence the effects seen in the absence of LPS nor the response of Kupffer cells triggered with 10 μg/ml of LPS. Finally, polyclonal antibodies to these cytokines did not prevent Kupffer cell activation by LPS as assessed in the hepatocyte coculture assay.[49] However, Figure 9 shows that cytokines could significantly alter Kupffer cell activation and LPS responsiveness. When Kupffer cells were preexposed to recombinant human

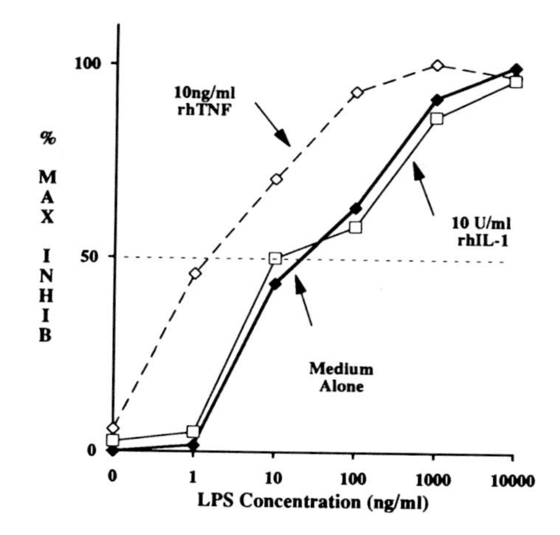

FIGURE 9. Effect of 24-h *in vitro* pretreatment with recombinant TNF or IL-1 on subsequent LPS-triggered, Kupffer cell-mediated inhibition of hepatocyte protein synthesis. No significant effect of IL-1 was observed. However, TNF pretreatment resulted in a significant ($p < 0.01$) leftward shift of the LPS dose-response curve.

TNF for 24 h, a significantly decreased threshold for Kupffer cell activation was seen. In the presence of TNF, as little as 1 ng/ml of LPS resulted in 50% maximal inhibition. Recombinant IL-1 added to Kupffer cell cultures for 24 h before subsequent LPS triggering was found to have no significant effect on the Kupffer cell LPS dose response. The importance of TNF for increasing Kupffer cell activation is supported indirectly by studies of portal vein infusion of TNF.[96] In these investigations, increased mortality was observed when TNF was administered via the portal vein compared to systemic arterial delivery.

The mechanism whereby TNF enhances Kupffer cell LPS responsiveness has not been investigated. In other macrophagelike cells, *in vitro* exposure to recombinant TNF-α has been shown to stimulate IL-1 synthesis and secretion.[97] TNF-α and TNF-β were also found to increase transcription of mRNA for IL-1, GM-CSF, and M-CSF.[98] TNF administration to animals, under appropriate conditions, has been associated with a high mortality which could be prevented with antibody against TNF.[99] *In vivo* exposure to recombinant TNF in low doses was found to desensitize animals to TNF lethality.[100] The mechanism responsible for this desensitization remains elusive, but two observations may provide some clues. First, it is known that TNF is a potent inducer of IL-6 synthesis.[101] More recently, it has been shown that IL-6 has significant inhibitory effects on LPS-induced TNF production *in vitro* and *in vivo*.[102] Second, it has been shown that treatment of monocytes with recombinant TNF-α or IL-1 markedly enhances PGE_2 synthesis in a dose-dependent manner.[97] The increased prostaglandin release by IL-1 and LPS could be inhibited by IFN-γ.[18,103]

TNF-α and PGE_2 may have reciprocal regulatory effects on macrophage-like cells. Low levels of PGE_2 augment TNF synthesis, while higher levels decrease TNF synthesis in a dose-dependent fashion.[104] The presence of PGE_2 appears to be required for the TNF signal to be processed by macrophages and has a positive influence on TNF production.[105] This role of PGE_2 appears to be mediated by changes in the cAMP/cGMP ratio. Increasing cGMP augmented TNF synthesis, whereas high levels of cAMP were inhibitory.[104] For example, sodium nitroprusside, which elevates intracellular cGMP, has been found to increase production of TNF.

Several cytokines have also been noted to synergize with IFN-γ and LPS for activation of macrophagelike cells. For example, TNF and IL-1 both augment IFN-γ-induced tumoricidal activity in macrophages. A synergistic stimulation of monocyte production of CSF was seen with IFN-γ and TNF-α, whereas neither mediator alone had any significant effect.[106] IL-2 was found to augment IFN-γ-induced production of TNF-α mRNA[88] and IL-3 increased production of both IL-1-α and IL-1-β.[89] IFN-γ production by macrophages may have a counterregulatory effect on cytokine production. *In vitro* treatment of peripheral blood mononuclear cells with IFN-γ markedly inhibited LPS-stimulated IL-1 production.[107] This inhibitory effect of IFN-γ was not mediated via induction of prostaglandin synthesis because no effects of cyclooxygenase inhibitors were noted.[108]

IL-1 levels have been shown to correlate positively with mortality[95] and decreased production has been shown in fatal sepsis as compared with survivors.[109] Production of IL-1 appears to be controlled by the second messengers, cAMP and protein kinase-c (PKC), as blockers to these molecules and their formation inhibits the production of IL-1. In this system, LPS and IFN-γ appear to act via the PKC pathway, though IFN-γ induction requires Ca^{++}.[110] It should be noted that the two subspecies of IL-1, alpha and beta, are controlled differently at both the translational and posttranslational levels.[111] IL-1 has been shown to increase the production of prostaglandins and collagenase,[112] CSF-1, IL-2,[24] plasminogen activating factor inhibitor,[113] and acute-phase proteins.[24,111] IL-1-β also has been shown to block hydrogen peroxide production.[94]

G. PLATELET ACTIVATING FACTOR (PAF) ALTERS KUPFFER CELL ACTIVATION STATE

PAF has been shown to stimulate hepatocyte gluconeogenesis in isolated perfused rat livers. PAF has no effect, however, on isolated hepatocytes in culture, in contrast to glucagon and vasopressin. This would suggest that PAF effects are indirect and are mediated by another cell group within the liver. Kupffer cells are noted to express a large number of PAF receptors, and it has been suggested that PAF-induced gluconeogenesis is mediated by Kupffer cells.[114] Preexposure of Kupffer cells to PAF downregulates PAF receptor expression, thus decreasing Kupffer cell PAF responsiveness.[115] This media-

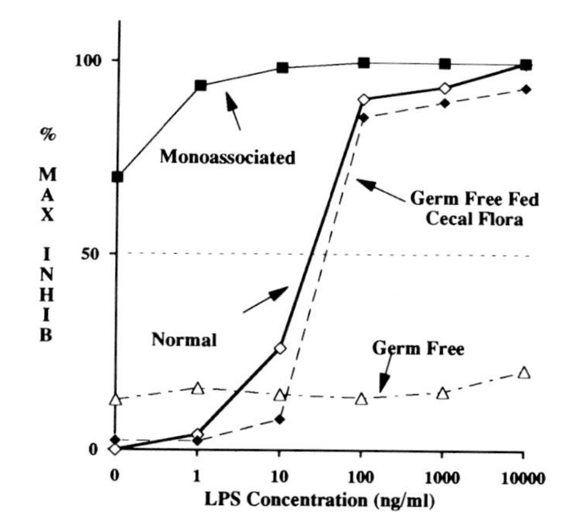

FIGURE 10. Schematic depiction of the effect of alterations in gut flora on subsequent Kupffer cell LPS dose-response relationships. Kupffer cells obtained from germfree animals did not respond to LPS *in vitro*. When the *in vivo* gastrointestinal flora was reconstituted by feeding animals normal cecal flora, *in vitro* Kupffer cell LPS responsiveness was restored. When experimental animals were monoassociated *in vivo* with a translocating strain of bacteria, *E. coli 25*, subsequent Kupffer cells LPS subsequent Kupffer cell LPS responsiveness was increased. (Adapted from Billiar, T. R., Maddaus, M. A., West, M. A., et al., *J. Surg. Res.*, 44, 397, 1988. With permission.)

tion is not blocked by indomethacin, indicating that products of cyclooxygenase are not involved. Gluconeogenesis has also been attributed to arachidonic acid metabolites, and may be stimulated by increased PGD_2 production.[116] PAF has been shown to stimulate the production of TNF, and this effect appears to be mediated via LTB_4.[117] TNF has also been shown to stimulate PAF production by macrophages, suggesting the potential for a positive feedback cycle. LTB_4 has also been shown to markedly enhance LPS-stimulated IL-1 production by macrophages and to increase the production of IL-2 and IFN-γ, indicating a strong upregulatory function.[115]

H. GUT FLORA-REGULATED KUPFFER CELL ACTIVATION

The effect of gut bacterial flora and bacterial translocation on regulation of Kupffer cell LPS responsiveness has also been investigated. Billiar et al.[68,71] obtained Kupffer cells from germfree animals and examined alterations in the LPS dose-response curve when compared to germfree animals whose gut flora was reconstituted by gavage feeding of normal cecal flora. Figure 10 shows that Kupffer cells from germfree animals were not activated by LPS for inhibition of hepatocyte protein synthesis. When similar animals had the gut flora restored, their Kupffer cells responded normally to LPS activation. Another group of littermates were monoassociated with a strain of *E. coli* that translocates into mesenteric lymph nodes with a high degree of frequency. Kupffer cells from these monoassociated animals had a significantly lowered threshold for LPS triggering of maximal inhibition in coculture.[68] In addition, Kupffer cells from animals that were monoassociated for 7 d produced significantly greater quantities of PGE_2 and IL-1 when triggered with LPS. There was no increase in the number of Kupffer cells recovered after 7 d of bacterial overgrowth, indicating that the Kupffer cells were functioning at a higher level than controls. The authors concluded that *in vivo* exposure to gastrointestinal bacteria is important for proper Kupffer cell function.[71]

I. REGULATION OF KUPFFER CELL ACTIVATION BY HYPOXIA

Another important stimulus which may have significant effects on Kupffer cell activation state is alterations in oxygen delivery.[118-120] The dual blood supply of the liver[121] makes a complete understanding of oxygen delivery difficult to assess or examine *in vivo*. Normally, the partial pressure of oxygen bathing Kupffer cells is considerably lower than arterial pO_2 because of the low oxygen tension in portal blood. Therefore, an attempt was made to examine the effects of alterations in oxygen delivery on Kupffer cell function by altering the atmospheric oxygen partial pressure *in vitro* in the presence or absence of activating concentrations of LPS.[65,66] All previous experiments utilizing the Kupffer cell: hepatocyte coculture system were performed in 95% air (20% O_2) and 5% CO_2. In an attempt to investigate the role of hypoxic stimuli on Kupffer cell functions, three different oxygen concentrations were studied: 20% O_2 (pO_2 = 120 torr), 5% O_2 (pO_2 = 40 torr), or 2.5% O_2 (pO_2 = 20 torr). These concentrations corresponded approximately to arterial pO_2, normal portal venous pO_2, and the pO_2 which might be present with decreased tissue perfusion in shock states, respectively.

Clinical transient hypotension (shock and decreased oxygen delivery) often precedes or is coincident with circulation of endotoxin or other inflammatory stimuli.[118,122] To stimulate this situation *in vitro,* the oxygen concentration was lowered during the coculture interval (24 to 48 h) preceding exposure to inflammatory stimuli (48 to 72 h). To eliminate any deleterious effects of lowered O_2 level during the initial plating, all cultured cells were plated under the normal atmospheric conditions (95% air, 5% CO_2) and allowed to recover for 24 h prior to lowering the oxygen concentration. Table 2 shows that exposure of Kupffer cells to 2.5% O_2 for 24 h resulted in significant inhibition of hepatocyte ^{3}H-leucine incorporation, even in the absence of LPS. Kupffer cell activation by LPS, following incubation in 5% O_2, was identical to that seen in 20% O_2. However, preexposure of Kupffer cells to 2.5% O_2 resulted in a significantly greater inhibition of hepatocyte protein synthesis. When the oxygen concentration was manipulated during the time interval utilized for LPS triggering of Kupffer cells, a very different response was noted. There was an oxygen concentration-dependent loss of Kupffer cell functions in the presence or absence of LPS.[66]

To further investigate this phenomenon, the effect of a 6-h exposure to lowered oxygen concentrations on Kupffer cell activation was examined using

TABLE 2
Effect of Alterations in Culture Oxygen (O_2) Concentration Prior to or During LPS Triggering on Kupffer Cell Activation (% Maximal Inhibition of Hepatocyte Protein Synthesis)

		% O_2 During hypoxic interval		
Hypoxic exposure	**LPS**	**20%**	**5%**	**2.5%**
24 h prior to LPS triggering	0	0 ± 5	50 ± 5^a	100 ± 10^a
	10 μg/ml	100 ± 3	94 ± 6^b	100 ± 8^b
During LPS triggering	0	0 ± 4	12 ± 5^c	10 ± 6^c
	10μg/ml	100 ± 7	70 ± 8^d	28 ± 5^d

[a] $p < 0.01$ vs. no LPS in 20% O_2.
[b] p = Not significant vs. 10 μg/ml LPS in 20% O_2.
[c] $p < 0.01$ vs. no LPS in 20% O_2.
[d] $p < 0.01$ vs. 10 μg/ml LPS in 20% O_2.

the LPS dose-response curve. Figure 11 shows a significant effect of transient hypoxia on the LPS dose-response curve for Kupffer cell activation in coculture. A progressive, marked shift of the LPS dose-response curve was seen after preexposure to 5 or 2% oxygen, respectively. This preexposure lowered the threshold concentration of LPS required to trigger inhibition of heptocyte protein synthesis. Thus, alterations in oxygen concentration appeared to regulate Kupffer cell activation. The ultimate effect was influenced by the degree of altered oxygenation as well as the timing relative to subsequent exposure to inflammatory stimuli. Under appropriate conditions, exposure to decreased oxygen alone was sufficient to activate Kupffer cells.[65]

Synthesis of acute-phase proteins by hepatocytes is controlled by cytokines produced by stimulated Kupffer cells.[123,124] IL-6 appears to be particularly important,[125] but IL-1[24] and TNF[123] also affect individual acute-phase proteins. Fibronectin is an opsonic acute-phase protein whose synthesis is increased in sepsis.[52,126] Repetitive low-dose exposure to LPS results in increased Kupffer cell fibronectin synthesis and decreased lethality.[52] Hepatic artery ligation in a porcine model was found to significantly decrease the production of fibronectin in response to intravenous injection of 10^7 of *E. coli* when compared with sham-operated animals.[127] Thus, the oxygen delivery to the liver may significantly influence synthesis of some acute-phase proteins, possibly via alteration in the functional state of Kupffer cells.

J. HEPATOCELLULAR REGULATION OF KUPFFER CELL ACTIVATION

Kupffer cell influence on hepatocytes is not a one-way pathway.[44,128,129] Hepatocytes release a heat-stable soluble substance (<10 kDa in size) that causes increased production of PGE_2 after 36 h of coculture. Furthermore, Kupffer cell production of IL-1 and TNF as well as tumor cytotoxicity were also increased in hepatocyte coculture vs. Kupffer cells cultured alone.[44]

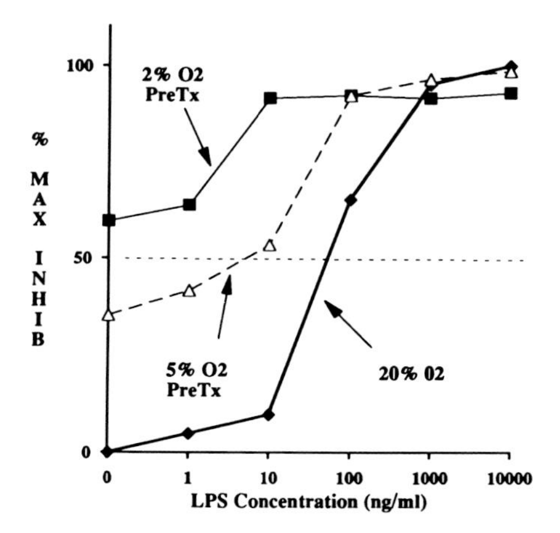

FIGURE 11. The effect of 6-h *in vitro* preexposure to altered oxygen concentrations on Kupffer cell LPS dose responsiveness 18 h later. Exposure to hypoxic culture conditions significantly lowered the threshold LPS concentration required for maximal LPS-triggered Kupffer cell activation, as assessed by inhibition of hepatocyte protein synthesis. Hepatocytes were obtained by enzymatic perfusion and plated in 20% O_2 and 5% CO_2 for 24 h. Following this, Kupffer cells were added to establish coculture, as previously described.[39] Parallel cocultures were performed with manipulation of the oxygen concentration for a 6-h interval. The oxygen concentrations examined were 20% O_2 ($\approx pO_2$ = 140 torr), 5% O_2 ($\approx pO_2$ = 35 torr), and 2% O_2 ($\approx pO_2$ = 14 torr). Following this, the cultures were all incubated in 20% O_2 for the remainder of the experiment. During the final 24-h *in vitro* period, the cultures were triggered with various concentrations of LPS to construct the LPS dose-response relationship. Kupffer cell inhibition of hepatocyte protein synthesis was assessed by measurement of ^{3}H-leucine incorporation into protein. Results were expressed here as percent maximal inhibition.

Laskin and Pilaro[128] found that Kupffer cell chemotaxis and superoxide production in response to PMA were augmented by coculture with hepatocytes. This effect was more pronounced if the hepatocytes were preexposed to acetaminophen. In addition, C-reactive protein, one of the acute-phase proteins synthesized by hepatocytes under the influence of various cytokines,[24,123] has been shown to be a potent activator of macrophages and Kupffer cells.[129]

K. EFFECT OF ANATOMIC FACTORS ON KUPFFER CELL ACTIVATION

Several deviations from normal Kupffer cell anatomy cause significant alterations in their function. McCuskey et al.[69] noted when looking at LPS tolerance that Kupffer cells exposed to LD_{10} of LPS caused a significantly decreased number of sinusoids with blood flow and a decrease in the number of Kupffer cells present. When these animals were subsequently exposed to a LD_{70} dose of LPS, their survival was greater. Though they initially attributed this improvement in survival to an increased activation state of the remaining Kupffer cells, they later concluded in a study comparing species sensitivity to endotoxin that there is a significant correlation between the LD_{50} for all species and the number of Kupffer cells in both the periportal and the centrilobular regions of the hepatic lobule.[70]

It is interesting in light of the correlation between Kupffer cell number and responsiveness to LPS that a significant decrease in reticuloendothelial phagocytosis was noted after hepatectomy.[130] In these studies, the degree of clearance dysfunction directly related to the extent of liver resection. Cecal ligation and puncture resulted in a significantly increased mortality in animals that had undergone 50% hepatectomy when compared to animals without resection. The importance of the role of decreased phagocytic function is supported by data demonstrating that Kupffer cell blockade resulted in decreased survival after infectious challenge, even though systemic immune function, measured by cell-mediated immunity and delayed-type hypersensitivity responses, was increased.[131]

IV. CONCLUSION

Kupffer cells appear to play a critical role in mediating the host response to serious infections. Kupffer cells are the primary line of defense for circulating microorganisms or other inflammatory stimuli. The huge mass of Kupffer cells within the liver, and their central location, underscore the importance of understanding how, when, and why these cells respond to septic stimuli. An ongoing question has been whether Kupffer cells, and other macrophagelike cells, are responding excessively or inadequately to repetitive septic stimuli. The studies summarized in this chapter show that Kupffer cells respond to a wide range of inflammatory stimuli. Furthermore, it is clear that many Kupffer cell functions are profoundly affected (regulated) by a myriad of other inflammatory factors relevant to sepsis. These factors include: LPS stimulation, arachidonic acid metabolites, cytokines, alterations in gut flora, oxygen delivery and partial pressure, lipid mediators, hepatocellular factors, and probably other factors which are presently not well characterized.

Regulation of Kupffer cell activation appears to occur via both local and systemic factors. On a systemic level, factors such as *in vivo* alterations in gut flora,[68] blood flow, or oxygen delivery[66] have been shown to influence Kupffer cells examined using *in vitro* assay systems. In addition, there is evidence that Kupffer cells may be influenced by several local or microenvironmental factors. Most importantly, Kupffer cell activation may be ''autoregulated'' by products released in the process of Kupffer cell activation.[50] Such factors may increase (TNF, IFN-γ) or decrease (PGE_2) Kupffer cell activation. Finally, local networks of intercellular communication exist between adjacent cells within the microenvironment of Kupffer cells. Thus, factors from nearby hepatocytes,[44] endothelial cells,[113,132] fibroblasts,[102,133] and probably other cells, have significant regulatory influences on Kupffer cell functions.

The major factors presently known to be responsible for up- or down-regulating Kupffer cell activation are summarized in Table 3. In general, IFN-

TABLE 3
Factors Regulating Kupffer Cell Activation

Function	Increase activation	Decrease activation
Cytokine synthesis	LPS	PGE_2
	IFN-γ	IL-4[a]
	IL-2[a]	IL-6[a]
	IL-3[a]	TGF-β[a]
Inhibition of hepatocyte protein synthesis	LPS	PGE_2
	IFN-γ	LPS pretreatment
	IL-2	Lack of GI flora
	TNF	Splenectomy
Arginine-dependent effector mechanism	Overgrowth of translocating bacteria	Hypoxia
	Hypoxic preexposure	TGF-β[a]
	Lymphokine supernatant	
Arachidonic acid metabolism	LPS	PGE_2
	IL-2	IFN-γ
	IL-1[a]	
Tumor cell cytolysis	LPS	PGE_2[a]
	IFN-γ	LPS pretreatment
	IL-2[a]	TGF-β[a]

[a] Study performed in macrophages; specific data unavailable for Kupffer cells.

γ and IL-2 appear to interact with LPS or bacteria to increase the activation state of Kupffer cells and most other macrophagelike cells. In contrast, PGE_2 has generally inhibitory, or downregulatory, effects on the Kupffer cell activation state. One exception to this generalization appears to be seen with respect to regulation of arachidonic acid metabolism. In this case, IFN-γ inhibits PGE_2 production, whereas IL-1 has stimulatory effects. These regulatory influences are common to all major functions associated with Kupffer cell activation. It has also been reported that alterations in gastrointestinal bacterial flora and splenectomy influence the Kupffer cell activation state. These effects may be due to induction of (or loss of) lymphocytic cytokines from the mesenteric lymph nodes or spleen, respectively. On the other hand, the mechanism could involve excessive LPS-like stimulation of Kupffer cells secondary to increased numbers of translocating bacteria or loss of reticuloendothelial phagocytic functions, respectively. We have found that O_2 concentration significantly affects the Kupffer cell activation state. Whether the alterations in O_2 result in stimulation or inhibition of Kupffer cell activation depends upon the timing of hypoxic exposure.

The regulatory influences on Kupffer cell, or macrophage, activation very likely also change significantly over time. At present, this area has not been adequately studied. We hypothesize that virtually all of the products of activated Kupffer cells, such as IL-1, TNF, IL-6, PGE_2, etc., have regulatory or counterregulatory influences on all other Kupffer cell functions. For example, loss of Kupffer cell LPS responsiveness appears to be secondary to

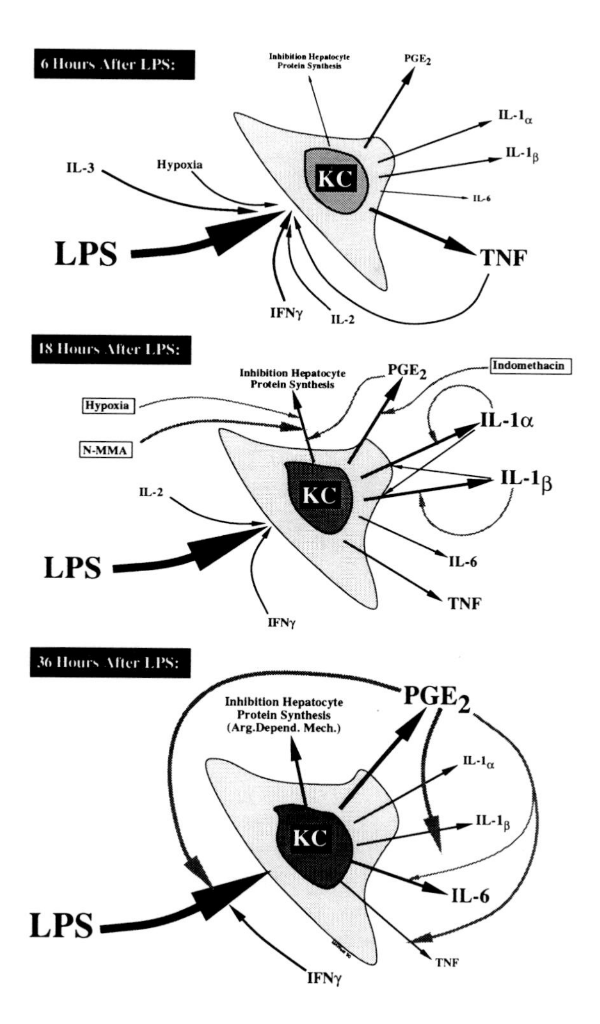

FIGURE 12. Hypothetical scheme depicting factors important in the regulation of Kupffer cell activation state over time. Negative (inhibitory) regulatory factors are shaded and intersect the Kupffer cell function which they influence. The approximate magnitude of individual Kupffer cell functions or regulatory factors are indicated by the size of the arrow and text.

LPS-stimulated production of PGE_2. In macrophages, it has been shown that production of IL-1 is inhibited by PGE_2 and by IL-1 itself. At the same time, IL-1-α can induce production of IL-1-β, and IL-1-β is capable of inducing IL-1-α synthesis. Furthermore, it is clear that Kupffer cells or other macrophagelike cells do not function autonomously, but rather interact continuously with adjacent cells. Thus, the Kupffer cell activation state is also influenced by complex cytokine networks, which are presently not well understood. In Figure 12, an attempt has been made to summarize some of the important regulatory loops which may control the Kupffer cell activation state.

Presently, it is impossible to predict the activation state of Kupffer cells in any clinical setting. Continued investigations, utilizing both *in vitro* and *in vivo* model systems, should produce major insights into the mechanisms of Kupffer cell regulation. More importantly, future discoveries will likely suggest new ways to intervene clinically to optimize the activation state of Kupffer cells and other macrophages so that patient mortality and morbidity are decreased.

REFERENCES

1. **Cerra, F. B., Siegel, J. H., Border, J. R., et al.,** The hepatic failure of sepsis: cellular versus substrate, *Surgery,* p. 409, 1979.
2. **Knaus, W. A. and Draper, E. A.,** Prognosis from combined organ system failure, *Ann. Surg.,* 202, 685, 1985.
3. **Carrico, C. J., Meakins, J. L., Fry, D. E., et al.,** Multiple organ failure syndrome, *Arch. Surg.,* 121, 196, 1986.
4. **Meakins, J. L., Hohn, D. C., Hunt, T. K., et al.,** Host defenses, in *Surgical Infectious Diseases,* R. L. Simmons and R. Howard, Eds., Appleton Century Crofts, New York, 1982.
5. **Jones, E. A. and Summerfield, J. A.,** Kupffer cells, in *The Liver: Pathobiology,* I. Aria, H. Popper, D. Schacter, and D. A., Shafritz, Eds., Raven Press, New York, 1982, 507.
6. **Gala, R. P., Sparks, R. S., and Golde, D. W.,** Bone marrow origin of hepatic macrophages (Kupffer cells), *Science,* 201, 937, 1978.
7. **Thomas, E. D., Ramberg, R. E., and Sale, G. E.,** Direct evidence for a bone marrow origin of the alveolar macrophage in man, *Science,* 192, 1016, 1976.
8. **West, M. A., Christou, N. V., Kasper, D., et al.,** Panel on macrophage: lymphocyte interactions in surgical sepsis; macrophage effector function in sepsis, *Arch. Surg.,* 122, 242, 1987.
9. **North, R. J.,** The concept of the activated macrophage, *J. Immunol.,* 121, 806, 1978.
10. **Karnovsky, M. L. and Lazdine, J. K.,** Biochemical criteria for activated macrophages, *J. Immunol.,* 121, 809, 1978.
11. **Rucco, R. and Meltzer, M. S.,** Macrophage activation for tumor cytotoxicity: Development of macrophage cytotoxic activity requires completion of a sequence of short-lived intermediary reactions, *J. Immunol.,* 121, 2035, 1978.
12. **Cohn, Z. A.,** The activation of mononuclear phagocytes: fact, fancy, and future, *J. Immunol.,* 121, 813, 1978.
13. **Crawford, R. M., Finbloom, D. S., Ohara, J., et al.,** B-Cell stimulatory factor-1 (interleukin-4) activates macrophages for increased tumoricidal activity and expression of Ia antigens, *J. Immunol.,* 139, 135, 1987.
14. **Nathan, C. F. and Root, R. K.,** Hydrogen peroxide release from mouse peritoneal macrophages: dependence on sequential activation and triggering, *J. Exp. Med.,* 146, 1648, 1977.

15. **Nathan, C. F., Silverstein, S. C., Brukner, C. H., et al.,** Extracellular cytolysis by activated macrophages and granulocytes. II. Hydrogen peroxidase a mediator of cytotoxicity, *J. Exp. Med.,* 149, 110, 1979.
16. **Johnson, W. J., Marion, P. A., Schreiber, R. D., et al.,** Sequential activation of murine mononuclear phagocytes for tumor cytolysis: differential expression of markers by macrophages in the several stages of development, *J. Immunol.,* 131, 1038, 1983.
17. **Pace, J. L. and Russel, S. W.,** Activation of mouse macrophages for tumor cell killing. I. Quantitative analysis of interactions between lymphokine and lipopolysaccharide, *J. Immunol.,* 126, 1863, 1981.
18. **Hamilton, T. A., Rigbee, J. E., Scott, W. A., et al.,** γ-Interferon enhances secretion of arachidonic acid metabolites from murine peritoneal macrophages stimulated with phorbol diesters, *J. Immunol.,* 134, 2631, 1985.
19. **Somers, S. D., Weiel, J. E., Hamilton, T. A., et al.,** Phorbol esters and calcium ionophore can prime murine peritoneal macrophages for tumor cell killing, *J. Immunol.,* 136, 4199, 1986.
20. **Tsunawaki, S. and Nathan, C. F.,** Enzymatic basis of macrophage activation. Kinetic analysis of superoxide production in lysates of resident and activated mouse peritoneal macrophages and granulocytes, *J. Biol. Chem.,* 259, 4305, 1984.
21. **Belosevic, M., Davis, C. E., Meltzer, M. S., et al.,** Regulation of activated macrophage antimicrobial activities: identification of lymphokines that cooperate with IFNγ for induction of resistance to infection, *J. Immunol.,* 141, 890, 1988.
22. **Bonney, R. J. and Humes, J. L.,** Physiological and pharmacological regulation of prostaglandin and leukotriene production by macrophages, *J. Leuk. Biol.,* 35, 1, 1984.
23. **Billiar, T. R., Bankey, P. E., Svingen, B. A., et al.,** Fatty acid intake and Kupffer cell function: fish oil alters eicosanoid and monokine production to endotoxin stimulation, *Surgery,* 104, 343, 1988.
24. **Dinarello, C. A.,** Interleukin-1 and the pathogenesis of the acute phase response, *N. Engl. J. Med.,* 311, 1413, 1984.
25. **Beutler, B., Greenwald, D., and Hulmes, J. D.,** Identity of tumor necrosis factor and the macrophage-secreted factor cachectin, *Nature,* 316, 552, 1985.
26. **Nathan, C. F.,** Secretory products of macrophages, *J. Clin. Invest.,* 79, 319, 1987.
27. **Russell, S. W. and Pace, J. L.,** Both the kind and magnitude of stimulus are important in overcoming the negative regulation of macrophage activation by PGE_2, *J. Leuk. Biol.,* 35, 291, 1984.
28. **Hibbs, J. B., Vavrin, Z., and Taintor, R. R.,** L-Arginine is required for expression of the activated macrophage effector mechanism causing selective metabolic inhibition in target cells, *J. Immunol.,* 138, 550, 1987.
29. **Bouwens, L. and Wisse, E,** Proliferation, kinetics, and fate of monocytes in liver during a zymosan-induced inflammation, *J. Leuk. Biol.,* 37, 531, 1985.
30. **Bouwens, L., Baekeland, M., and Wisse, E.,** Importance of local proliferation in the expanding Kupffer cell population of rat liver after zymosan stimulation and partial hepatectomy, *Hepatology,* 4, 213, 1984.

31. **Diesselhoff-Den Dulk, M. M. C., Crofton, R. W., and Van Furth, R.,** Origin and kinetics of Kupffer cells during an acute inflammatory response, *Immunology,* 37, 7, 1979.
32. **Nolan, J. P.,** Endotoxin, reticuloendothelial function, and liver injury, *Hepatology,* 1, 458, 1981.
33. **Largen, M. T. and Tannenbaum, C. S.,** LPS regulation of specific protein synthesis in murine-peritoneal macrophages, *J. Immunol.,* 136, 988, 1986.
34. **Smith, E. M.,** Hormonal activities of lymphokines, monokines, and other cytokines, *Prog. Allergy,* 43, 121, 1988.
35. **Dinarello, C. A. and Mier, J. W.,** Lymphokines, *N. Engl. J. Med.,* 317, 940, 1987.
36. **Wardle, E. N.,** Interferon gamma: actions and importance, *Br. J. Hosp. Pharm.,* p. 446, 1987.
37. **Zahlten, R. N., Hagler, H. K., Nejeter, M. E., et al.,** Morphological characterization of Kupffer cells and endothelial cells of rat liver isolated by counterflow elutriation, *Gastroenterology,* 75, 80, 1978.
38. **Munthe-Kass, A. C., Berg, T., Seglen, P. O., et al.,** Mass isolation of rat Kupffer cells, *J. Exp. Med.,* 141, 1, 1975.
39. **West, M. A., Keller, G. A., Hyland, B. J., et al.,** Hepatocyte function in sepsis: Kupffer cells mediate a biphasic protein synthesis response in hepatocytes after exposure to endotoxin or killed E. coli, *Surgery,* 98, 388, 1985.
40. **West, M. A., Billiar, T. R., Curran, R. D., et al.,** Evidence that rat Kupffer cells stimulate and inhibit hepatocyte protein synthesis in vitro by different mechanisms, *Gastroenterology,* 96, 1572, 1989.
41. **Billiar, T. R., Curran, R. D., West, M. A., et al.,** Toxic L-arginine metabolites produced by endotoxin-activated Kupffer cells induce hepatocyte death, *Arch. Surg.,* 124, 1416, 1989.
42. **Billiar, T. R., Curran, R. D., Stuehr, D., West, M. A., et al.,** An L-arginine dependent mechanism mediates Kupffer cell inhibition of hepatocyte protein synthesis *in vitro, J. Exp. Med.,* 169, 1467, 1989.
43. **Bankey, P., Carlson, A., Ortiz, M., Singh, R., et al.,** Tumor necrosis factor production by Kupffer cells requires protein kinase-C activation, *J. Surg. Res.,* 49, 256, 1990.
44. **Billiar, T. R., Lysz, T. W., Curran, R. D., et al.,** Hepatocyte modulation of Kupffer cell prostaglandin E2 production *in vitro, J. Leuk. Biol.,* 47, 304, 1990..
45. **Bhatnagar, R., Schirmer, R., Ernst, M. et al.,** Superoxide release by zymosan-stimulated rat Kupffer cells in vitro, *Eur. J. Biochem.,* 119, 171, 1981.
46. **Laskin, D. L., Sirak, A. A., Pilaro, A. M., et al.,** Functional and biochemical properties of rat Kupffer cells and peritoneal macrophages, *J. Leuk. Biol.,* 44, 71, 1988.
47. **Roh, M. S., Wang, L., Oyedeji, C., et al.,** Human Kupffer cells are cytotoxic against human colon adenocarcinoma, *Surgery,* 108, 400, 1990.
48. **West, M. A., Keller, G. A., Hyland, B. J., et al.,** Further characterizations of Kupffer cell/macrophage mediated alterations in hepatocyte protein synthesis, *Surgery,* 100, 416, 1986.
49. **West, M. A., Billiar, T. R., Mazuski, J. E., et al.,** Endotoxin modulation of hepatcyte secreted and cellular protein synthesis is mediated by Kupffer cells, *Arch. Surg.,* 123, 1400, 1988.

50. **West, M. A., Billiar, T. R., Hyland, B. J., et al.,** Regulation of Kupffer cell mediated alterations in hepatocyte protein synthesis in vitro, *Cur. Surg.*, 44, 467, 1987.
51. **West, M. A., Keller, G. A., Cerra, F. B., et al.,** Killed E. coli stimulate macrophage-mediated alterations in hepatocellular function during in vitro coculture, *Infect. Immun.*, 49, 563, 1985.
52. **Vincent, P. A., Cho, E., and Saba, T. M.,** Effect of repetitive low-dose endotoxin on liver parenchymal and Kupffer cell fibronectin release, *Hepatology,* 9, 1989, 1989.
53. **Curran, R. D., Billiar, T. R., Stuehr, D. J., et al.,** Multiple cytokines are required to induce hepatocyte nitric oxide production and inhibit total protein synthesis, *Ann. Surg.*, 212, 462, 1990.
54. **Keller, G. A., West, M. A., Harty, J. T., et al.,** Modulation of hepatocyte protein synthesis by endotoxin activated Kupffer cells. III. Evidence for the role of a monokine similar but not identical to interleukin-1, *Ann. Surg.*, 203, 436, 1985.
55. **Billiar, T. R., Curran, R. D., Ferrari, F. K., et al.,** Kupffer cell: hepatocyte cocultures release nitric oxide in response to bacterial endotoxin, *J. Surg. Res.*, 48, 349, 1990.
56. **Maier, R. V. and Hahne, G. B.,** Potential for endotoxin-activated Kupffer cells to induce microvascular thrombosis, *Arch. Surg.*, 119, 62, 1984.
57. **Fong, Y., Moldawer, L. L., Marano, M., et al.,** Endotoxemia elicits increased B2-IFN/IL-6 in man, *J. Immunol.*, 142, 2321, 1989.
58. **Fong, Y., Tracey, K. J., Moldawer, L. L., et al.,** Antibodies to cachectin/tumor necrosis factor reduce interleukin-1β and interleukin-6 appearance during lethal bacteremia, *J. Exp. Med.*, 170, 1627, 1989.
59. **Decker, K. and Birmelin, M.,** Ca mediates phagocytosis-evoked eicosanoid synthesis in Kupffer cells, in *Prostaglandin and Membrane Ion Transport,* P. Braquet, Ed., Raven Press, New York, 1984, 113.
60. **Casteleijn, E., Kuiper, J., Van Rooij, H. C. J., et al.,** Prostaglandin D_2 mediates the stimulation of glycogenolysis in the liver by phorbol ester, *Biochem. J.*, 250, 77, 1988.
61. **LePay, D. A., Nathan, C. F., Steinman, R. M., et al.,** Murine Kupffer cells. Mononuclear phagocytes deficient in the generation of reactive oxygen intermediates, *J. Exp. Med.*, 161, 1079, 1985.
62. **Decker, T., Kiderlen, A., and Lohmann-Matthes, M. L.,** Liver macrophages (Kupffer cells) as cytotoxic effector cells in extracellular and intercellular cytotoxicity, *Infect. Immun.*, 50, 358, 1985.
63. **Stukart, M. J., Rijnsent, A., and Roos, E.,** Induction of tumoricidal activity in isolated rat liver macrophages by liposomes containing recombinant rat gamma-interferon supplemented with lipopolysacchride or muramyldipeptide, *Cancer Res.*, 47, 3880, 1987.
64. **Keller, F., Wild, M. T., and Kirn, A.,** In vitro cytostatic properties of unactivated rat Kupffer cells, *J. Leuk. Biol.*, 35, 467, 1984.
65. **West, M. A., Knighton, D. R., Hyland, B. J., et al.,** Hypoxic preexposure alters Kupffer cell modulation of hepatocyte function, *Hepatology,* 8, 1985.
66. **West, M. A., Knighton, D. R., Hyland, B. J., et al.,** Hypoxia alters hepatocyte: Kupffer cell interactions, *Surg. Forum,* 36, 75, 1985.

67. **Billiar, T. R., West, M. A., Hyland, B. J., et al.,** Splenectomy alters Kupffer cell response to endotoxin, *Arch. Surg.*, 121, 327, 1988.
68. **Billiar, T. R., Maddaus, M. A., West, M. A., et al.,** Gram negative intestinal overgrowth in vivo augments the in vitro response of Kupffer cells to endotoxin, *Ann. Surg.*, 208, 532, 1988.
69. **McCuskey, R. S., McCuskey, P. A., Urbaschek, R., et al.,** Kupffer cell function in host defense, *Rev. Infect. Dis.*, 9, S616, 1987.
70. **McCuskey, R. S., McCuskey, P. A., Urbaschek, R., et al.,** Species differences in Kupffer cells and endotoxin sensitivity, *Infect. Immun.*, 45, 278, 1984.
71. **Billiar, T. R., Maddaus, M. A., West, M. A., et al.,** The role of intestinal flora in the interactions between liver nonparenchymal cells and hepatocytes, *J. Surg. Res.*, 44, 397, 1988.
72. **Scammon, J. P., Zawacki, J. K., McMurrich, J., et al.,** Glycosidases of rat Kupffer cells, hepatocytes, and peritoneal macrophages, *Biochim. Biophys. Acta,* 404, 281, 1975.
73. **Taffet, S. M., Pace, J. L., and Russel, S. W.,** Lymphokine maintains macrophage activation for tumor cell killing by interfering with the negative regulatory effects of prostaglandin E_2, *J. Immunol.*, 127, 121, 1981.
74. **Billiar, T. R.,** Personal communication, 1988.
75. **Zimmer, T. and Jones, P. P.,** Combined effects of tumor necrosis factor-prostaglandin E_2, and corticosterone on induced Ia expresison on murine macrophages, *J. Immunol.*, 145, 1167, 1990.
76. **Wahl, L. M., Corcoran, M. E., Mergenhagen, S. E., et al.,** Inhibition of phospholipase activity in human monocytes by IFNγ blocks endogenous prostaglandin-E_2 dependent collagenase production, *J. Immunol.*, 144, 3518, 1990.
77. **Boraschi, D., Censini, S., Bartalini, M., et al.,** Interferon inhibits prostaglandin biosynthesis in macrophages: effects on arachidonic acid metabolism, *J. Immunol.*, 132, 1987, 1984.
78. **Reider, H., Ramadori, G., and Meyer zum Buschenfelde, K. H.,** Guinea pig Kupffer cells can be activated in vitro to an enhanced superoxide response. II. Involvement of eicosanoids, *Hepatology,* 7, 345, 1988.
79. **Chen, L., Suzuki, Y., and Wheelock, E.,** Interferon-gamma synergizes with tumor necrosis factor in interleukin-1 and requires the presence of both monokines to induce antitumor cytotoxic activity in macrophages, *J. Immunol.*, 139, 3, 4096, 1987.
80. **Stuehr, D. J. and Marletta, M. A.,** Induction of nitrite/nitrate synthesis in murine macrophages by BCG infection, lymphokines, or interferon-γ, *J. Immunol.*, 139, 518, 1987.
81. **Ding, A. H., Nathan, C. F., and Stuehr, D. J.,** Release of reactive nitrogen intermediates and reactive oxygen intermediates from mouse peritoneal macrophages, *J. Immunol.*, 141, 2407, 1988.
82. **Drapier, J. C. and Hibbs, J. B.,** Differentiation of murine macrophages to express nonspecific cytotoxicity for tumor cells results in L-arginine-dependent inhibition of mitochondrial iron-sulfur enzymes in the macrophage effector cells, *J. Immunol.*, 140, 2829, 1988.
83. **Curran, R. D., Billiar, T. R., West, M. A., et al.,** Effects of interleukin-2 on Kupffer cell activation and response to endotoxin, *Arch. Surg.*, 123, 1373, 1988.

84. **Hancock, W. W., Muller, W. A., and Cotran, R. S.,** Interleukin-2 receptors are expressed by alveolar macrophages during pulmonary sarcoidosis and are inducible by lymphokine treatment of normal lung macrophages, blood monocytes, and monocytic cell lines, *J. Immunol.,* 138, 185, 1987.
85. **Herrmann, F., Cannistra, S. A., Levine, H., et al.,** Expression of interleukin-2 receptors and binding of interleukin-2 by gamma interferon-induced human leukemic and normal monocytic cells, *J. Exp. Med.,* 162, 1111, 1985.
86. **Numerof, R. P., Aronson, F. R., and Mier, J. W.,** IL-2 stimulates the production of IL-1α and IL-1β by human peripheral blood mononuclear cells, *J. Immunol.,* 141, 4250, 1988.
87. **Tilden, A. B. and Dunlap, N. E.,** Interleukin-2 augmentation of interleukin-1 and prostaglandin E_2 production, *J. Leuk. Biol.,* 45, 474, 1989.
88. **Belosevic, M., Finbloom, D. S., Meltzer, M. S., et al.** IL-2, a cofactor for induction of activated macrophage resistance to infection, *J. Immunol.,* 145, 831, 1990.
89. **Frendl, G., Fenton, M. J., and Belkler, D. I.,** Regulation of macrophage activation by IL-3. II. IL-3 and lipopolysaccharide act synergistically in the regulation of IL-1 expression, *J. Immunol.,* 144, 3400, 1990.
90. **Essner, R., Rhoades, K., McBride, W. H., et al.,** IL-4 downregulates IL-1 and TNF gene expression in human monocytes, *J. Immunol.,* 142, 3857, 1989.
91. **Hart, P. H., Vitti, G. F., Burgess, D. R., et al.,** Potential anti-inflammatory effects of interleukin-4: suppression of human monocyte tumor necrosis factor-α, interleukin-1 and prostaglandin E_2, *Proc. Natl. Acad. Sci. U.S.A.,* 86, 3803, 1989.
92. **Abrahamson, S. L. and Gallin, J. I.,** IL-4 inhibits superoxide production by human mononuclear phagocytes, *J. Immunol.,* 144, 625, 1990.
93. **Sporn, M. B., Roberta, A. B., Wakefield, L. M., et al.,** Transforming growth factor-beta: biological function and chemical structure, *Science,* 233, 532, 1986.
94. **Tsunawaki, S., Sporn, M., Ding, A., et al.,** Deactivation of macrophages by transforming growth factor-beta, *Nature,* 334, 260, 1988.
95. **Cilliari, E., Dieli, M., Maltese, E., et al.,** Enhancement of macrophage IL-1 production by Leishmania major infection in vitro and its inhibition by IFN, *J. Immunol.,* 143, 2001, 1989.
96. **Kahky, M. P., Daniel, C. O., Cruz, A. B., et al.,** Portal infusion of tumor necrosis factor increases mortality in rats, *J. Surg. Res.,* 49, 138, 1990.
97. **Dianrello, C. A., Cannon, J. G., Wolff, S. M., et al.,** Tumor necrosis factor (cachectin) is an endogenous pyrogen and induces interleukin-1, *J. Exp. Med.,* 163, 1433, 1986.
98. **Kaushansky, K., Broudy, V. C., Harlan, J. M., et al.,** Tumor necrosis factor and tumor necrosis factor-beta (lymphotoxin) stimulate the production of granulocyte-macrophage colony stimulating factor, macrophage colony-stimulating factor, and IL-1 in vitro, *J. Immunol.,* 141, 3410, 1988.
99. **Buetler, B., Milsark, I. W., and Cerami, A. C.,** Passive immunization against cachectin/tumor necrosis factor protects mice from the lethal effects of endotoxin, *Science,* 229, 869, 1985.
100. **Wallach, D., Holtmann, H., Engelmann, H., et al.,** Sensitization and desensitization to the lethal effects of tumor necrosis factor and IL-1, *J. Immunol.,* 140, 2994, 1988.

101. **Defillippi, P., Puopart, P., Tavernier, J., et al.,** Induction and regulation of mRNA encoding 26 kD protein in human cell lines treated with recombinant human necrosis factor, *Proc. Natl. Acad. Sci. U.S.A.*, 845, 4557, 1987.
102. **Aderka, D., Le, J., and Vilcer, J.,** IL-6 inhibits lipopolysaccharide-induced tumor necrosis factor production in cultured human monocytes, U937 cells, and in mice, *J. Immunol.*, 143, 3517, 1989.
103. **Browning, J. L. and Ribolini, A.,** Interferon blocks interleukin-1 induced prostaglandin release from human peripheral monocytes, *J. Immunol.*, 138, 2857, 1987.
104. **Renz, H., Gong, J. H., Schmidt, A., et al.,** Release of tumor necrosis factor-α from macrophages: enhancement and suppression are dose-dependently related by prostaglandin E_2 and cyclic nucleotides, *J. Immunol.*, 141, 2388, 1988.
105. **Lehmann, V., Benninghoff, B., and Droge, W.,** Tumor necrosis factor-induced activation of peritoneal macrophages is regulated by prostaglandin-E_2 and cAMP, *J. Immunol.*, 141, 587, 1988.
106. **Li Lu, D. W., Graham, C. D., Waheed, A., et al.,** Enhancement of release from MHC class II antigen-positive monocytes of hematopoietic colony stimulating factors SCF-1 and G-CSF by recombinant human tumor necrosis factor-alpha: synergism with recombinant human interferon-gamma, *Blood,* 72, 34, 1988.
107. **Brandwein, S.,** Regulation of interleukin 1 production by mouse peritoneal macrophages, *J. Biol. Chem.*, 261, 8624, 1986.
108. **Ghezzi, P. and Dinarello, C. A.,** IL-1 induces IL-1. III. Specific inhibition of IL-1 production by IFN-γ, *J. Immunol.*, 140, 4238, 1988.
109. **Luger, A., Graf, H., Schwarz, H. P., et al.,** Decreased serum interleukin-1 activity and monocyte interleukin-1 production in patients with fatal sepsis, *Crit. Care Med.*, 14, 458, 1986.
110. **Kovacs, E. J., Brock, B., Varesio, L., et al.,** IL-2 induction of IL-1 beta mRNA expression in monocytes. Regulation by agents that block second messenger pathways, *J. Immunol.*, 143, 3532, 1989.
111. **Turner, M., Chantry, D., Buchan, G., et al.,** Regulation of expression of human IL-1 alpha and IL-1 beta genes, *J. Immunol.*, 143, 3556, 1989.
112. **Mizel, S. B., Dayer, J. M., Krane, S. M., et al.,** Stimulation of rheumatoid synovial cell collagenase and prostaglandin production by partially purified lymphocyte-activating factor (interleukin-1), *Proc. Natl. Acad. Sci. U.S.A.*, p. 78, 1981.
113. **Nachman, R. L., Hajjar, K. A., Silverstein, R. L., et al.,** Interleukin-1 induces endothelial cell synthesis of plasminogen activator inhibitor, *J. Exp. Med.*, 163, 1595, 1986.
114. **Rola-Pleszczynski, M. and Lemaire, I.,** Leukotrienes augment interleukin-1 production by human monocytes, *J. Immunol.*, 135, 3958, 1985.
115. **Chao, W., Liu, H., Hanahan, D. J., et al.,** Regulation of platelet-activating factor receptor in rat Kupffer cells, *J. Biol Chem.*, 264, 20448, 1989.
116. **Casteleijn, E. Kuiper, J., Van Rooij, H. C. J., et al.,** Hormonal control of glycogenolysis in parenchymal liver cells by Kupffer and endothelial cells, *J. Biol. Chem.*, 263, 2699, 1988.
117. **Dubois, C., Bissonette, E., and Rola-Pleszczynski, M.,** Platelet activating factor (PAF) enhances tumor necrosis factor production by alveolar macrophages, *J. Immunol.*, 143, 964, 1989.

118. **Cerra, F. B., Border, J. R., McMenamy, R. H., et al.,** Multiple system organ failure, in *Surgical Infectious Diseases,* R. L. Simmons and R. Howard, Eds., Appleton Century Crofts, New York, 1982.
119. **Cerra, F. B., Siegel, J. H., Border, J. R., et al.,** Correlations between metabolic, cardiopulmonary measurements in patients after trauma, general surgery and sepsis, *J. Trauma,* 19, 621, 1979.
120. **Knighton, D. R., Hunt, T. K., et al.,** Oxygen tension regulates the expression of angiogenesis factor by macrophages, *Science,* 221, 1283, 1983.
121. **Lautt, W. W. and Greenway, C. V.,** Conceptual review of the hepatic vascular bed, *Hepatology,* 7, 952, 1987.
122. **Eiseman, B., Beart, R., and Norton, L.,** Multiple organ failure, *Surg. Gynecol. Obstet.,* 144, 323, 1977.
123. **Perlmutter, D. H., Dinarello, C. A., Punsal, P. I., et al.,** Cachectin/tumor necrosis factor regulates hepatic acute phase gene expression, *J. Clin. Invest.,* 78, 1349, 1986.
124. **Baumann, H., Jahreis, G. P., Sauder, D. N., et al.,** Human keratinocytes and monocytes release factors which regulate the synthesis of major acute phase proteins in hepatic cells from man, rat and mouse, *J. Biol. Chem.,* 259, 7331, 1984.
125. **Marinkovic, S., Jahreis, G. P., Wong, G. G., et al.,** IL-6 modulates the synthesis of a specific set of acute phase proteins in vivo, *J. Immunol.,* 142, 808, 1989.
126. **Richards, P. S. and Saba, T. M.,** Alterations of fibronectin and reticuloendothelial phagocytic function during adapation to experimental shock, *Circ. Shock,* 10, 189, 1983.
127. **Heaney, M. L. and Marini, C. P.,** Personal communication, 1990.
128. **Laskin, D. L. and Pilaro, A. M.,** Potential role of activated macrophages in acetaminophen hepatotoxicity. I. Isolation and characterization of activated macrophages from rat liver, *Toxicol. Appl. Pharmacol.,* 86, 204, 1986.
129. **Zahedi, K. and Mortenson, R. F.,** Macrophage tumoricidal activity induced by human C-reactive protein, *Cancer Res.,* 46, 5077, 1986.
130. **Vo, N. M. and Chi, D. S.,** Effect of hepatectomy on the reticuloendothelial system of septic rats, *J. Trauma,* 28, 852, 1988.
131. **Callery, M. P., Kamei, T., and Flye, W.,** Kupffer cell blockade increases mortality during intra-abdominal sepsis despite improving systemic immunity, *Arch. Surg.,* 125, 36, 1990.
132. **Stuehr, D. J., Gross, S. S., and Sakuma, I., et al.,** Activated murine macrophages secrete a metabolite of arginine with the bioactivity of endothelium-derived relaxing factor and the chemical reactivity of nitric oxide, *J. Exp. Med.,* 169, 1011, 1989.
133. **Van Damme, J., Opdenakker, G., Simpson, R. J., et al.,** Identification of the human 26-kd protein interferon $beta_2$ (IFN-$beta_2$), as a B-cell hybridoma/plasmacytoma growth factor induced by interleukin-1 and tumor necrosis factor, *J. Exp. Med.,* 165, 914, 1987.

Chapter 11

THE ROLE OF KUPFFER CELLS IN CONTROL OF ACUTE-PHASE PROTEIN SYNTHESIS

P.H. Kispert

TABLE OF CONTENTS

ISBN 0-8493-6109-5

I. INTRODUCTION

Kupffer cells (KC) account for 90% of the reticuloendothelial system and are of critical importance in the clearance of bacteria and endotoxins in the portal and systemic circulation. In the past, KC have been studied to clarify their phagocytic properties. It is clear, however, that KC function in a role far beyond that of a solely phagocytic cell. The control of hepatocyte (HC) protein synthesis by macrophage products has been clarified by many investigators over the last 15 years. The majority of these studies have involved the addition of monocyte supernatants or cytokines to HC to clarify their role in altering HC protein synthesis. Little attention has been given to interactions between cells as they might occur *in vivo*. KC and HC are in intimate physical contact in the liver. This histologic arrangement places KC in a unique position to regulate HC protein synthesis. Our laboratory has been particularly interested in examining how KC interact with HC to alter protein synthesis. In the remainder of this chapter, we will review our experience with KC-HC interactions *in vitro* and *in vivo,* as well as the factors that alter KC interleukin-6 (IL-6) production.

II. FACTORS AFFECTING HC ACUTE-PHASE PROTEIN SYNTHESIS

The synthesis of acute-phase proteins (APP) by HC is an important part of the response of an organism to injury and infection, and is one component of the acute-phase response. The systemic acute-phase response is characterized by fever, leukocytosis, elevated sedimentation rate, elevated ACTH and glucocorticoid levels, complement activation, depressions in serum zinc and iron, a negative nitrogen balance, and changes in hepatic acute protein synthesis.[1,4,5] The hepatic APP have been defined as those proteins whose plasma concentration increases or decreases by 25% after an inflammatory stimulus.[4] Hepatic APP synthesis appears to be an important component of this response because of the protective functions some of these proteins possess. The synthesis of many proteins by the liver is dramatically increased (positive APP), while the synthesis of others is depressed (negative APP), reflecting changes in the patterns of gene expression for proteins in HC after inflammatory stimuli. Total HC protein synthesis changes little in experimental acute inflammation, while the synthesis of specific APP changes greatly, indicative of a change in the pattern of protein synthesis.[2] The importance of APP can be appreciated by examining their functions in the context of injury or inflammation. The acute-phase response, including protein synthesis, is a relatively nonspecific reaction to stressful stimuli. Injury is complicated by infection, hemorrhage, and inflammation. The APP act to limit tissue injury, control infection, and arrest hemorrhage. Hemostatic proteins (fibrinogen), opsonins (C-reactive protein and fibronectin),[18,22] hemoglobin-binding proteins (hepatoglobin), complement components (C3),[19,38] antiproteases (alpha-2-macroglobulin, alpha-1-antitrypsin, alpha-1-acid glycoprotein, and thiostatin),[2,23] and lipopolysaccharide (LPS)-binding protein[61-63] are synthesized at an increased rate, while the synthesis of carrier proteins such as albumin and transferrin is depressed.[15,16] This reprioritization of protein synthesis in favor of APP may permit amino acids normally used in the synthesis of carrier proteins to be shunted into the synthesis of proteins needed in increased amounts during periods of stress. The failure of APP synthesis results in increased morbidity and mortality.

The control of APP synthesis has recently been comprehensively reviewed by Baumann.[14] It was originally recognized that crude leukocyte supernatants were capable of inducing APP synthesis when applied to cultured HC.[12,17,79] Among the leukocytes, macrophage supernatants were the most active in stimulating HC APP synthesis. Supernatants from plated peripheral blood mononuclear cells and KC were noted to contain HC stimulating factor (HSF) activity.[69] HSF was recognized as a family of glycoproteins differing in their degree of glycosylation, capable of inducing APP synthesis in HC.[28] Similarly, when APP are secreted, they also vary in their degree of glycosylation.[13] HSF was initially thought to be IL-1. It soon became apparent, however, that IL-1 applied to isolated HC would induce only a small fraction of the APP response. HSF was subsequently identified as a class of interferon (IFN), being identical with several known factors, including B2 IFN and plasmacytoma growth factor.[34,36] Mediators distinct from IL-6 have also been demonstrated to possess IL-6-like activity.[32] IFN-α, IFN-γ, or INF-β, possessed no HSF activity. The name IL-6 was proposed for this mediator. IL-6 is the major inducer of the acute-phase response in human and rat liver cells.[7-11] The role of IL-6 in controlling APP synthesis has been extensively reviewed in a recent symposium.[75] HSF was noted to be capable of inducing the synthesis of many APP such as fibrinogen and alpha-2-macroglobulin, and depressing the synthesis applied to HC.[36] Recombinant human IL-6 applied to rat HC is capable of inducing nearly the entire spectrum of APP synthesis. Many cytokines, including IL-6, are postulated to play roles in human disease.[20]

Cytokines other than IL-6 have been implicated in control of APP synthesis, principally IL-1 and tumor necrosis factor (TNF). Injection of IL-1 alone into an animal results in changes only for the negative acute-phase reactant mRNA.[2] It was initially believed that IL-1 was responsible for controlling APP synthesis. *In vitro* experiments revealed that IL-1 or TNF were capable of inducing only a small fraction of APP synthesis in isolated HC culture in humans and rats. Treatment of rat HC monolayers with TNF and IL-1 fails to augment the synthesis of positive APP when applied alone for most proteins. IL-1 synergizes with IL-6 to augment alpha-1-acid glycoprotein synthesis. IL-1 was capable of enhancing IL-6 production by fibroblasts.[6] Several authors have described a depression of albumin synthesis through the addition of TNF or IL-1. Our studies have failed to consistently demonstrate this depression of albumin synthesis using IL-1 and TNF, but have consistently demonstrated an IL-6-dependent depression in albumin synthesis. It appears that IL-6 is primarily responsible for augmentation or depression of APP synthesis. IL-1 has also been described by some authors as inhibitory to the effects of IL-6 on inducing alpha-2-macroglobulin or fibrinogen synthesis. We have noted that a combination of cytokines, including IL-1, TNF, and IFN-γ antagonize the effects of IL-6 on HC fibrinogen production. IFN-γ itself has been demonstrated to induce the synthesis of some complement components under certain circumstances.[12] Complement production in some tumor cell lines is increased by IL-1.[38]

IL-6 has been shown to act at a transcriptional and translational level. The synthesis of mRNA for some APP (fibrinogen, alpha-2-macroglobulin) is dramatically increased after IL-6 treatment and results in augmented APP synthesis.[66] Increased gene transcription, increased message RNA half-life, or increased rate of translation of genes for APP will increase their synthesis. There is some evidence that other cytokines may have an enhancing effect on the stabilization of mRNA with resultant increases in APP synthesis. Changes in intracellular mRNA concentrations may be brought about by a change in the rate of gene transcription or the rate of degradation of the mRNA. Negative APP, such as albumin and its mRNA, exhibited a large decrease in transcriptional rates in the first 24 h after acute inflammation. This was followed by a fall in the mRNA levels. A change in mRNA stability is important for the positive APP. Alpha-2-macroglobulin mRNA rises strongly in response to tissue injury, with only small changes in transcriptional activity. Prolongation of the half-life may increase protein synthesis.[2]

Hormonal influences have been demonstrated to have profound effects on APP synthesis in some systems, both *in vitro* and *in vivo*. Glucocorticoids have been demonstrated to be essential for rat HC response to IL-6 *in vivo* and *in vitro*. Maximal APP synthesis occurs only with simultaneous exposure to IL-6 and glucocorticoids. Dexamethasone has classically been the steroid hormone most commonly used in these studies.[25,26] *In vivo*, simultaneous administration of IL-6 and dexamethasone is necessary to induce a maximal acute-phase response. Other hormones, including estrogen, catecholamines, and thyroid hormone, have minimal effects on APP synthesis.

III. KC:HC INTERACTIONS IN THE CONTROL OF APP SYNTHESIS

We have demonstrated that KC can provide the signal for APP synthesis *in vivo* and *in vitro*.[73] KC supernatants in the presence of glucocorticoids have previously been shown to be capable of inducing alpha-2-macroglobulin synthesis.[48,49,64,67] We have previously demonstrated that LPS-treated KC induce a significant depression in total HC protein synthesis,[50-58] thereby demonstrating that KC could lead to alterations in HC protein metabolism. This inhibition in protein synthesis can be reversed by the addition of dexamethasone.[24] We have developed and characterized a KC:HC coculture model to examine the possible interactions controlling APP synthesis. The close association of KC and HC in coculture approximates the interactions that normally exist in the liver parenchyma. We have examined several facets of this response, including the numbers of KC needed to alter HC APP synthesis (KC:HC ratios), the response of the coculture model to LPS and other cytokines, and the effect of glucocorticoids on this response. KC have previously been demonstrated to produce factors that can augment HC fibrinogen synthesis. Fibrinogen-degradation products are capable of activating KC to induce HC fibrinogen synthesis.[21]

The synthesis of APP by HC has been characterized in our KC:HC model. HC which are initially cultured alone are cultured in dexamethasone. Failure to initially expose HC to dexamethasone dramatically depressed basal and IL-6-stimulated fibrinogen and albumin synthesis as well as the HC response to exogenous IL-6. After 24 h, KC were added to HC at a KC:HC ratio of 0 (no KC), 0.1, 0.5, 1.0, 2.0, and 4.0 in the absence of added dexamethasone. We postulated that dexamethasone would inhibit KC responses to LPS and would inhibit any interactions with HC. The coculture system was then treated with *Escherichia coli* endotoxin (type 0111:B4) 100 ng/ml for 24 h and the supernatant harvested and assayed for albumin, fibrinogen, alpha-1-acid glycoprotein, and IL-6 (Figure 1A). In response to LPS, fibrinogen synthesis was augmented and albumin synthesis depressed as the KC:HC ratio was increased. Supernatants were assayed for IL-6 using proliferation of the B9 hybridoma cell line. IL-6 levels in the culture supernatants increased as the KC:HC ratio increased and in response to LPS (Figure 2). Interestingly, the IL-6 levels in cocultures not treated with LPS were only slightly augmented as KC numbers increased, yet increases in fibrinogen synthesis were still observed, suggesting that a non-IL-6 factor or nonimmunoreactive IL-6 was responsible for the augmented fibrinogen synthesis. The KC:HC ratio experiments as described above were repeated in the presence of 10^{-7} *M* dexamethasone. In these experiments, KC were plated in 10^{-7} *M* dexamethasone. We had anticipated that the addition of dexamethasone to the coculture would inhibit HC IL-6 secretion and fail to augment APP production. To the contrary, coculture in the presence of dexamethasone with or without LPS resulted in even greater increases in APP synthesis than if dexamethasone was excluded from the culture after the addition of KC. This suggests that KC were not inhibited significantly or that HC responsiveness was significantly increased (Figure 1B). It has been demonstrated that adherence of monocytes alone

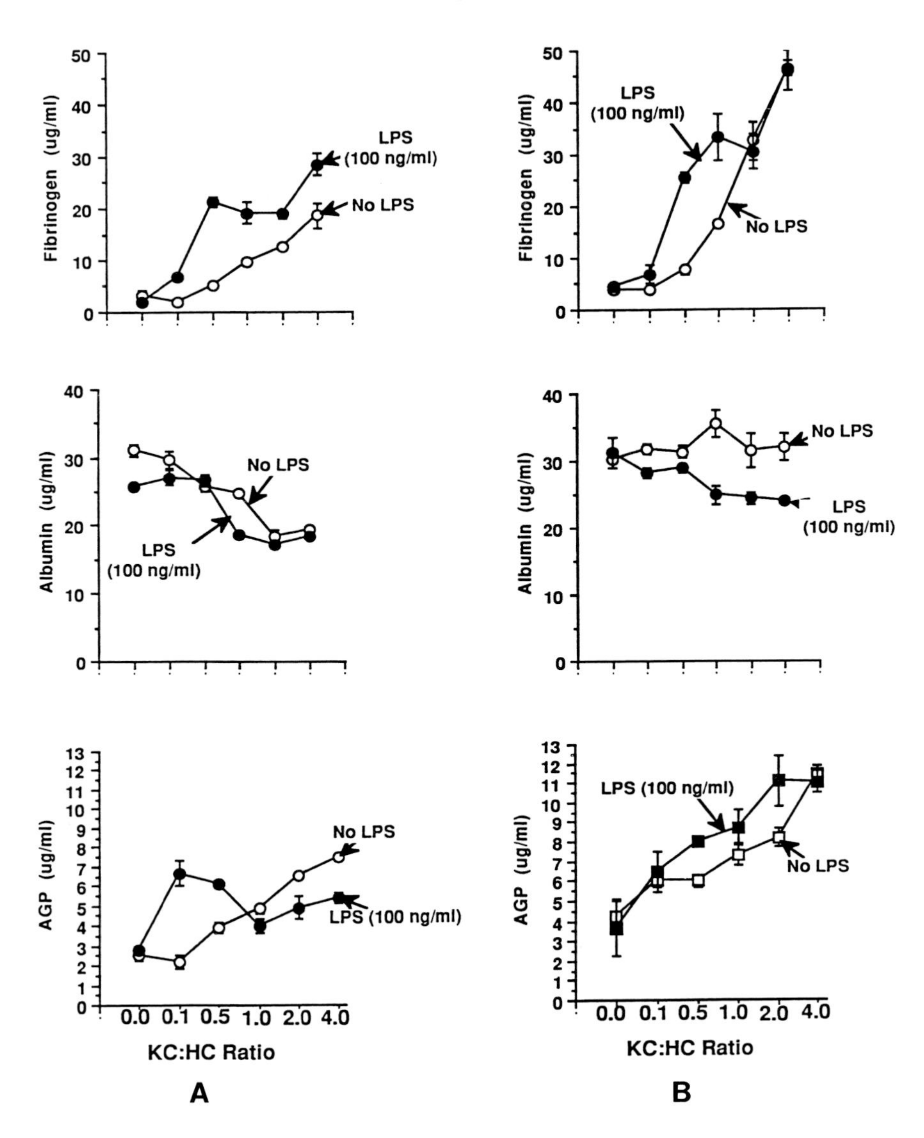

FIGURE 1. Effect of KC:HC ratios on fibrinogen, albumin, and alpha-1-acid glycoprotein (AGP) synthesis. HC were initially plated in dexamethasone. The HC cultures were washed and increasing numbers of KC were added to the culture without added dexamethasone. The supernatants were harvested 24 h later and assayed for fibrinogen, albumin, and AGP (Figure 1A). The addition of increasing numbers of KC increased fibrinogen and AGP synthesis and depressed albumin synthesis. LPS addition (100 ng/ml) increased fibrinogen and AGP at low KC:HC ratios. The experiment was repeated with KC being added in the presence of 10^{-7} *M* dexamethasone (Figure 1B). Dexamethasone augmented the KC-induced synthesis of fibrinogen and AGP with or without dexamethasone.

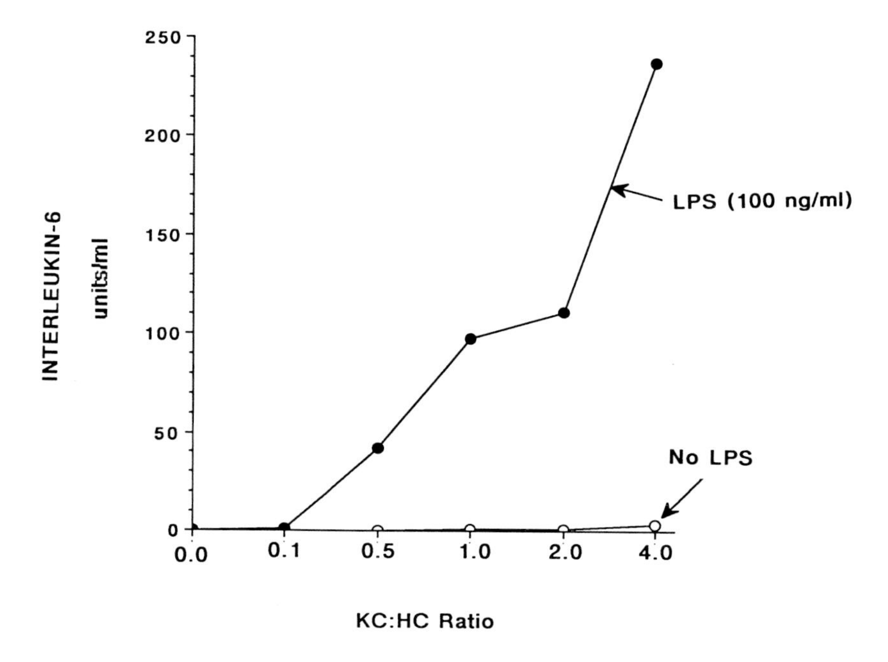

FIGURE 2. IL-6 levels in culture supernatants at varying KC:HC ratios. As the KC:HC ratio increased, basal IL-6 production increased slightly, but LPS-stimulated IL-6 production increased greatly.

increases IL-6 production and mRNA expression for IL-6. IL-6 may have been associated and failed to be released into the culture supernatants, accounting for the low supernatant IL-6 levels at a time when APP synthesis was still augmented. KC-associated IL-6 could theoretically be presented to the HC without release into the supernatant, possibly accounting for the disparity between IL-6 levels and APP synthesis.

The endotoxin sensitivity of this coculture system has been examined. KC were cocultured with HC at a ratio of 0.5:1 and exposed to increasing concentrations of *E. coli* endotoxin in the presence or absence of dexamethasone (Figure 3). The increased fibrinogen production was greater in cocultures where dexamethasone was present throughout, again suggesting that KC responsiveness to LPS was maintained in the presence of dexamethasone. As LPS levels were increased above 10 ng/ml, fibrinogen synthesis was increased and albumin synthesis depressed. The response was maximal at 1000 ng/ml. It was occasionally noted that at LPS concentrations of 1000 ng/ml, less fibrinogen was produced in coculture in spite of increased IL-6 synthesis, suggesting that the maximal HC response to IL-6 was inhibited. We have also examined the effect of TNF and IL-1 added to the coculture model. IL-1 or TNF alone failed to induce APP synthesis compared to KC:HC coculture

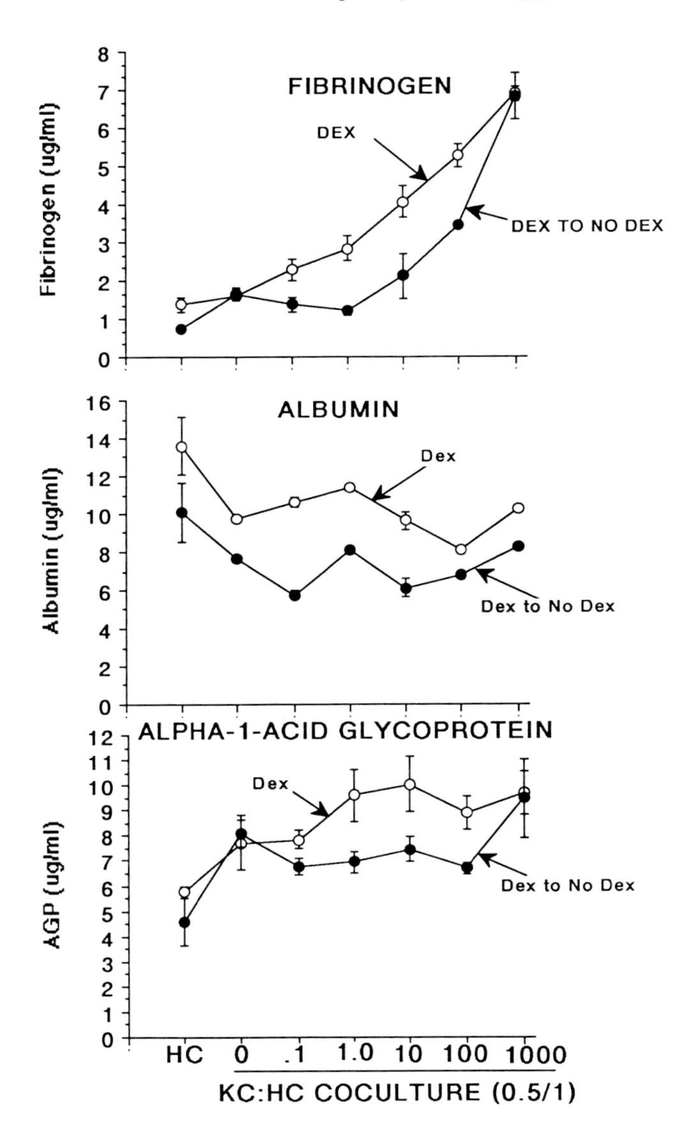

FIGURE 3. Effect of dexamethasone on APP synthesis in coculture. HC were initially plated in dexamethasone. KC were added at a KC:HC ratio of 0.5:1 in the presence or absence of 10^{-7} *M* dexamethasone, and LPS was added into the cocultures. Albumin, fibrinogen, and AGP were measured after 24 h of culture. The addition of KC with dexamethasone increased the production of AGP, fibrinogen, and albumin compared to KC added without dexamethasone. Increasing LPS levels augmented fibrinogen production in a dose-dependent fashion, but had lesser effects on albumin and AGP.

without added cytokines or LPS, despite the fact that both are capable of inducing IL-6 synthesis when injected into animals.

The induction of HC APP synthesis is a coordinated response involving cytokine (mainly IL-6) production and augmented adrenal glucocorticoid secretion. This demonstrates an example of the endocrine-immune axis. Classically, it has been accepted that glucocorticoids inhibit many macrophage functions, including cytokine production. However, under conditions in which APP occurs, such as injury and septic states, glucocorticoid levels are markedly elevated. We were interested in studying the interaction between dexamethasone, KC, and HC in culture. We would expect that under these circumstances, the acute-phase response might be inhibited because of inhibitory effects on macrophage IL-6 production by elevated glucocorticoid levels. We investigated this question by evaluating the effect that dexamethasone, a prototype glucocorticoid, had on APP response in coculture. HC were initially cultured without dexamethasone. After HC adherence to the plates, the media was changed and replaced with varying concentrations of dexamethasone (0 to 10^{-6} *M*). KC were then added to the coculture (KC:HC 0.5:1) and stimulated with LPS. We found that the basal production of fibrinogen, albumin, and alpha-1-acid glycoprotein by HC without added KC was markedly enhanced as the concentration of dexamethasone increased from 0 to 10^{-6} *M* (Figure 4). Similarly, we found that KC-stimulated fibrinogen and albumin synthesis in coculture without LPS was also increased at increasing levels of dexamethasone. When LPS was added to the cocultures, a similar finding was seen. Albumin production was depressed by KC with or without LPS at all dexamethasone concentrations, but LPS failed to further depress albumin production at higher dexamethasone levels (Figure 5). At the same time, IL-6 levels in the culture media were depressed as dexamethasone levels increased (Figure 6). To investigate this phenomenon further, we evaluated the effect of rhIL-6 on the induction of fibrinogen and albumin production after the cells had been cultured in varying concentrations of dexamethasone from the time of their initial plating (Figure 7). It can be seen that as the dexamethasone concentration in which the cells were cultured decreased, the IL-6-induced fibrinogen production also decreased. Basal albumin secretion was greatly depressed and IL-6 continued to inhibit albumin synthesis. As other authors have shown, it appears that dexamethasone upregulates HC responsiveness to IL-6.

This phenomenon similarly occurs in the coculture model, where lower peak levels of IL-6 are associated with the highest levels of fibrinogen synthesis by HC. The interaction of the endocrine and adrenal axis can easily be appreciated. The role of other cytokines and hormones in control of the APP synthesis remains to be investigated, but several of these have been studied by others. IL-4 has been shown to depress LPS-induced macrophage IL-6 production and therefore possibly influences APP synthesis.[76] Transforming

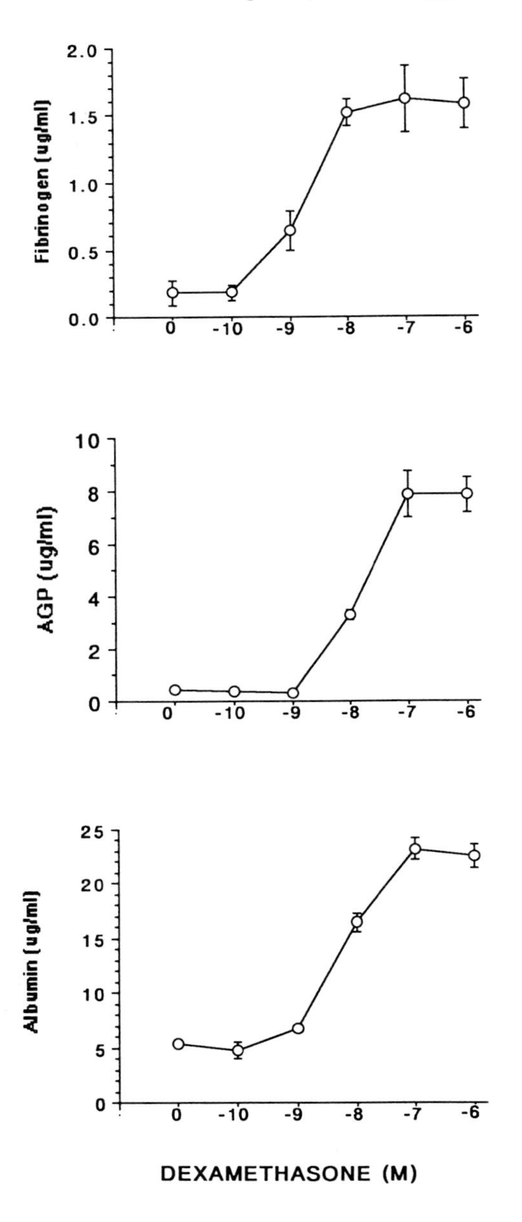

FIGURE 4. Dexamethasone alters basal synthesis of APP. HC were cultured from the time of plating in varying concentrations of dexamethasone; albumin, fibrinogen, and AGP were measured 24 h later. Increasing levels of dexamethasone augmented basal synthesis of all three proteins, and reached maximum at 10^{-7} *M* dexamethasone.

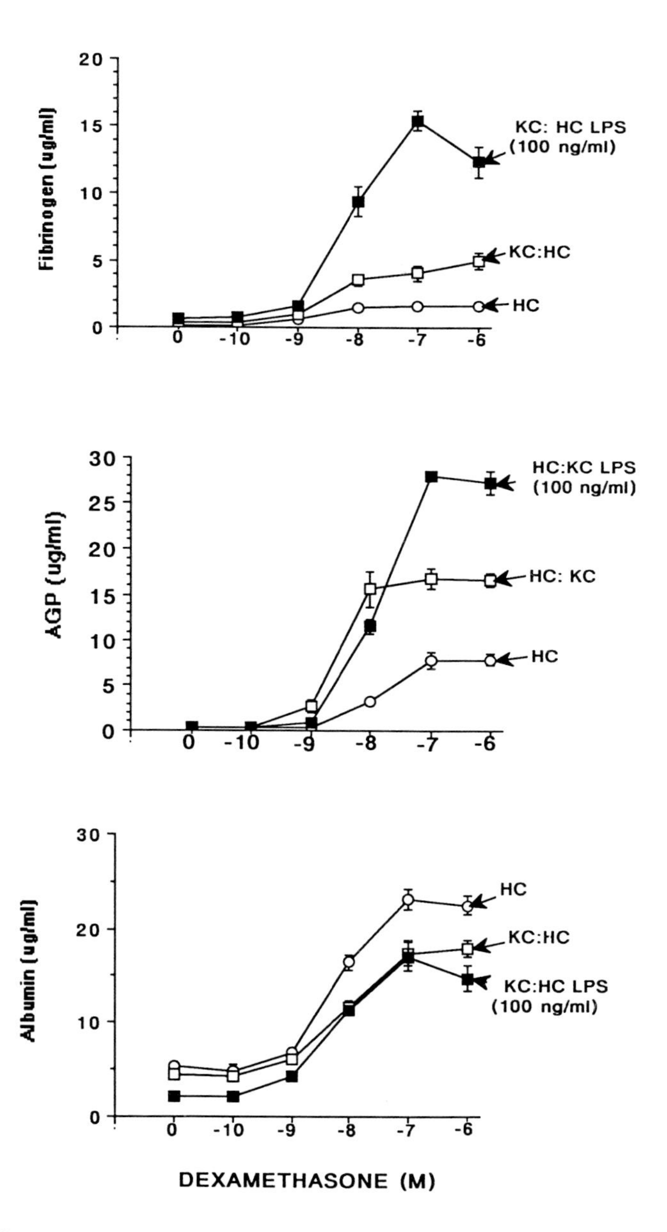

FIGURE 5. Interactions between KC, HC, and dexamethasone in the control of APP synthesis. HC were plated in varying concentrations of dexamethasone and KC were added at the same level of dexamethasone at a KC:HC ratio of 0.5:1. The cultured cells were treated with or without LPS 100 ng/ml for 24 h, and fibrinogen, AGP, and albumin production was determined. As the dexamethasone increased, the synthesis of fibrinogen and AGP increased as KC were added, and increased further with the addition of LPS. Albumin synthesis decreased with the addition of KC compared to HC alone and was further depressed by LPS at low levels of dexamethasone.

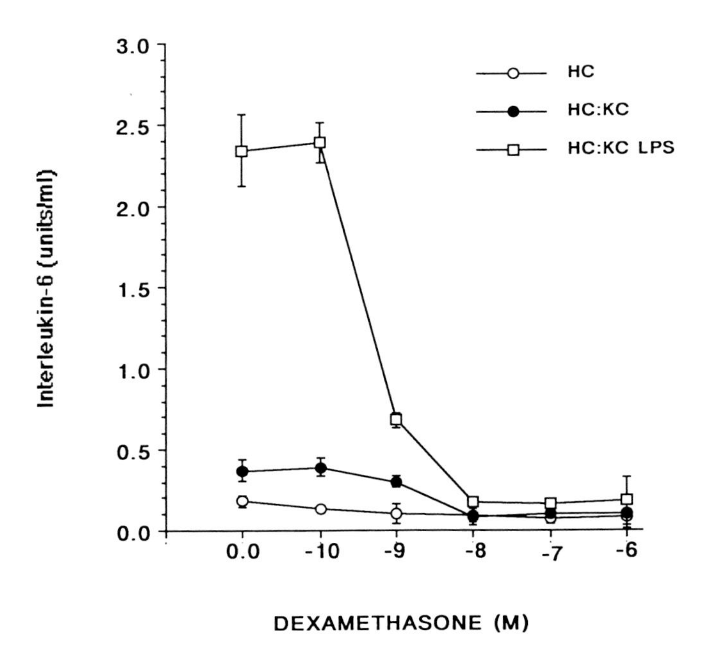

FIGURE 6. IL-6 levels in the cocultures described in Figure 5. Virtually no IL-6 was detectable in the cultures without KC. KC alone produced low levels of IL-6. When LPS was added to KC, large increases in culture IL-6 levels were seen at dexamethasone levels of 0 and 10^{-10} *M*. At dexamethasone levels of less than 10^{-9} *M*, IL-6 levels dropped to those seen with KC alone, demonstrating an inhibition of IL-6 synthesis as the dexamethasone increased.

growth factor-beta has been demonstrated to inhibit CRP production in hepatoma cell lines.[77] Glucagon rapidly enhances transport of amino acids into HC and may thereby accelerate HC responsiveness to IL-6.[78]

The interaction between KC and HC in control of APP synthesis may be bidirectional. We have demonstrated that coculture of KC and HC results in augmented APP synthesis. Recent data have suggested that a circulating protein (LPS binding protein [LBP]) binds to LPS in the plasma to form an LPS-LBP complex. This LPS-LBP complex binds to macrophage CD-14 receptors, resulting in increased macrophage TNF production compared to exposure to LPS without LBP. LBP appears to amplify the response of macrophage cytokine production to LPS. There is preliminary evidence that LBP is an acute-phase reactant produced by HC. LBP released from HC may bind gut-derived LPS and amplify KC responses to small levels of circulating LPS.

IV. FACTORS AFFECTING KC IL-6 PRODUCTION

The stimuli responsible for activating macrophages to increase production of IL-6 are important because of the preeminent role of IL-6 in controlling

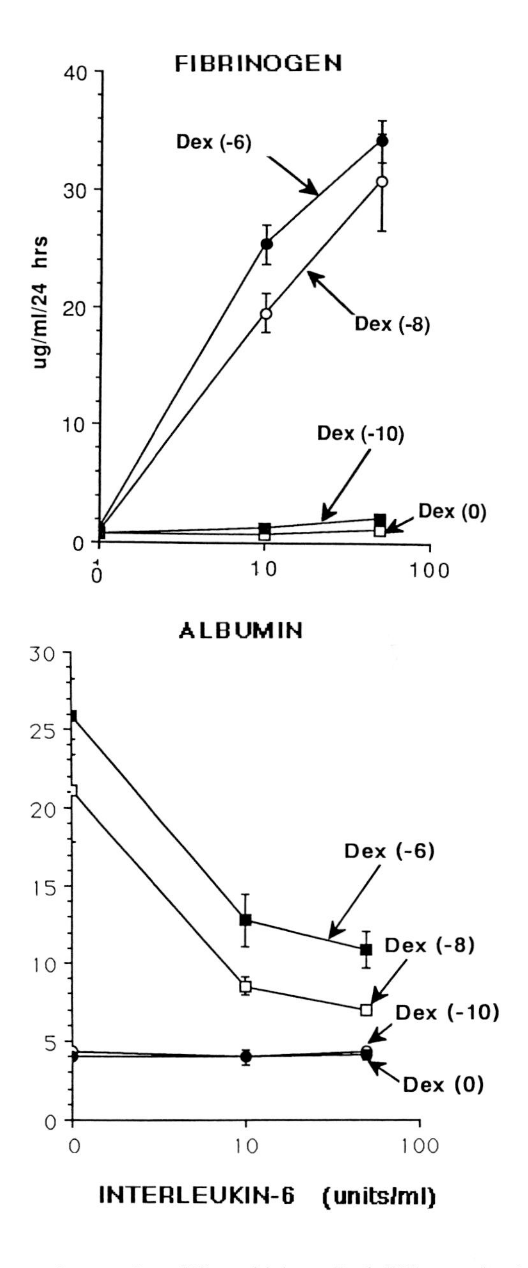

FIGURE 7. Dexamethasone alters HC sensitivity to IL-6. HC were plated at dexamethasone levels from 0 to 10^{-6} *M* and exposed to increasing levels of IL-6. Fibrinogen synthesis was greatly increased at dexamethasone levels of 10^{-6} and 10^{-8} *M*. At lower levels, HC fibrinogen synthesis was not responsive to added IL-6. Basal albumin synthesis was maximal at dexamethasone levels of 10^{-6} and 10^{-8} *M*, and was inhibited by the addition of IL-6. Albumin synthesis was low and unaffected by IL-6 at 0 and 10^{-10} *M* dexamethasone. This clearly demonstrates that dexamethasone upregulated HC responsiveness to IL-6.

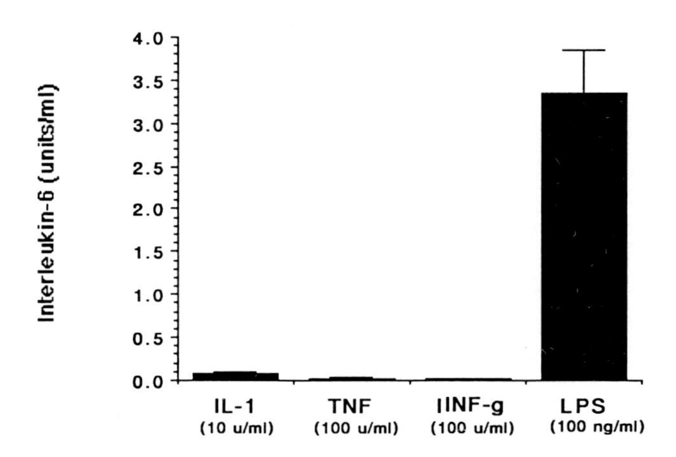

FIGURE 8. Cytokines fail to augment cell IL-6 synthesis. Elutriation-purified KC were plated for 24 h and exposed to recombinant human IL-1 (10 U/ml), recombinant murine TNF (100 U/ml), rat IFN-γ (100 U/ml), or *E. coli* endotoxin (100 ng/ml) for 24 h. Only endotoxin augmented KC IL-6 production.

APP synthesis. IL-6 is a ubiquitous cytokine produced by macrophages,[29] endothelial cells,[33,59,60] T cells,[31] fibroblasts,[43] and keratinocytes[68] in response to various stimuli. IL-6 synthesis by fibroblasts and endothelial cells can be preferentially augmented by IL-1 and TNF, while macrophage populations are activated by endotoxin and to a lesser degree by IL-1.[35] We have investigated the stimuli that act on the KC to alter IL-6 synthesis. It appears that IL-6 and IL-1 are regulated in different ways. Monocytes cultured for 24 h with LPS lose their ability to produce IL-1, while synthesis of IL-6 remains unaffected.[39] We have similarly demonstrated that KC IL-6 production is nearly linear to 36 h after LPS exposure. It also appears that T lymphocyte products may indirectly affect APP synthesis through modulation of macrophage responses. IL-2-treated mononuclear cell supernatants are capable of augmenting C3 synthesis and depressing albumin synthesis.[40] IFN-γ, another well-known lymphocyte product, is able to upregulate cytokine synthesis, including IL-6.

We have examined the role that various cytokines and hormones have on KC IL-6 synthesis. The role of TNF, IL-1, IFN-γ, and LPS on KC IL-6 production (Figure 8) was examined first. IFN-γ has been shown to augment cytokine and prostanoid production by human monocytes.[41] Elutriation-purified KC alone failed to increase IL-6 synthesis following high doses of TNF, IL-1, and IFN-γ. In our system, only endotoxin was capable of augmenting IL-6 synthesis (Figure 8). IL-1, murine TNF, and IFN-γ added without endotoxin failed to enhance IL-6 production. IL-6 synthesis was initially increased at endotoxin levels between 1 to 10 ng/ml. Below 1 ng/ml, no augmentation in IL-6 production cold be noted. Synthesis increased most rapidly between

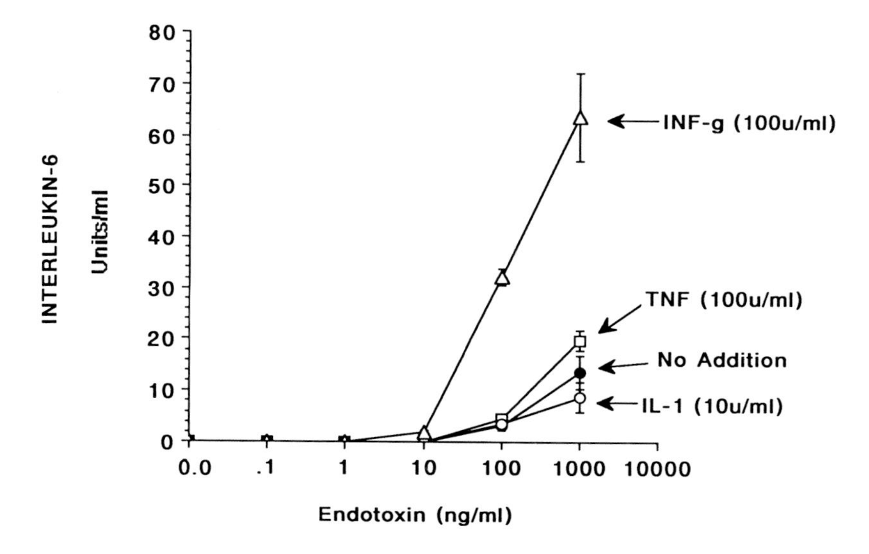

FIGURE 9. Effect of preincubation of KC with recombinant cytokines on Il-6 synthesis. KC were plated and cultured for 24 h. TNF, IL-1, and IFN-γ were added for 8 h, then the cultures were washed and increasing concentrations of endotoxin added. Preincubation with TNF and IL-1 failed to increase IL-6 production to levels above those seen with LPS addition alone. IFN-γ preincubation greatly increased IL-6 production in response to endotoxin.

10 to 100 ng/ml and continued to increase slowly to 1000 ng/ml. Higher LPS concentrations failed to further augment IL-6 production.

We next examined the role that preincubation of cytokines with KC might have on IL-6 production in response to added LPS (Figure 9). KC were cultured and preincubated for 8 h in the presence of TNF, IL-1, IFN-γ, or no cytokines. Preincubation with TNF or IL-1 did not augment IL-6 production in response to LPS. However, pretreatment with rat IFN-γ resulted in large increases in IL-6 production compared with LPS alone or LPS in combination with TNF or IL-1. The effect of IFN-γ was first seen at LPS levels of 10 ng/ml. Pretreatment of KC with varying concentrations of IFN-γ resulted in a dose-dependent increase in IL-6 production as IFN-γ concentration increased from 0.1 to 100 U/ml (Figure 10). *In vivo,* infusion of TNF led to increases in IL-6 levels that were further augmented by simultaneous IFN-γ infusion. IFN-γ alone failed to augment IL-6 levels in patients.[42] IL-1 and TNF have been demonstrated to induce IL-6 production by fibroblasts.[30]

Because KC and macrophages are exposed to systemic hormones during times of stress and infection, and glucocorticoid levels are markedly elevated during these periods, the effects of dexamethasone on KC cultured with or without LPS were also examined. We noted that IL-6 production was markedly inhibited as the dexamethasone concentration increased from 10^{-10} to 10^{-8} *M* (Figure 11). This finding of dexamethasone inhibition of IL-6 production confirmed our previous findings with the coculture model.

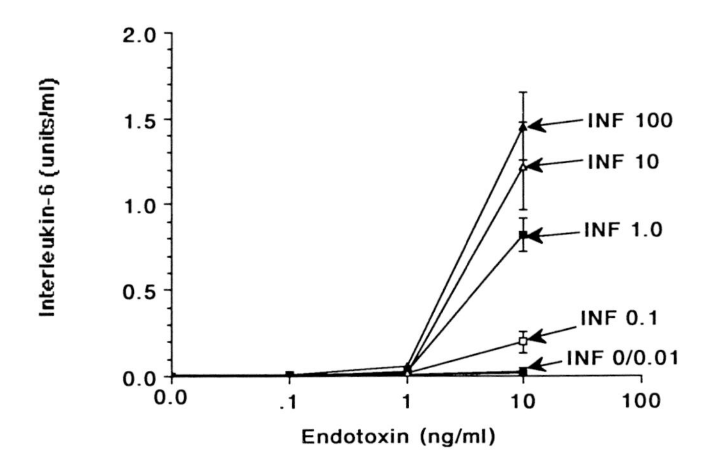

FIGURE 10. Dose-dependent effects of IFN-γ on KC IL-6 productioN. As described in Figure 9, plated KC were preincubated for 8 h in concentrations of IFN-γ from 0 to 100 U/ml. As the preincubation concentration of IFN-γ was increased, IL-6 production increased, demonstrating an increased KC sensitivity to endotoxin.

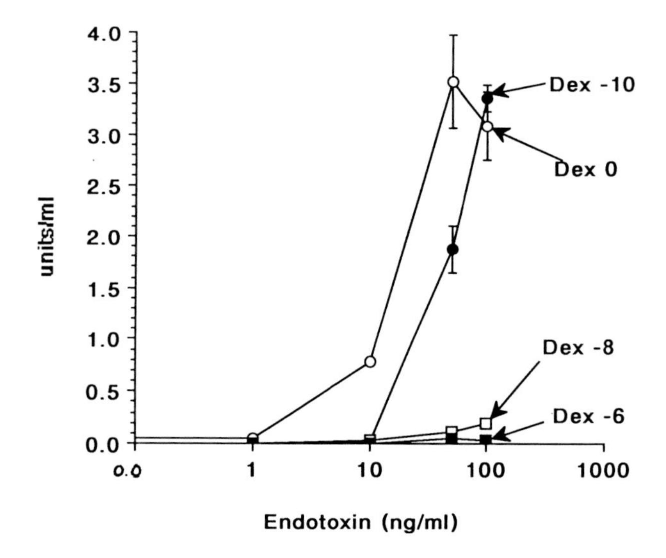

FIGURE 11. Effects of dexamethasone on KC IL-6 production. KC were plated in varying concentrations of dexamethasone and stimulated with endotoxin. As the dexamethasone increased, the sensitivity of the KC to endotoxin decreased and IL-6 production decreased.

V. *IN VIVO* STUDIES OF IL-6 AND APP SYNTHESIS

The origin of the signal for the induction of APP synthesis is not clear. It is apparent, however, that experimental inflammation leads to large changes in APP synthesis.[65,71] As mentioned, multiple cell types, including macrophages, endothelial cells, and fibroblasts are able to synthesize IL-6. The source of IL-6 synthesis for the induction of HC APP synthesis may be from several sources, including the wound.[74] It has been demonstrated that fluid obtained from subcutaneously implanted polyvinyl chloride sponges in the absence of bacteria or endotoxin results in the accumulation of large amounts of IL-6 in the wound. The wound fluid is capable of augmenting APP synthesis when applied to HC cultures. It has been shown that infusion of endotoxin results in rapid increases in serum IL-6 levels.[44] Large increases in serum IL-6 levels are seen in patients with septic shock;[45] however, the role of IL-6 as a proximal mediator of septic shock is being investigated. Antibodies to IL-6 in mice have been shown to be protective from lethal *E. coli* infusions.[72] In addition, infusion of TNF induces IL-6 production and APP synthesis *in vivo*.[27,46] When endotoxin is administered to experimental animals, peak serum TNF levels always precede the appearance of IL-6. Blockage of TNF with anti-TNF antibodies decreases the IL-6 response, demonstrating that IL-6 secretion is at least partially dependent on TNF. IL-2 infusion increases IL-6 levels, but significantly less than with TNF in humans[46] and in mice.[47] The inflammatory cells in the wound, systemic macrophages, fibroblasts, and endothelial cells are all capable of supporting IL-6 synthesis. Possible sources of IL-6 for induction of APP synthesis by the liver after stress or infection include the cells in the wound releasing IL-6 that then circulate to the liver, the KC being activated and releasing IL-6 that stimulate the HC directly, or the wound releasing a factor that stimulates the KC which then stimulate the liver cell. It has been shown that up to 80% of injected IL-6 is rapidly reversibly bound to HC. Therefore, the liver could continuously extract low levels of IL-6 from the plasma to augment APP synthesis.

We have tried to clarify the role that the KC plays in the control of APP synthesis by examining the role of septic and nonseptic inflammatory stimuli in inducing IL-6 production by liver nonparenchymal cells, which are approximately 40% KC.[52] We postulated that the measurement of fibrinogen synthesis by rat HC would be indicative of the IL-6 effect on HC APP synthesis and that IL-6 spontaneously secreted by cultured nonparenchymal cells isolated from these animals would be indicative of *in vivo* IL-6 production. Sprague-Dawley rats were injected with LPS intraperitoneally or with turpentine subcutaneously. Turpentine was selected as a sterile inflammatory stimulus and LPS was chosen to stimulate a systemic infection and Gram-negative bacteremia. HC and nonparenchymal cells were harvested from these animals and plated separately. They were allowed to spontaneously secrete

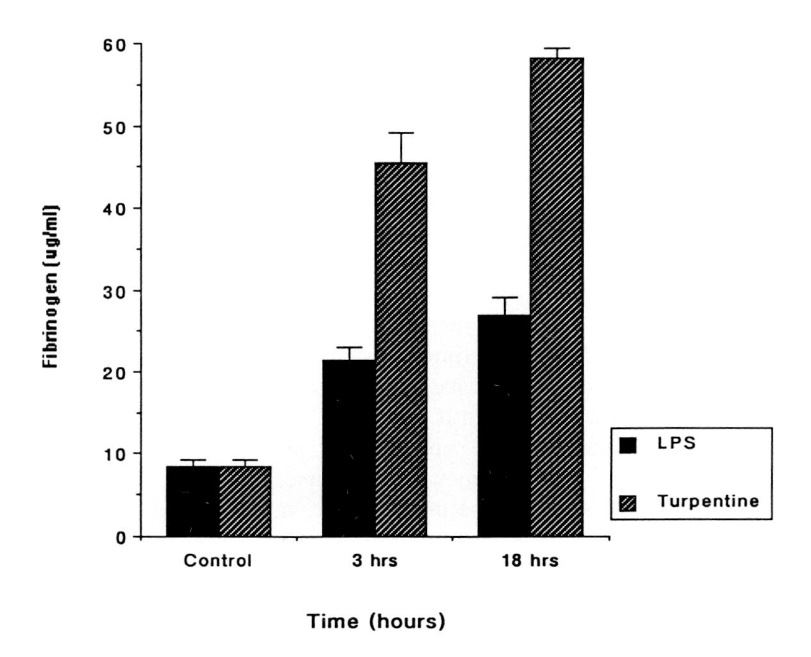

FIGURE 12. Effects of endotoxin and turpentine treatment of rats on HC fibrinogen production. Rats were treated with either *E. coli* endotoxin (LPS) intraperitoneally or turpentine subcutaneously. At 3 and 18 h after treatment, the livers were removed and the HC isolated and allowed to secrete in cultures for an additional 24 h. Control rats were untreated. Both turpentine and LPS treatment greatly increased fibrinogen production. Turpentine was a more effective stimulant for HC fibrinogen production.

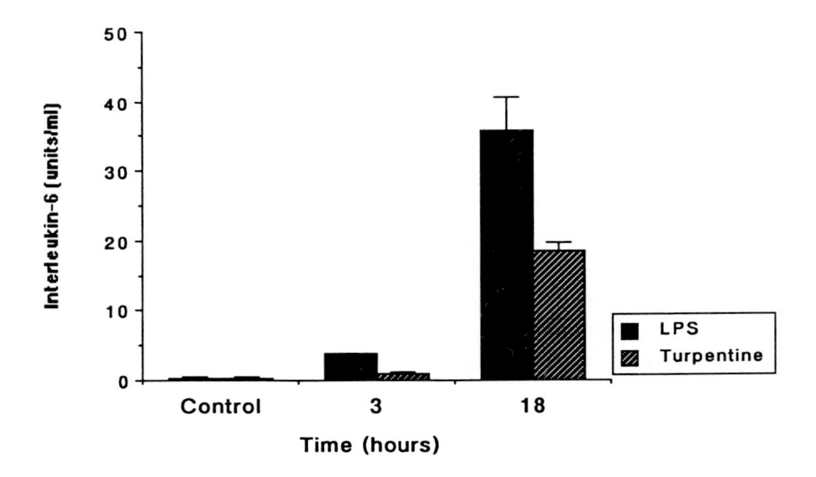

FIGURE 13. Effects of LPS and turpentine treatment of rats on hepatic nonparenchymal cell IL-6 synthesis. Nonparenchymal cells from LPS-treated animals produced more IL-6 than nonparenchymal cells from turpentine-treated animals at 3 and 18 h after treatment.

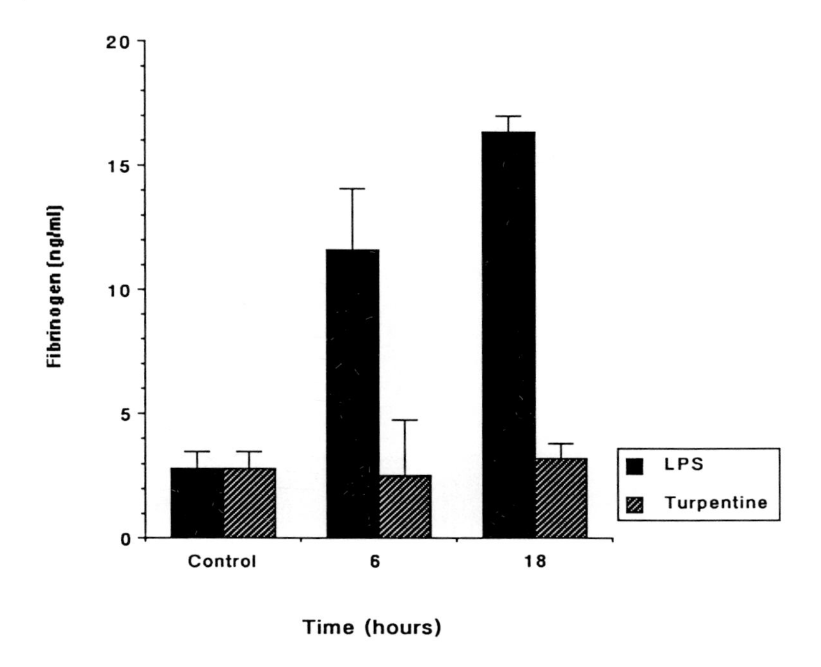

FIGURE 14. Effects of nonparenchymal cells supernatants on normal HC fibrinogen synthesis. Plated nonparenchymal cells from turpentine- and LPS-treated rats were isolated, plated, and allowed to secrete without stimulation for 24 h. The supernatants were harvested and plated at a 50% concentration on normal HC. After an additional 24 h the supernatants were harvested and assayed for fibrinogen. Nonparenchymal cells supernatants from LPS-treated animals increased normal HC fibrinogen synthesis, while supernatants from turpentine-treated animals did not.

fibrinogen and IL-6 for 24 h. HC from both turpentine- and LPS-treated rats secreted significantly increased amounts of fibrinogen compared to saline-treated controls (Figure 12). Much more fibrinogen was secreted by HC isolated from turpentine-treated rats than LPS-treated rats. Nonparenchymal cells from the livers of turpentine-treated animals did not produce more IL-6 than control animals at 3 h after treatment, but small increases were seen at 18 h. Nonparenchymal cells from LPS-treated animals produced much more IL-6 than controls or turpentine-treated animals at 3 and 18 h (Figure 13). These findings were confirmed by placing supernatants from nonparenchymal cells of LPS- or turpentine-treated animals on HC isolated from normal animals. Supernatants from nonparenchymal cells of LPS-treated animals, but not turpentine-treated animals increased fibrinogen synthesis by normal HC (Figure 14). This confirmed the low IL-6 production by nonparenchymal cells from the turpentine-treated animals and the high IL-6 production by the LPS-treated animals. In addition, we measured the serum levels of IL-6 at various times after LPS and turpentine administration (Figure 15). LPS administration resulted in rapid and large increases in IL-6 at 3 h that rapidly

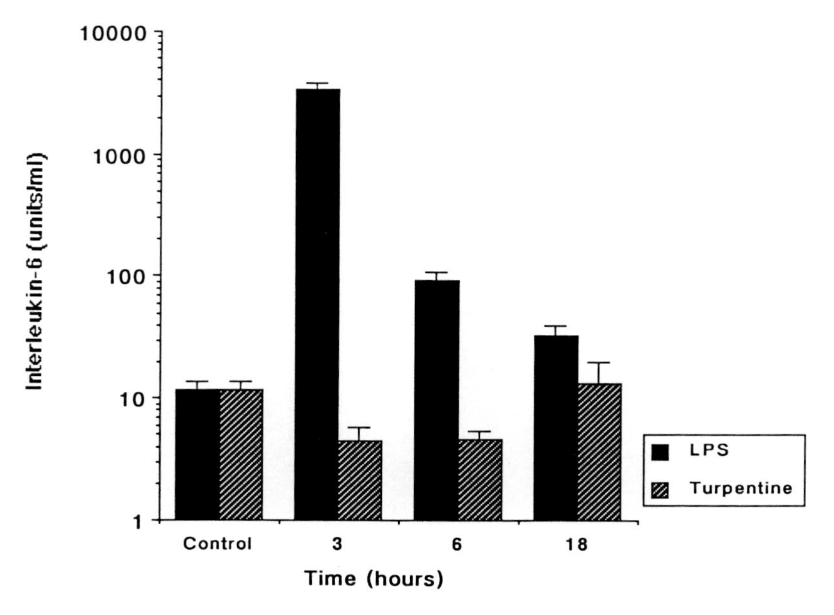

FIGURE 15. Serum IL-6 levels in LPS- and turpentine-treated animals. LPS administration resulted in large increases in serum IL-6 that peaked at 3 h and returned toward normal by 18 h. Subcutaneous turpentine failed to significantly increase serum IL-6 level.

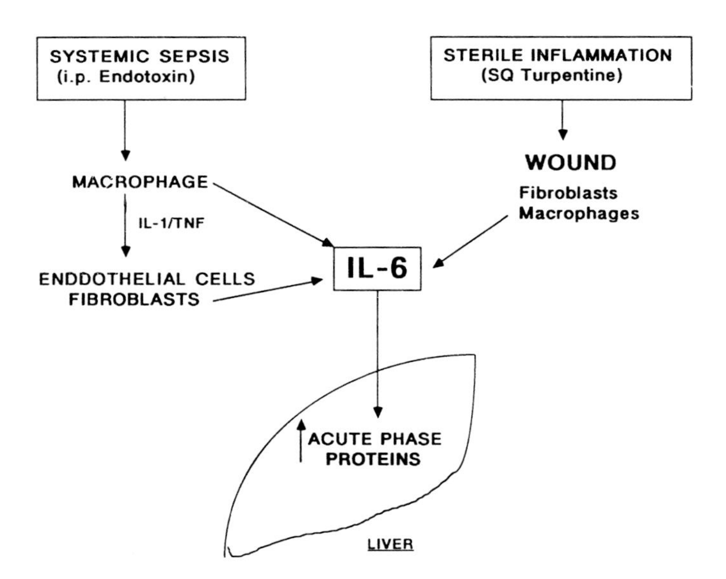

FIGURE 16. Hypothesis for the control of HC APP synthesis. Local inflammation may increase IL-6 synthesis and release by cells in the wound. HC would bind circulating IL-6 to initiate APP synthesis. Systemic sepsis would result in diffuse macrophage activation with IL-6 release. In addition, macrophages may synthesize IL-1 and TNF, which could activate fibroblasts and endothelial cells to produce IL-6, which would lead to increased APP synthesis.

declined. Turpentine injection in the hind limb failed to significantly alter serum IL-6 levels.

These data would suggest that KC can act as a source of IL-6 to stimulate HC APP synthesis in response to endotoxemia. However, a different source of IL-6, most likely the wounded tissue, is likely to be responsible for augmented HC APP synthesis in noninfectious inflammatory stimuli. Figure 16 demonstrates a proposed hypothesis to explain how the control of APP synthesis might differ in systemic septic (intraperitoneal LPS) and nonseptic inflammation (turpentine). Endotoxemia probably results in systemic macrophage activation with rapid and large increases in serum IL-6 leading to the initiation of APP response. Late activation of KC by septic and nonseptic stimuli also contributes to APP synthesis. In addition, other cytokines such as TNF and IL-1, which are released by macrophages, may subsequently activate endothelial cells and fibroblasts to release IL-6. Turpentine likely results in large increases in wound levels of IL-6, which then "spill over" into the systemic circulation at low levels to be extracted by the liver cell and induce the APP response. The failure to detect elevated circulating IL-6 levels in turpentine-treated animals is vexing. We suspect that the levels are slightly elevated above normal, but that we are unable to detect this in our assay. The high affinity of HC for Il-6 may allow the HC to bind very low levels of circulating IL-6 and initiate the acute-phase response.

REFERENCES

1. **Heinrich, P. C., Castell, J. V., and Andus, T.,** Interleukin-6 and the acute phase response, *Biochem. J.*, 265, 621, 1990.
2. **Schreiber, G., Tsykin, A., Alred, A. R., Thomas, T., Fung, W., Dickson, P. W., Cole, T., Birch, H., DeJong, F. A., and Millard, J.,** The acute phase response in the rodent, *Ann. N.Y. Acad. Sci.*, 557, 61, 1989.
3. **Kushner, I., Ganapathi, M., and Schultz, D.,** The acute phase response is mediated by heterogeneous mechanisms, *Ann. N.Y. Acad. Sci.*, 557, 19, 1989.
4. **Kushner, I.,** The phenomenon of the acute phase response, *Ann. N.Y. Acad. Sci.*, 389, 39, 1982.
5. **Dinarello, C. A.,** Interleukin-1 and the pathogenesis of the acute phase response, *N. Engl. J. Med.*, 311, 1413, 1984.
6. **Kohase, M., May, L. T., Tamm, I., Vilcek, J., and Sehgal, P. B.,** A cytokine network in human diploid fibroblasts: interaction of beta-interferons, tumor necrosis factor, platelet-derived growth factor and interleukin-1, *Mol. Cell Biol.*, 7, 273, 1987.

7. **Andus, T., Geiger, T., Hirano, H., Northoff, H., Ganteh, U., Bauer, J., Kishimoto, T., and Heinrich, P. C.,** Recombinant human B cell stimulatory factor (BSF-2/IFN-B2) regulates fibrinogen and albumin mRNA levels in Fao-9 cells, *FEBS Lett.*, 221, 18, 1987.
8. **Andus, T., Geiger, T., Hirano, T., Kishimoto, T., and Heinrich, P. C.,** Action of recombinant human interleukin-6, interleukin-1B, and tumor necrosis factor alpha on the mRNA induction of acute phase proteins, *Eur. J. Immunol.*, 18, 739, 1988.
9. **Baumann, H., Richards, C., and Gauldie, J.,** Interaction among hepatocyte-stimulating factors, interleukin-1, and glucocorticoids for regulation of acute phase plasma protein in human hepatoma (HepG2) cells, *J. Immunol.*, 139, 4122, 1987.
10. **Baumann, H. and Eberhard-Mueller, U.,** Synthesis of hemopexin and cystein protease inhibitor is coordinately regulated by HSF-III and interferon B2 in rat hepatoma cells, *Biochem. Biophys. Res. Commun.*, 146, 1218, 1987.
11. **May, L. T., Ghrayeb, J., Santhanam, U., Tatter, S. B., Sthoeger, Z., Helfgott, C., Chiorazzi, N., Greininger, G., and Sehgal, P. B.,** Synthesis and secretion of multiple forms of B2 interferon/B cell differentiation factor 2/hepatocyte stimulating factor by human fibroblasts and monocytes, *J. Biol. Chem.*, 263, 7760, 1988.
12. **Miura, M., Prentice, U., Schneider, M., and Perlmutter, D. H.,** Synthesis and regulation of the two human C4 complement genes in stable transfected mouse fibroblasts, *J. Biol. Chem.*, 262, 7298, 1987.
13. **Koj, A., Gauldie, J., Regoeczi, E., Sauder, D. N., and Sweeney, G., D.,** The acute phase response of cultured rat hepatocytes: system characterization and the effect of human cytokines, *Biochem. J.*, 224, 505, 1984.
14. **Baumann, H.,** Hepatic acute phase reaction *in vivo* and *in vitro, In Vitro Cell Dev. Biol.*, 25, 115, 1989.
15. **Koj, A.,** Acute phase reactants, in *Structure and Function of Plasma Proteins*, Vol. 1, A. C. Allison, Ed., Plenum Press, New York, 1974, 73.
16. **Koj, A.,** Biological functions of acute phase proteins and the cytokines involved in them induces synthesis, in *Modulation of Liver Cell Expression,* P. C. Heinrich, H. Popper, and D. Keppler, Eds., MTP Press, Lancaster, PA, 1987, 331.
17. **Rupp, R. G. and Fuller, G. M.,** Comparison of albumin and fibrinogen biosynthesis in stimulated rats and cultured fetal rat hepatocytes, *Biochem. Biophys. Res. Commun.*, 88, 327, 1979.
18. **Gotschlich, E. L.,** C-Reactive protein, *Ann. N.Y. Acad. Sci.*, 557, 9, 1989.
19. **Morris, K. M., Aden, D. P., Knowles, B. B., and Colten, H. R.,** Complement biosynthesis by the human hepatoma-derived cell line HepG2, *J. Clin. Invest.*, 70, 906, 1982.
20. **Fong, Y., Moldawar, L. L., Shires, G. T., and Lowry, S. F.,** The biological characteristics of cytokines and their implication in surgical injury, *Surg. Gyn. Obstet.*, 170, 363, 1990.
21. **Fuller, G. M. and Ritchie, D. G.,** A regulatory pathway for fibrinogen biosynthesis involving an indirect feedback loop, *Ann. N.Y. Acad. Sci.*, 389, 303, 1982.

22. **Lanser, M. E. and Brown, G. E.,** Stimulation of rat hepatocyte fibronectin productions by monocyte-conditions medium is due to interleukin-6, *J. Exp. Med.,* 170, 1781, 1989.
23. **Urban, J., Chan, D., and Schreiber, G.,** A rat serum glycoprotein whose synthesis increases greatly during inflammation, *J. Biol. Chem.* 254, 10565, 1979.
24. **Keller, G. A., West, M. A., Cerra, F. B., and Simmons, R. L.,** Macrophage-mediated modulation of hepatocyte protein synthesis: effect of dexamethasone, *Arch. Surg.,* 121, 1199, 1986.
25. **Grieninger, G., Hertzberg, K. M., and Pindyk, J.,** Fibrinogen synthesis in serum-free cultures: stimulation by glucocorticoids, *Proc. Natl. Acad. Sci. U.S.A.,* 75, 5506, 1978.
26. **Baumann, H. and Held, W. A.,** Biosynthesis and hormone-regulated expression of secretory glycoproteins in rat liver and hepatoma cells: effect of glucocorticoid and inflammation, *J. Biol. Chem.,* 256, 10145, 1981.
27. **Fong, Y., Tracey, K. J., Moldawer, L. L., Hesse, D. G., Monogue, K. B., Kenney, J. S., Lee, A. T., Kuo, G. C., Allison, A. C., Lowry, S. F., and Cerami, A.,** Antibodies to cachectin/tumor necrosis factor reduce interleukin-1B and interleukin-6 appearance during lethal bacteremia, *J. Exp. Med.,* 170, 1627, 1989.
28. **Baumann, H., Onoraato, V., Gauldie, J., and Jahreis, G. P.,** Distinct sets of acute phase plasma proteins are stimulated by separate human hepatocyte-stimulating factors and monokines in rat hepatoma cells, *J. Biol. Chem.,* 262, 9756, 1987.
29. **Sanceau, J., Falcoff, R., Zilberstein, A., Beranger, F., Lebeau, J., Revel, M., and Vaquero,** Interferon-B2 (BSF-2) mRNA is expressed in human monocytes, *J. Interferon Res.,* 8, 473, 1988.
30. **Van Damme, J., Opdenakker, G., Simpson, R. J., Rubira, M. R., Cayphas, S., Vink, A., Billau, A., and Van Snick, J.,** Identification of the human 26-kD protein, interferon B2 (IFN-B2), as a B cell hybridoma/plasmacytoma growth factor induced by interleukin-1 and tumor necrosis factor, *J. Exp. Med.,* 165, 914, 1987.
31. **Horii, Y., Muraguchi, A., Suematsu, S., Matsuda, T., Yoshizaki, K., Hirano, T., and Kishimoto, T.,** Regulation of BSF-2/IL-6 production by human mononuclear cells: macrophage-dependent synthesis of BSF/IL-6 by T cells, *J. Immunol.,* 141, 1529, 1988.
32. **Baumann, H., Won, K. A., and Jahreis, G. P.,** Human hepatocyte-stimulating factor-III and interleukin-6 are structurally and immunologically distinct but regulate the production of the same acute phase protein, *J. Biol. Chem.,* 264, 8046, 1989.
33. **Sironi, M., Breviario, F., Proserpio, P., Biondi, A., Vecchi, A., Van Damme, J., Dejana, E., and Mantovani, A.,** IL-1 stimulates IL-6 production in endothelial cells, *J. Immunol.,* 142, 549, 1989.
34. **Gauldie, J., Richards, C., Harnish, D., Lansdorp, P., and Baumann, H.,** Interferon B2/B-cell stimulatory factor type 2 shares identity with monocyte-derived hepatocyte-stimulating factor and regulates the major acute phase protein response in liver cells, *Proc. Natl. Acad. Sci. U.S.A.,* 84, 7251, 1987.

35. **Navarro, S., Debili, N., Bernaudin, J. F., Vainchenker, W., and Doly, J.,** Regulation of the expression of IL-6 in human monocytes, *J. Immunol.,* 142, 4339, 1989.
36. **Fuller, G. M., Otto, J. M., Woloske, M., McGary, C. T., and Adams, M. A.,** The effects of hepatocyte stimulating factor on fibrinogen biosynthesis in hepatocyte monolayers, *J. Cell. Biol.,* 101, 1481, 1985.
37. **May, L. T., Ghrayeb, J., Santhanam, U., Tatter, S. B., Sthoeger, Z., Helfgott, D. C., Chiorazzi, N., Grieninger, G., and Sehgal, P. B.,** Synthesis and secretion of multiple forms of B2-interferon/B-cell differentiation factor 2/ hepatocyte-stimulating factor by human fibroblasts and monocytes, *J. Biol. Chem.,* 263, 7760, 1986.
38. **Buesher, H. U., Fallon, R. J., and Colten, H. R.,** Macrophage membrane interleukin 1 regulates the expression of acute phase proteins in human hepatoma Hep 3B cells, *J. Immunol.,* 139, 1896, 1987.
39. **Northoff, H., Andus, T., Tran-Thi, T., Bauer, J., Decker, K., Kubanek, B., and Heinrich, P. C.,** The inflammation mediators interleukin-1 and hepatocyte-stimulating factors are differently regulated in human monocytes, *Eur. J. Immunol.,* 17, 707, 1987.
40. **Mier, J. W., Dinarello, C. A., Atkins, M. B., Punsal, P. I., and Perlmutter, D. H.,** Regultion of hepatic acute phase protein synthesis by products of interleukin 2 (IL-2)-stimulated human peripheral blood mononuclear cells, *J. Immunol.,* 139, 1268, 1987.
41. **Hart, P. H., Whitty, G. A., Piccoli, D. A., and Hamilton, J. A.,** Control by IFN-gamma and PGE2 of TNF alpha and IL-1 production by human monocytes, *Immunology,* 66, 376, 1989.
42. **Brouckaert, P., Spriggs, D. R., Demetri, G., Kufe, D. W., and Fiers, W.,** Circulating interleukin-6 during a continuous infusion of tumor necrosis factor and interferon gamma, *J. Exp. Med.,* 169, 2257, 1989.
43. **Katz, Y. and Strunk, R. C.,** IL-1 and tumor necrosis factor: similarities and differences in stimulation of expression of alternative pathway of complement and IFN-B2/IL-6 genes in human fibroblasts, *J. Immunol.,* 142, 3862, 1989.
44. **Fong, Y., Moldawer, L. L., Marano, M., Wei, H., Tatter, S. B., Clarick, R. H., Santhanam, U., Sherris, D., May, L. T., Sehgal, P. B., and Lowry, S.,** Endotoxemia elicits increased circulating B2-IFN/IL-6 in man, *J. Immunol.,* 142, 2321, 1989.
45. **Waage, A., Brandtzaeg, P., Halstensen, A., Kierulf, P., and Expevik, T.,** The complex pattern of cytokines in serum from patients with meningococcal septic shock, *J. Exp. Med.,* 169, 333, 1989.
46. **Jablone, D. M., Mule, J. J., McIntosh, J. K., Sehgal, P. B., May, L. T., Huang, C. M., Rosenberg, S. A., and Lotze, M. T.,** IL-6/IFN-B2 as a circulating hormone: induction by cytokine administration in humans, *J. Immunol.,* 142, 1542, 1989.
47. **McIntosh, J. K., Jablons, D. M., Mule, J. J., Mordan, R. P., Rudikoff, S., Lotze, M. T., and Rosenberg, S. A.,** *In vivo* induction of IL-6 by administration of exogenous cytokines and detection of *de novo* serum levels of IL-6 in tumor bearing mice, *J. Immunol.,* 143, 162, 1989.

48. **Kurakowa, S., Ishibashi, H., Hayashida, K., Tsuchiya, Y., Hirata, Y., Sakaki, Y., Okubo, H., and Niho, Y.,** Kupffer cell stimulation of alpha-2-macroglobulin synthesis in rat hepatocytes and the role of glucocorticoids, *Cell Struct. Function,* 12, 35, 1987.
49. **Bauer, J., Birmelin, M., Northoff, G. H., Northemann, W., Traan-Thi, T., Ueberberg, H., Decker, K., and Heinrich, P. C.** Induction of rat alpha-2-macroglobulin *in vivo* and in hepatocyte primary cultures: synergistic action of glucocorticoids and a Kupffer cell-derived factor, *FEBS Lett.,* 177, 87, 1981.
50. **Keller, G. A., West, M. A., Cerra, F. B., and Simmons, R. L.,** Macrophage mediated modulation of hepatic function in multiple system failure, *J. Surg. Res.,* 39, 555, 1985.
51. **West, M. A., Keller, G. A., Hyland, B. J., Cerra, F. B., and Simmons, R. L.,** Further characterization of Kupffer cell/macrophage-mediated alterations in hepatocyte protein synthesis, *Surgery,* 100, 416, 1986.
52. **Keller, G. A., West, M. A., Cerra, F. B., and Simmons, R. L.,** Macrophage-mediated modulation of hepatocyte protein synthesis: effect of dexamethasone, *Arch. Surg.,* 121, 1199, 1986.
53. **Keller, G. A., West, M. A., Cerra, F. B., and Simmons, R. L.,** Multiple systems organ failure: modulation of hepatocyte protein synthesis by endotoxin activated Kupffer cells, *Ann. Surg.,* 201, 88, 1984.
54. **West, M. A., Keller, G. A., Hyland, B. J., Cerra, F. B., and Simons, R. L.,** Hepatocyte function is sepsis: Kupffer cells mediate a biphasic protein synthesis response in hepatocytes after exposure to endotoxin or killed *Escherichia coli, Surgery,* 98, 388, 1985.
55. **Keller, G. A., West, M. A., Harty, J. T., Cerra, F. B., and Simmons, R. L.,** Modulation of hepatocyte protein synthesis during co-cultivation with macrophage-rich peritoneal cells *in vitro, Arch. Surg.,* 120, 180, 1985.
56. **Keller, G. A., West, M. A., Harty, J. T., Wilkes, L. A., Cerra, F. B., and Simmons, R. L.,** Modulation of hepatocytes protein synthesis by endotoxin activated Kupffer cells. III. Evidence for the role of a monokine similar to but not identical with interleukin-1, *Ann. Surg.,* 201, 436, 1984.
57. **West, M. A., Keller, G. A., Cerra, F. B., and Simmons, R. L.,** Killed *Escherichia coli* stimulates macrophage mediated alterations in hepatocellular function during *in vitro* co-culture: a mechanism of altered liver function in sepsis, *Infect. Immun.,* 49, 563, 1985.
58. **Keller, G. A., West, M. A., Wilkes, L. A., Cerra, F. B., and Simmons, R. L.,** Modulation of hepatocyte protein synthesis by endotoxin activated Kupffer cells. II. Mediation by soluble transferrable factors, *Ann. Surg.,* 201, 429, 1985.
59. **Jirik, F. R., Podor, T. J., Hirano, T., Kishimoto, T., Loskutoff, D. J., Carson, D. A., and Lotze, M.,** Bacterial lipopolysaccharide and inflammatory mediators augment IL-6 secretion by human endothelial cells, *J. Immunol.,* 142, 144, 1989.
60. **Shalaby, M. R., Waage, A., and Espevik, T.,** Cytokine regulation of interleukin-6 production by human endothelial cells, *Cell. Immunol.,* 121, 372, 1989.

61. **Schumann, R. R., Leong, S. R., Flaggs, G. W., Gray, P. W., Wright, S. D., Mathison, J. C., Tobias, P. S., and Ulevitch, R. J.,** Structure and function of human lipopolysaccharide binding protein, *Science,* 249, 1429, 1990.
62. **Wright, S. D., Ramos, R. A., Tobias, P. S., Ulevitch, R. J., and Mathison, J. C.,** CD 14, a receptor for complexes of lipopolysaccharide (LPS) and LPS binding protein, *Science,* 249, 1431, 1990.
63. **Ramadori, G., Bushenfelde, K. M., Tobias, P. S., Mathison, J. C., and Ulevitch, R. J.,** Biosynthesis of lipopolysaccharide-binding protein in rabbit hepatocytes, *Pathobiology,* 58, 89, 1990.
64. **Hirata, Y., Ishibashi, H., Kimura, H., Hayashida, K., Nagano, M., and Okubu, H.,** Alpha-2-macroglobulin secretion enhanced in rat hepatocytes by partially characterized factor from Kupffer cells, *Inflammation,* 9, 201, 1985.
65. **Okubu, H., Miyanaga, O., Nagano, M., Ishibashi, H., Kudo, J., Ikuta, T., and Shibata, K.,** Purification and immunological determination of alpha-2-macroglobulin in serum from injured rats, *Biochem. Biophys. Acta,* 668, 257, 1981.
66. **Gehring, M. R., Shiels, B. R., Northemann, W., de Bruijn, M. H. L., Kan, C. C., Chain, A. C., Noonan, D. J., and Fey, G.,** Sequence of rat liver alpha-2-macroglobulin and acute phase control of its messenger RNA, *J. Biol. Chem.,* 262, 446, 1987.
67. **Sanders, K. D. and Fuller, G. M.,** Kupffer cell regulation of fibrinogen synthesis in hepatocytes, *Thromb. Res.,* 32, 133, 1983.
68. **Baumann, H., Jahreis, G. P., Sauder, D. N., and Koj, A.,** Human keratinocytes and monocytes release factors which regulate the synthesis of major acute phase plasma proteins in hepatic cells from man, rat, and mouse, *J. Biol. Chem.* 259, 7331, 1984.
69. **Darlington, G. J., Wilson, D. R., and Lachman, L. B.,** Monocyte-conditioned medium, interleukin-1, and tumor necrosis factor stimulate the acute phase response in human hepatoma cells *in vitro, J. Cell. Biol.,* 103, 787, 1986.
70. **Rupp, R. G. and Fuller, G. M.,** The effects of leukocytic and serum factors on fibrinogen biosynthesis in cultured hepatocytes, *Exp. Cell. Res.,* 118, 23, 1979.
71. **Sevaljevic, L., Glibetic, M., Poznanociv, Petrovic, Matic, S., and Pantelic,** Thermal injury-induced expression of acute phase proteins in rat liver, *Burns,* 14, 280, 1988.
72. **Starnes, H. F., Pearce, M. K., Tewari, A., Yim, J. H., Zou, J., and Abrans, J. S.,** Anti-IL-6 monoclonal antibodies protect against lethal *Escherichia coli* infection and lethal tumor necrosis factor alpha challenge in mice, *J. Immunol.,* 145, 4185, 1990.
73. **Kispert, P. H., Curran, R., Billiar, T., and Simmons, R. L.,** Kupffer cells can provide the signal for hepatocyte acute phase protein synthesis, *Surg. Forum,* 41, 80, 1990.
74. **Ford, H. R., Hoffman, R. A., Tweardy, D. J., Kispert, P., Wang, S., and Simmons, R. L.,** Production of interleukin-6 (IL-6) within the rejecting allograft coincides with cytotoxic T lymphocyte development, *Transplantation,* in press.

75. **Sehgal, P., Grienger, G., and Tosato, G.,** Regulation of the acute phase and immune response: interleukin-6, *N.Y. Acad. Sci.,* 557, 1, 1989.
76. **Lee, J. D., Swisher, S. G., Minehart, E. H., McBride, W. H., and Economou, J. S.,** Interleukin-4 downregulates interleukin-6 production in human peripheral blood mononuclear cells, *J. Leuk. Biol.,* 47, 475, 1990.
77. **Taylor, A. W., Ku, N. O., and Mortensen, R. F.,** Regulation of cytokine induced human C-reactive protein production by transforming growth factor beta, *J. Immunol.,* 145, 2705, 1990.
78. **Krebs, H. A., Lund, P., and Stubbs, M.,** Interrelations between gluconeogenesis and urea synthesis, in *Gluconeogenesis: Its Regulation in Mammalian Species,* R. W. Hanson and M. A. Mehlman, Eds., John Wiley & Sons, New York, 1976, 269.

Chapter 12

THE CHANGES IN HEPATIC LIPID AND CARBOHYDRATE METABOLISM IN SEPSIS*

T.W. Lysz and G.W. Machiedo

TABLE OF CONTENTS

* Supported in part by NIH grant EY 05437 (TWL) and Veterans Administration Merit Review Award.

ISBN 0-8493-6109-5

I. INTRODUCTION

The care of the severely septic patient is often centered on the repair of the injury and the treatment of infection. These patients often die with an acquired hepatic insufficiency despite the use of appropriate antibiotics and careful patient monitoring. The hypermetabolism generally associated with sepsis is the most common reason for admission to the surgical intensive care unit with between 47 and 85% of the patients succumbing to the syndrome. Care of these patients consumes the greatest amount of hospital resources on a surgical service.[1,2]

Sepsis and its accompanying hypermetabolic syndrome have long been associated with marked changes in the vascular, metabolic, endocrine, and immunological systems. Of these changes, the etiology of the metabolic and endocrine alterations are the least clearly understood. Clinically, the study of the metabolic events that herald sepsis has proven difficult due to the vast number of factors that impact on whether the patient recovers from the bacteremia or progresses to organ failure. The complexity of these metabolic events, which appear to occur in a number of different organ systems simultaneously, has stimulated research in animal models and cell culture in an attempt to elucidate the regulatory factors that precipitate or accompany the presence of the bacteremia and endotoxemia.

The clinical picture of the late phase of organ failure is one that includes liver failure. This process seems to begin some time before clinical recognition and may be the result of a common pathway for the onset of different organ failure syndromes.[3] Due to the pivotal role of liver cells in the maintenance of glucose and fatty acid metabolism, information on how hepatocytes and Kupffer cells communicate with one another to modulate these functions may help in understanding some of the metabolic events that precede organ failure and death. The use of coculture techniques has already been applied to examine the changes that occur in liver protein synthesis.[4-6] However, few studies have examined the effect of sepsis on fatty acid metabolism in these two different cell populations.

In this chapter, the properties of the Kupffer cell and hepatocyte are reviewed with emphasis upon the contribution of these cells in carbohydrate and lipid metabolism during sepsis. We will discuss how these two cells might act in concert to support the energy requirements needed to sustain the defensive and survival functions utilized by the host to combat sepsis. Finally, we propose that as part of the acute-phase response of sepsis, a "remodeling" of Kupffer cell fatty acid content occurs following exposure to endotoxin which is modified by the presence of the hepatocytes or possibly by a substance elaborated by hepatocytes.

II. CARBOHYDRATE METABOLISM IN THE LIVER

A. GLUCOSE METABOLISM

The early phases of sepsis are generally characterized by hyperglycemia.[7] This state is probably due to an elevated glucose production coupled with a diminished peripheral tissue utilization. Sympathetic activity associated with the release of hormones from the pituitary-adrenal axis mobilizes glycogen and stimulates glucose production while generally enhancing plasma insulin levels.[8] As sepsis proceeds, a hypoglycemia may result as glycogen stores are depleted during stress. Gluconeogenesis then becomes important for the maintenance of adequate plasma glucose levels. Since many acute compensatory physiological functions depend on glucose as fuel (e.g., cells involved in modulating inflammation and wound repair, red blood cells, and the central nervous system), conditions which undermine the transition from glycogenolysis to gluconeogenesis will alter the host's ability to respond to sepsis.[9,10]

It is generally agreed that in septic and trauma patients glucose turnover is increased and that gluconeogenesis is enhanced and not easily suppressed by the administration of exogenous glucose.[11,12] The mass flow of glucose is increased and the peripheral cellular uptake is normal. Although the total glucose oxidation is increased relative to an equivalent stage of starvation, the fraction of calories expended from glucose is reduced.[13] The elevated hepatic glucose production is maintained in sepsis despite the hyperglycemia characteristic of severely ill patients.[14] The lack of suppression of gluconeogenesis during sepsis has been attributed, in part, to the production of lactate by hypoxic tissues and the release of glycerol from adipose tissues which induces the gluconeogenic state.[7] Elevated plasma glucose levels persist despite high levels of insulin. This apparent insulin resistance has been attributed to the presence of glucocorticoids, although the mechanism of the response has not been fully elucidated.[15]

B. DO MACROPHAGE-DERIVED PRODUCTS INFLUENCE GLUCOSE METABOLISM?

The ability of products synthesized by the macrophage in response to an invading organism to modify glucose metabolism is still unresolved. Interleukin-1 (IL-1) and tumor necrosis factor (TNF) have received considerable interest recently due to their production during inflammatory states and their ability to markedly affect protein metabolism.[16] Infusion of IL-1 had no effect on carbohydrate metabolism in one study,[11] while others have reported an increased gluconeogenesis and proteolysis following IL-1 administration.[17] Bagley et al.[18] have reported that an infusion of recombinant TNF increased glucose metabolism, suggesting a role for TNF in the elevated carbohydrate metabolism observed during sepsis. However, TNF administration also produces a sustained increase in the plasma insulin concentration and an elevation of circulating catecholamine levels, suggesting that the effect of TNF may be a consequence of the action of the cytokine on other systems that require glucose for energy. This latter possibility has recently been studied by Meszarus et al.,[19] who demonstrated that glucose uptake *in vivo* is markedly elevated in the liver, spleen, skin, and lung following TNF infusion into nonseptic animals. Thus, during infection, the TNF-stimulated macrophages located in these tissues increase their requirement for glucose. The sustained requirement for glucose by these cells increases lactate production, which in turn supports an elevated carbohydrate output by the liver via gluconeogenesis, since total hepatic blood flow is maintained in the hypermetabolic state.[20]

C. KUPFFER CELL-HEPATOCYTE INTERACTIONS IN MODIFYING GLUCOSE METABOLISM BY PLATELET ACTIVATING FACTOR

Due to the complex array of inflammatory responses to tissue invasion provoked by endotoxin in the host, a number of studies have begun to investigate how the communication between cells within a tissue contributes to the observed response elicited by infection. However, the anatomic organization of the cells in the tissue under study may be as vital as the cells themselves in attempting to determine how the action of a mediator may impact upon the host. Recently, Kuiper et al.[21,22] and Shukla et al.[23] suggested that cooperation between the liver nonparenchymal cells and the hepatocyte could induce glucogenolysis in the perfused liver following the administration of platelet activating factor (PAF, or 1-0-alkyl-2-acetyl-sn-glycerol-3 phosphocholine). PAF, a biologically active phospholipid, has potent effects on inducing the aggregation of platelets,[24,25] exerting its physiological response probably by stimulating inositol turnover and mobilizing intracellular calcium via the formation of phosphatidylinositol-4,5-biphosphates.[26,27] PAF increases glucose production in the isolated perfused liver,[28,29] but does not affect oxygen consumption or glycogenolysis in isolated hepatocytes.[30] The effect of PAF in the perfused liver has previously been postulated to be due to an ischemia produced by the ability of PAF to constrict the hepatic sinusoid, which promotes the glycolytic response.[31,32]

However, Kuiper et al.[22] have suggested that the generation of prostaglandin D_2 (PGD_2) by the Kupffer cell (or possibly the endothelial cell) in response to administered PAF may be responsible for the glycolytic effect in freshly isolated hepatocytes. Evidence to support this apparent cooperation between different cells was: (1) exogenously added PGD_2 produced a dose-related enhanced glycogenolysis when measured in either isolated hepatocytes or perfused liver; (2) indomethacin, an inhibitor of PGD_2 production, eliminated the PAF response in the perfused rat liver, suggesting that the action of PAF on glucose output was secondary to inhibition of prostaglandin synthesis; (3) Kupffer cells can synthesize PAF in response to phorbol ester, diacylglycerol, and endotoxin, which then stimulates PGD_2 production.

While an attractive hypothesis, attempts to replicate the effect of PGD_2 on glucose metabolism in cultured hepatocytes have been unsuccessful.[29]

However, differences in the methods for assay and the utilization of cultured hepatocytes rather than freshly prepared hepatocytes may partially account for the discrepancy. The utilization of cultured hepatocytes in the latter study produces a more stable glucose base line release than the freshly isolated cell preparations and allows for the recovery of various liver functions.[32] In addition, the concentration of PAF required to stimulate glycogenolysis and vasoconstriction in the perfused liver has been reported to be between 10 to 30 times lower than that required for PAF to induce the synthesis of Kupffer cell PGD_2.[33] The inability to inhibit the glycogenolytic response in the perfused liver by low-dose indomethacin has also been reported.[32] Thus, it is uncertain whether or not the PAF response to glycogenolysis in the isolated perfused liver is indirect via the production of PGD_2 from the Kupffer cell.

III. LIPID METABOLISM IN THE LIVER

A. THE EFFECT OF SEPSIS ON FATTY ACID METABOLISM

Lipids constitute 80% of the energy reserves of the body. In times of stress, the mobilization and utilization of fat stores permits the preservation of protein critical for cellular function and the maintenance of oncotic gradients. As an alternative fuel source, fat metabolism also permits the sparing of carbohydrates for utilization by the central nervous and immune systems. The importance of lipids as fuel during infection is reflected by the use of lipid as nutritional therapy in septic patients.

With the onset of sepsis, lipolysis is increased due to the stimulation of sympathetic innervation that increases the levels of adrenaline, glucagon, and cortisol.[35] Adipose tissue and liver assume primary roles in the storage and processing of the lipid for fuel. In the short term, sepsis produces a pronounced increase in plasma free fatty acid concentration that is primarily caused by an elevated rate of appearance of the lipid from adipose tissue. The increased rate of free fatty acid mobilization from the adipose tissue is accompanied by an elevated rate of fatty acid oxidation. This latter finding may be an important feature of the altered metabolism seen in sepsis, since the accelerated lipolysis occurs despite the prevailing hyperglycemia and an elevated plasma insulin level that normally curtail lipid catabolism.[36,37] However, there is little correlation between plasma free fatty acid levels and the severity of sepsis.[38] This lack of correlation may be due to the reduced blood flow in the peripheral tissue seen during severe sepsis which limits the availability of albumin carrier needed for the transport of the fatty acid, and a lactic acidosis that results from the resulting systemic hypoxia which encourages reesterification of free fatty acids.[39,40]

The albumin-bound free fatty acids are only one source of fatty acid that could be utilized by the peripheral tissue for fuel. A second important source of fatty acid is the triglyceride fatty acid found in lipoproteins. Wolfe et al.[36] have reported that the increased mobilization of free fatty acid from the adipose tissue during sepsis was responsible for the elevated triglyceride production by the liver. Normally, fatty acids are converted by the liver into triglyceride and secreted by the hepatocyte as a component of very low density lipoprotein (VLDL), which is processed by the extrahepatic tissues to support oxidative metabolism.

In sepsis, a hypertriglyceridemia probably occurs through an increased hepatic synthesis of triglyceride and a decreased clearance of triglyceride-rich lipoprotein from the systemic circulation.[41,42] The uptake of free fatty acid by the liver does not involve an enzymatic mechanism, although the rate of uptake is directly proportional to their plasma concentration.[43] However, the plasma levels of free fatty acid are often low in septic patients, and this suggests that profound changes in the intracellular fat metabolism occur in these individuals.[44,45] The VLDL triglyceride, once in circulation, must first undergo an enzymatic hydrolysis by lipoprotein lipase (LPL) before the fatty acid can be utilized by the peripheral tissues. LPL is located on the luminal surface of the capillary endothelial cells[46] and modification of the activity of the enzyme has been proposed to be one mechanism that may allow for the triglycerides to be directed to tissues requiring fatty acid as an energy source.[11] Indeed, LPL activity has been reported to be selectively increased in the heart and skeletal muscle while decreased in adipose tissue.[47] It is unclear, however, whether or not LPL suppression plays a significant role in the course of human sepsis and whether or not a differential inhibition of LPL activity in the different tissues during sepsis can actually occur. In Gram-negative sepsis and endotoxemia, the observed hypertriglyceridemia is believed to be due in large measure to the enhanced macrophage production of the monokines IL-1 and TNF, which have been shown to suppress LPL activity.[48,49]

B. LIPID MEDIATORS IN SEPSIS: THE EICOSANOIDS AND PLATELET ACTIVATING FACTOR

A large number of compounds acting as mediators of the inflammatory response in sepsis are derivatives of either the fatty acids (e.g., arachidonic acid) or phospholipid (e.g., PAF) of tissue lipids. Both classes of lipid mediators are extremely potent substances that have a variety of important biological effects contributing to cellular injury. The most important of these compounds are the eicosanoids, derivatives of arachidonic acid (a 20 carbon polyunsaturated fatty acid) that is found esterified at the second position in membrane phospholipids and, to a smaller degree, in neutral lipids. Eicosanoids are produced following the release of arachidonic acid from phospholipids by the stimulation of phospholipase activity due to hypoxia, mechanical damage, or the occupation of cell-surface receptors by chemical mediators.[50] The free fatty acid, normally low in the cell, can be either quickly reesterified into the plasma phospholipids via an acyl-CoA transferase[51] or can be utilized by the cyclooxygenase or the lipoxygenase pathways to produce the prostaglandins, leukotrienes, and related hydroxy fatty acids (for review see References 50, 52). Prostaglandin production can be selectively inhibited by nonsteroidal anti-inflammatory agents such as aspirin. Manipulation of the diet by feeding omega-3 fatty acids which compete for the second position of the glycerol backbone of the phospholipid and which are either poor substrates for eicosanoid synthesis or yield products which are not as biologically active, is another means of limiting eicosanoid production.[53]

PAF is another lipid mediator that has already been implicated in the gluconeogenic response produced during sepsis (see above). PAF, also known as acetyl glyceryl ether phosphorylcholine, exists as a phospholipid with either a C16 or a C18 fatty acid chain in the 1 position.[54] PAF, or its precursor lyso-PAF, is a constituent of cell membranes and, interestingly, can release arachidonic acid from its phospholipid matrix.

The eicosanoids have been increasingly implicated as significant pathophysiologic mediators of endotoxic shock.[55,56] Inhibition of prostaglandin synthesis increases the survival of endotoxic dogs and alters the pathophysiologic responses observed during the course of endotoxemia.[57-59] Rats are significantly more resistant to the lethal effects of endotoxin when maintained on an essential fatty acid-deficient diet.[60] Such diets deplete the membrane phospholipids of precursor arachidonic and linoleic acids. Supplementing the fatty acid-deficient diet with arachidonic acid can restore the sensitivity of the rat to the lethal effects of endotoxin, which can be reversed by using an inhibitor of thromboxane (TxA_2) synthesis.[61,62] These results imply that the prostaglandins, in particular TxA_2, which is normally associated with platelet aggregation, contribute to the mortality in the rat model. Elevated levels of TxA_2 have also been measured in a human population of nonsurvivors of sepsis,[63] but these results could not be substantiated.[64]

PAF is a particularly potent inducer of capillary fluid leakage with a potent thrombogenic (i.e., platelet aggregation) action. Like TxA_2, PAF is very labile, both having half-lives of approximately 30 s. PAF can also be measured in the body fluids during endotoxic shock. Its biological significance in sepsis is as yet unknown, although a picture is emerging that suggests that PAF acts synergistically with TNF and endotoxin to promote leukotriene and TxA_2 biosynthesis in shock.[65]

C. HEPATIC LIPID MEDIATOR SYNTHESIS AND DEGRADATION DURING SEPSIS

In the liver, the Kupffer cell and, to a lesser extent, the endothelial cells are thought to be the principle cell types involved in prostanoid production in response to sepsis and other stimuli.[66] PGD_2 and PGE_2 are the most abundant prostaglandins produced by cultured Kupffer cells,[21] while leukotriene biosynthesis has also been demonstrated in mast cells, blood-borne elements (except red blood cells), and possibly hepatocytes.[67] Inflammatory cells infiltrating the liver under pathological conditions may also contribute to leukotriene production. The hepatocytes also have the capacity to catabolize the prostanoids from both extrahepatic and intrahepatic production sites[68] by inactivation of the postanoids via 15-dehydrogenase activity followed by 13-reductase reduction of the double bond at carbons 13 and 14.[15,69] Further catabolism of the inactive products occurs principally by beta and omega oxidation.[70,71] Leukotrienes such as LTC_4, LTD_4, and LTE_4 are rapidly eliminated from the vascular space and are taken up by the liver and kidney.[72] The hepatocellular system is most effective for LTE_4, but can also transport LTD_4 and LTC_4. Hepatocytes can quickly release the ingested leukotrienes after uptake and partial metabolism, and thus assist in the rapid hepatobiliary elimination of circulating cysteinyl leukotrienes.[72] Endotoxin reduces the biliary elimination of cysteinyl leukotrienes, and their presence may be related to the severe intrahepatic cholestasis often associated with sepsis and severe trauma by promoting edema around the bile ducts.

D. IS HEPATIC DYSFUNCTION IN SEPSIS DUE TO PRODUCTS RELEASED FROM PHAGOCYTES OR TO ALTERATIONS IN THE HEPATOCYTE LIPID METABOLISM?

Abnormalities of hepatic function exist in surgical or trauma patients without any direct hepatic injury, especially when the posttraumatic period is complicated by sepsis.[3,73] While a direct effect due to endotoxin seems unlikely, based on recently published results,[74] two proposals have evolved that may help explain how hepatic dysfunction occurs with sepsis. One hypothesis suggests that the extracellular release of proteases and oxygen metabolites, by-products of the phagocytic activity of the resident Kupffer cells and especially marginating polymorphonuclear leukocytes, promotes hepatocellular injury.[74,75] The second hypothesis suggests that impairments of arachidonic acid metabolism and inositol turnover resulting from the down-regulation of the alpha-1 and vasopressin receptors on the hepatocytes significantly interfere with the calcium-dependent mechanisms in the liver during chronic sepsis. Spitzer et al.[76] and Rodriguez de Turco and Spitzer[77-79] have demonstrated that an altered membrane phospholipid and changes in triglyceride content result from chronic nonlethal endotoxemia, which is believed to be linked to changes in phosphoinositide turnover and calcium mobilization in hepatocytes.[76-79] Impairments in the calcium-signaling process may ultimately be a significant consequence of endotoxemia in the hepatocyte and underlie some of the metabolic changes measured in chronic sepsis. It is unclear whether the continuous presence of the endotoxin per se is the inciting factor in the bacteremia or if other inflammatory mediators produced in response to sepsis cause the effects on calcium mobilization. Nevertheless, these data suggest that changes in the signaling mechanisms of the cell may play a significant role in cell response to infection.

E. HEPATOCYTE MODULATION OF KUPFFER CELL LIPID METABOLISM IN CULTURE

The anatomic position of the Kupffer cell and hepatocyte suggests that the Kupffer cell responds to circulating stimuli and provides signals to the neighboring parenchymal cells during normal conditions and sepsis. The Kupffer cell, as a macrophage, secretes a number of activation products (e.g., interleukins, TNF, eicosanoids) which have been reported to induce the synthesis of acute-phase proteins from hepatocytes.[80,81] While the hepatocyte is not believed to release significant quantities of the arachidonic acid metabolites, its inherent capacity to degrade the prostaglandins has been proposed as a means of regulating prostanoid levels in the liver microenvironment. Modifying arachidonic acid metabolism, particularly PGE_2 and PGD_2, could then influence proinflammatory monokine production. PGE_2 is recognized as a potent inhibitor of macrophage function in assays of cytokine release[82] and tumor cytotoxicity.[83]

We have recently demonstrated that the local control of endotoxin-stimulated Kupffer cell prostaglandin synthesis may also be mediated by a substance(s) elaborated by the hepatocyte. Addition of a cell-free hepatocyte supernatant significantly increased LPS-stimulated Kupffer cell PGE_2 production[84] (Figure 1). In contrast, cocultures of heptocytes and Kupffer cells consistently reduced the prostaglandin levels in the culture media following endotoxin challenge. The ability to increase eicosanoid production appears specific for endotoxin, since neither calcium ionophore nor phorbol ester could mimic the response on Kupffer cell PGE_2 synthesis. The activity in the hepatocyte supernatant may be distinct from most hepatocyte-derived factors, since the substance was obtained from unstimulated hepatocytes and was not active when the hepatocyte supernatant was added without endotoxin. Additionally, the hepatocyte supernatant was synergistic with only endotoxin and not phorbol ester or the calcium ionophore A23187, suggesting that neither changes in intracellular calcium nor the activation of a protein kinase C are involved in the increased Kupffer cell PGE_2 production.

Proof of the existence of the unique hepatocyte factor was also derived from experiments that demonstrate: (1) the activity of the hepatocyte supernatant can be eliminated by boiling for 5 min, and (2) partial purification of the factor using a Sephacryl® S-200 and DEAE-Affi-Blue affinity chromatography (Figures 2 and 3). The hepatocyte factor shares some of the characteristics (e.g., molecular weight, tissue origin) of the lipopolysaccharide (LPS)-binding protein which reportedly binds to the endotoxin and enhances the interaction of the toxin to the macrophage, probably through interaction with a receptor on the cell surface.[86,88]

The hepatocyte supernatant also produces changes in the labeled arachidonic acid distribution between the phospholipid and neutral lipid fractions in the LPS-stimulated Kupffer cell (Table 1).[85] Thus, the factor released by the hepatocytes appears to dramatically affect both the structural (phospho-

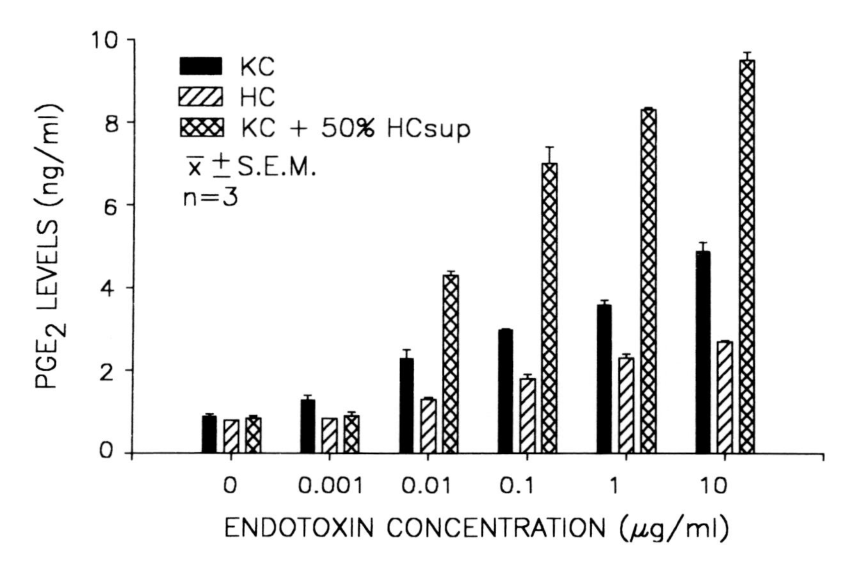

FIGURE 1. Endotoxin concentration-dependent effects on Kupffer cell PGE_2 production. Effect of hepatocyte (HC) supernatant. Kupffer Cells (KC: 0.5×10^6 cells per milliliter) were either cultured alone (black bars), with HC (1×10^6 cells per milliliter) (hatched bars), or with 50% HC supernatant (cross-hatched bars) for 18 h. LPS in concentrations from 0.001 to 10 µg/ml were added to the cell cultures and the culture supernatants collected after 24 h for analysis of PGE_2 by radioimmunoassay. Results are expressed as the mean of three different experiments, each performed in triplicate. (From Billiar, T. R., Lysz, T. W., Curran, R. D., Bentz, B. G., Machiedo, G. W., and Simmons, R. L., *J. Leuk. Biol.*, 47, 304, 1990. With permission.)

lipid) and functional (arachidonic acid) lipids in the LPS-challenged Kupffer cell. Specifically, the presence of the hepatocyte supernatant promotes changes in the labeled arachidonic acid distribution into the neutral lipid fraction with a corresponding decreased distribution of label into the major phospholipids, phosphatidylethanolamine and phosphatidylcholine. Altered membrane phospholipid composition and changes in triglyceride content have been reported in hepatocytes isolated from rats after chronic nonlethal endotoxemia.[78,79] These results suggest that the action of the hepatocyte supernatant on the Kupffer cell constitutes an acute lipid response to the presence of endotoxin and may be part of the attempt of the liver to compensate for the changes normally observed with sepsis.

F. KUPFFER CELL MODULATION OF HEPATOCYTE LIPID METABOLISM FOLLOWING ENDOTOXIN CHALLENGE

Lipid metabolism is affected in a variety of ways during infection, depending on the causative agents. This interaction is further modulated by the nutritional status of the host and the severity of the infection. The most profound effects occur with a Gram-negative infection and principally involve

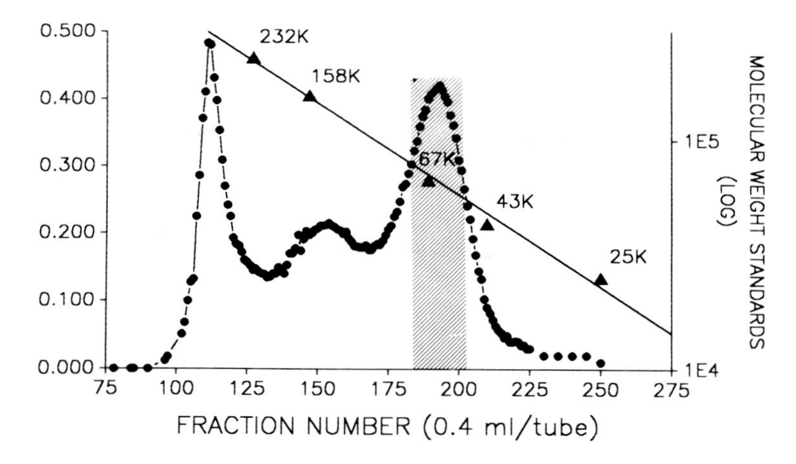

FIGURE 2. Sephacryl® S-200 column chromatography of lyophilized rat hepatocyte culture media. Approximately 500 cc of unstimulated rat hepatocyte culture supernatant was filtered/dialyzed using an Amicon ultrafiltration device with a PM10 filter at 4°C and lyophilized. The lyophilized sample was dissolved in 2 ml phosphate-buffered saline, pH 7.2, centrifuged to remove undissolved particulate matter, and added to a Sephacryl® S-200 column (47 × 2.4 cm). Fractions (0.4 ml) were collected at a flow rate of 0.25 ml/min and read at 280 nm. The peaks were collected, dialyzed, and lyophilized. Shaded area denotes the region demonstrating a net increased Kupffer cell PGE_2 production after incubation with 10 μg/ml LPS ± lyophilized hepatocyte supernatant. The column was initially standardized using known molecular weight standards.

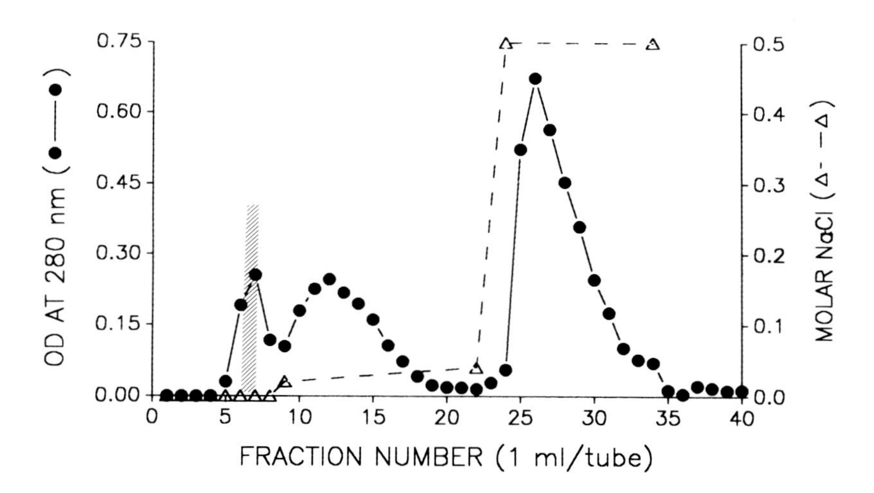

FIGURE 3. DEAE-Affi-Blue column chromatography of the 67 kd fraction obtained from Sephacryl® S-200. Fractions 180 to 205 from the Sephacryl® S-200 column were combined, dialyzed against distilled water, concentrated, and lyophilized. The fraction demonstrating an increased net PGE_2 production (shaded area) was recovered in the void volume following elution with 0.01 *M* phosphate buffer, pH 7.3. Albumin avidly binds to the DEAE-Affi-Blue and is eluted with 0.5 *M* NaCl.

TABLE 1
Effect of Hepatocyte Supernatant on Percent [^{14}C] Arachidonic Acid Incorporation into Kupffer Cells

Lipid fraction	KC	KC + LPS	KC + HC Sup	KC + HC Sup + LPS
PS/PI	10.1 ± 0.9	11.9 ± 0.9	10.4 ± 0.3	10.7 ± 0.6
PC	30.2 ± 1.5	30.2 ± 1.5	34.0 ± 3.6	25.2 ± 2.8[a]
PE	15.2 ± 2.6	25.8 ± 2.4[b]	20.2 ± 3.2	14.0 ± 1.4[a]
NL	19.6 ± 2.1	14.5 ± 2.9[a]	19.7 ± 2.1	26.2 ± 2.1[a]

Note: The amount of 1-[^{14}C]-arachidonate incorporated into phosphatidyl serine/inositol (PS/PI), phosphatidyl ethanolamine (PE), phosphatidyl choline (PC), and neutral lipid (NL) was determined in 0.5 × 10^6 Kupffer cells in the presence of 50% hepatocyte supernatant (HC Sup) ± 10 μg LPS per milliliter. Average dpm recovered from the KC = 32,165; KC + LPS = 68,297; KC + HC Sup = 59,916; KC + LPS + HC Sup = 66,369. Standard error was between 10 and 15% of the mean.

[a] $p < 0.05$.
[b] $p < 0.01$ control vs. LPS treatment

Reprinted with permission.

free fatty acids and increased plasma level of triglycerides. The etiology of the hypertriglyceridemia that is often measured during sepsis is likely to be due to a number of different cellular responses other than a decreased lipoprotein lipase activity. Endotoxin has a major influence on several hepatic functions,[89] thus the moderating effect of LPS on the triglyceride synthesis and the induction of hypertriglyceridemia should not be dismissed.

We have begun to assess what factors may contribute to the fate of radiolabeled oleic acid, a monoenoic saturated fatty acid, and arachidonic acid, a polyunsaturated long-chain fatty acid, in cultured hepatocytes after incubation with either endotoxin-stimulated Kupffer cell supernatant or the cytokines IL-1, IL-6, and TNF. In our studies, the hepatocyte responds to the presence of Kupffer cell supernatant stimulated with LPS by significantly reducing the distribution of both labeled oleic acid into the triglyceride fraction found in the hepatocyte supernatant while increasing the extracellular levels of labeled free fatty acid (Figure 4). Similar changes were measured when the hepatocytes were incubated with labeled arachidonic acid. No changes were detected in the phospholipid fraction in the hepatocyte. These results agree with those of Rodrigez de Turco and Spitzer,[77] who reported that the arachidonic acid-associated triglyceride fraction accounted for approximately 50 to 60% of the total labeled arachidonate esterified into glycerolipids and probably reflects a small, but metabolically very active acyl-storage site. Our results suggest a possible decreased efficiency of esterifying lipids by the

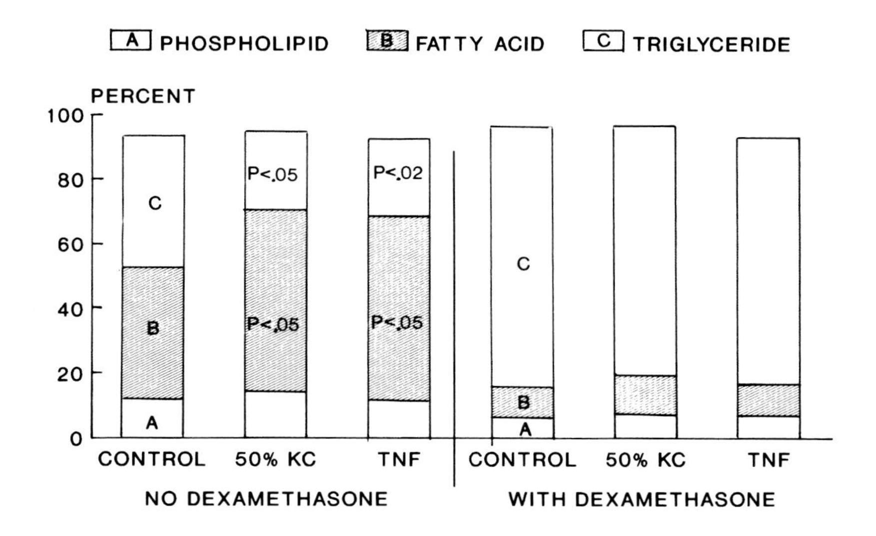

FIGURE 4. Percent radiolabeled oleic acid distribution in the hepatocyte (HC) culture media after incubating the cells with LPS-stimulated Kupffer cell (KC) supernatant (Sup) and tumor necrosis factor (TNF): Effect of dexamethasone. HC were incubated with either 500 U TNF or 50% LPS-stimulated KC Sup for 18 h, followed by a 24-h incubation with 1-[^{14}C]-oleic acid. The HC Sup was collected and the lipids extracted and separated using thin-layer chromatography. Values represent the means of three separate experiments, each performed in duplicate. Average total dpm recovered from the HC Sup without dexamethasone = 367,302; HC + 50% KC Sup = 283,776; HC + TNF = 415,543. Total dpm of HC with dexamethasone = 647,202; HC + 50% KC Sup = 682,141; HC + TNF = 511,965. The balance of the 1 million dpm added to the cultures was found in the cells.

acyl-CoA synthetase system that effects both the utilization of metabolic lipids (oleic acid) and structural fatty acids (arachidonate).

The addition of TNF, but not IL-1 or IL-6, was capable of mimicking the response observed with the endotoxin-stimulated Kupffer cell supernatant (Figure 4). Addition of dexamethasone to the cultures eliminated the inhibitory action of both the LPS-stimulated Kupffer cell supernatant and TNF. Thus, in the absence of "stress hormones", TNF may act to simultaneously limit the production of triglyceride in the liver while elevating the plasma triglyceride content by inhibiting LPL activity. The presence of steroids during sepsis, however, appears to effectively blunt the effect of TNF on the arachidonic acid and oleic acid incorporation into triglyceride.

IV. SUMMARY

The liver plays a pivotal role in the maintenance of glucose and fatty acid metabolism during sepsis. Information of how the hepatocyte and Kupffer cell communicate with one another to modulate these functions may help in our understanding of some of the metabolic events that accompany infection and the hypermetabolic syndrome. Currently, the link between the observed changes in metabolism between liver cells is thought to reside in the chemical mediator system which includes the interleukins, eicosanoids, and PAF, with the flow of information generally occurring from Kupffer cell to hepatocyte. We proposed that a *bidirectional* flow of information exists between the hepatocyte and the Kupffer cell with the cross talk serving to modulate the levels of chemical mediators produced following endotoxin challenge. In the long term, an altered membrane phospholipid composition and a modified fatty acid neutral lipid content may serve to support specific metabolic pathways that may be involved in the structural and functional integrity of organs and organ systems. The redistribution of fatty acid in the glycerolipids of the hepatocyte and Kupffer cell by LPS may thus represent an "acute-phase response" analogous to the hepatic acute-phase protein production observed in sepsis.

REFERENCES

1. **Machiedo, G. W., LoVerme, P. J., McGovern, P. J., and Blackwood, J. M.,** Patterns of mortality in a surgical intensive care unit, *Surg. Gyn. Obstet.*, 152, 757, 1981.
2. **Madoff, R. D., Sharpe, S. M., Fath, J. J., Simmons, R. L., and Cerra, F. B.,** Prolonged surgical intensive care, *Arch. Surg.*, 120, 698, 1985.
3. **Cerra, F. B., Siegel, J. H., Border, J., and Coleman, B.,** The hepatic failure of sepsis: cellular vs. substrate, *Surgery*, 86, 409, 1979.
4. **Casteleijn, E., Kuiper, J., Rooij, H. C. J., and Koster, J.,** Conditioned media of Kupffer and endothelial liver cells influence protein phosphorylation in parenchymal liver cells: involvement of prostaglandins, *Biochem. J.*, 252, 601, 1988.
5. **Billiar, T. R., Curran, R. D., Stuehr, D. J., West, M. A., Bentz, B. G., and Simmons, R. L.,** An L-arginine-dependent mechanism mediates Kupffer cell inhibition of hepatocyte protein synthesis *in vitro*, *J. Exp. Med.*, 169, 1467, 1989.
6. **West, M. A., Billiar, T. R., Curran, R. D., Hyland, B. J., and Simmons, R. L.,** Kupffer cell-mediated alterations of hepatocyte protein synthesis: evidence for separate mechanism in increases and decreases of protein synthesis, *Gastroenterology*, 196, 1572, 1989.
7. **Wolfe, R. R., Miller, H. I., and Spitzer, J. J.,** Glucose and lactate metabolism in burn and shock, *Am. J. Physiol.*, 232, 415, 1977.
8. **Rennie, M. J. and Edwards, R. H. T.,** Carbohydrate metabolism of skeletal muscle and its disorders, in *Carbohydrate Metabolism and Its Disorders*, F. Dickens, P. J. Randle, and W. Whelan, Eds., Academic Press, London, 1968, chap. 3.
9. **Clowes, G. H. A., Randal, H. R., and Cha, C. J.,** Amino acid and energy metabolism in septic and traumatized patients, *J. Paren. Enter. Nutr.*, 4, 195, 1980.

10. **Ryan, N. T.,** Metabolic adaptations for energy production during trauma and sepsis, *Surg. Clin. N. Am.,* 56, 999, 1976.
11. **Spitzer, J. J., Bagby, G. J., Meszaros, K., and Lang, C. H.,** Alterations in lipid and carbohydrate metabolism in sepsis, *J. Paren. Enter. Nutr.,* 8, 53S, 1988.
12. **Imamura, M., Clowes, G. H. A., Blackburn, G. L., O'Donnell, T. F., Trerice, M., Bhimjee, Y., and Ryan, N. T.,** Liver metabolism and gluconeogenesis in trauma and sepsis, *Surgery,* 77, 868, 1975.
13. **Burke, J. F., Wolfe, R., and Mullaney, J.,** Glucose requirements following burn injury. Parameters of optimal glucose infusion and possible hepatic and respiratory abnormalities following excessive glucose uptake, *Ann. Surg.,* 190, 274, 1979.
14. **Long, C., Kinney, J., and Greiger, J.,** Nonsuppressibility of gluconeogenesis by glucose in septic patients, *Metabolism,* 25, 193, 1976.
15. **Barton, R. N. and Passingham, B. J.,** Evidence for a role of glucocorticoids in the development of insulin resistance after ischemic limb injury in the rat, *J. Endocrinol.,* 86, 363, 1986.
16. **Clowes, G. H. A., George, B. C., Villee, C. A., and Saravis, C. A.,** Muscle proteolysis induced by a circulating peptide in patients with sepsis or trauma, *N. Engl. J. Med.* 308, 545, 1983.
17. **Alberti, K. G., Batsone, G. F., Foster, K., and Johnston, D. G.,** Relative role of various hormones in mediating the metabolic response to injury, *J. Paren. Enter. Nutr.,* 4, 141, 1980.
18. **Bagley, G. J., Lang, C. H., Hargove, D. M., Thompson, J. J., Wilson, L. A., and Spitzer J. J.,** Glucose kinetics in rats infused with endotoxin-induced monokines or tumor necrosis factor, *Circ. Shock,* 24, 111, 1988.
19. **Meszarus, K., Lang, C. H., and Bagby, G. J.,** Tumor necrosis factor increases in vivo glucose utilization of macrophage rich tissues, *Biochem. Biophys. Res. Commun.,* 149, 1, 1987.
20. **Lang, C. H., Bagby, G. J., and Ferguson, J. L.,** Cardiac output and the redistribution or organ blood flow in hypermetabolic sepsis, *Am. J. Physiol.,* 246, R331, 1984.
21. **Kuiper, J., Zijlstra, F. J., Kamps, J. A. A. M., and van Berkel, T. J. C.,** Identification of prostaglandin D2 as the major eicosanoid from liver endothelial and Kupffer cells, *Biochim. Biophys. Acta,* 959, 143, 1988.
22. **Kuiper, J., DeRoike, Y. B., Zijlstra, F. J., Van Waas, M. P., and van Berkel, T. J. C.,** The induction of glycogenolysis in the perfused liver by platelet activating factor is mediated by prostaglandin D2 from Kupffer cells, *Biochem. Biophys. Res. Commun.,* 157, 1288, 1988.
23. **Shukla, S. D., Buxton, D. B., Olson, M. S., and Hanahan, D. J.,** Acetyl glyceryl ether phosphorylcholine. A potent activator of hepatic phosphoinositide metabolism and glycogenolysis, *J. Biol. Chem.,* 258, 10212, 1983.
24. **Buxton, D. B., Shukla, S. D., Hanihan, J. J., and Olsen, M. S.,** Stimulation of hepatic glycogenolysis by acetylglyceryl ether phosphorylcholine, *J. Biol. Chem.,* 259, 1468, 1986.
25. **Hanahan, D. J., Demopoulos, C. A., Liegr, J., and Pinchard, R. N.,** Identification of platelet-activating factor isolated from rabbit basophils as acetyl glycerylether phosphorylcholine, *J. Biol. Chem.,* 255, 5514, 1980.

26. **McManus, L. M., Hanahan, D. J., and Pinchard, R. N.,** Human platelet stimulation by acetyl glyceryl ether phosphorylcholine, *J. Clin. Invest.,* 67, 903, 1981.
27. **Shukla, S. D. and Hanahan, D. J.,** AGEPC (platelet activating factor) induced stimulation of rabbit platelets: effects on phosphatidylinositol, di- and triphosphoinositides and phosphatidic acid metabolism, *Biochem. Biophys. Res. Commun.,* 106, 697, 1982.
28. **Billah, M. M. and Lapetina, E. G.,** Platelet-activating factor stimulates metabolism of phosphoinositides in horse platelets: possible relationship to a Ca^{2+} mobilization during stimulation, *Proc. Natl. Acad. Sci. U.S.A.,* 80, 965, 1983.
29. **Hems, D. A. and Whitton, P. D.,** Control of hepatic glycogenolysis, *Physiol. Rev.,* 60, 1, 1980.
30. **Fisher, R. A., Shulka, S. D., Debuysere, M. S., Hanahan, D. S., and Olson, M., S.,** The effect of acetylglyceryl ether phosphorylcholine on gluconeogenesis and phosphatidylinositol 4,5, bisphosphate metabolism in rat hepatocytes, *J. Biol. Chem.,* 259, 8685, 1984.
31. **Okumara, T. and Saito, K.,** Effect of prostaglandins on glycogenesis and glycogenolysis in primary cultures of rat hepatocytes — a role of prostaglandin D2 in the liver, *Prostaglandins,* 39, 525, 1990.
32. **Tanaka, K., Sato, M., Tomita, Y., and Ichikara, A.,** Biochemical studies on liver function in primary cultures of hepatocytes in adult rats, *J. Biochem. (Tokyo),* 84, 987, 1978.
33. **Fisher, R. A., Sharma, R. V., and Bhalla, R. C.,** Platelet-activating factor increases inositol phosphate production and cytosolic free Ca^{2+} concentrations in cultured rat Kupffer cells, *FEBS Lett.,* 251, 22, 1989.
34. **LaPoint, D. S. and Olson, M. S.,** Vasoactive and glycogenolytic effects of AGEPC in the perfused rat liver are not mediated by cyclooxygenase derived metabolites, *J. Biol. Chem.,* 107, 858a, 1988.
35. **Frayn, K. N.,** Hormonal control of metabolism in trauma and sepsis, *Clin. Endocrinol.,* 24, 577, 1986.
36. **Wolfe, R. R., Shaw, J. H. F., and Durkot, M. J.,** Effect of sepsis on VLDL kinetics: responses in basal state and during glucose infusion, *Am. J. Physiol.,* 248, E732 1985.
37. **Nanni, G., Seigel, J. H., Coleman, B., Fader, P., and Castiglione, R.,** Increased lipid fuel dependence in the critically ill septic patient, *J. Trauma,* 24, 14, 1984.
38. **Spitzer, J. J.,** Lipid metabolism in endotoxin shock, *Circ. Shock,* 1, (Suppl.), 69, 1979.
39. **Miller, H. I., Issekutz, B., and Paul, P.,** Effect of lactic acid on plasma free fatty acid in pancreatectomized dogs, *Am. J. Physiol.,* 207, 1226, 1964.
40. **Stoner, H. B. and Matthews, J.,** Studies in the mechanism of shock. Fat mobilization after injury, *Br. J. Exp. Pathol.,* 48, 58, 1967.
41. **Kaufamm, R. L., Matson, C. R., and Beisel, W. R.,** Hypertriglyceridemia produced by endotoxin: role of impaired triglyceride disposal mechanisms, *J. Infect. Dis.,* 133, 548, 1976.
42. **Guckian, J. C.,** Role of metabolism in the pathogenesis of bacteremia due to diplococcus pneumoniae in rabbits, *J. Infect. Dis.,* 127, 1, 1973.
43. **Vranic, M.,** Turnover of free fatty acid and triglyceride: an overview, *Fed. Proc. Fed. Am. Soc. Exp. Biol.,* 34, 368, 1975.

44. **Frayn, K. N., Little, R. A., Stoner, H. B., and Galaski, C. S. B.,** Metabolic control in non-septic patients with musculo-skeletal injuries, *Injury,* 16, 73, 1984.
45. **Frayn, F. N.,** Substrate turnover after injury, *Br. Med. Bull.,* 41, 232, 1985.
46. **Fielding, C. J. and Havel, R. J.,** Lipoprotein lipase, *Arch. Pathol. Lab. Med.,* 101, 225, 1977.
47. **Bagby, G. J. and Pekala, P. H.,** Lipoprotein lipase in trauma and sepsis, in *Lipoprotein Lipase,* J. Borensztajn, Ed., Evener, Chicago, 1987, 247.
48. **Pekala, P. H., Kawakami, M., Angus, C. W., Lane, M. D., and Cerami, A.,** Selective inhibition of synthesis of enzymes for de novo fatty acid biosynthesis by an endotoxin-induced mediator from exudate cells, *Proc. Natl. Acad. Sci. U.S.A.,* 80, 2743, 1983.
49. **Kawakami, M. and Cerami, A.,** Studies of endotoxin-induced decrease in lipoprotein lipase activity, *J. Exp. Med.,* 154, 631, 1981.
50. **Smith, W. L.,** The eicosanoids and their biochemical mechanisms of action, *Biochem. J.,* 259, 315, 1989.
51. **Irvine, R. W.,** How is the level of free arachidonic acid controlled in mammalian cells?, *Biochem. J.,* 204, 3, 1982.
52. **Lands, W. E. M. and Samuelsson, B.,** Phospholipid precursors of prostaglandins, *Biochim. Biophys. Acta,* 164, 426, 1968.
53. **Magrum, P. I. and Johnson, M.,** Modulation of prostaglandin synthesis in rat peritoneal macrophages with omega-3 fatty acids, *Lipids,* 18, 514, 1983.
54. **Hannahan, D. J. and Kumar, R.,** Platelet activating factor: chemical and biochemical characteristics, *Prog. Lipid Res.,* 26, 1, 1987.
55. **Fletcher, J. R.,** The role of prostaglandins in sepsis, *Scand. J. Infect. Dis. (Suppl.),* 31, 55, 1982.
56. **Lefer, A. M.,** Role of prostaglandins and thromboxanes in shock states, in *Handbook of Shock and Trauma,* B. M. Altura, A. M. Lefer, and W. Schumer, Eds., Raven Press, New York, 1982, 355.
57. **Halushka, P. V., Wise, W. C., and Cook, J. A.,** Protective effects of aspirin in endotoxin shock, *J. Pharm. Exp. Ther.,* 218, 464, 1981.
58. **Fletcher, F. R. and Ramwell, P. W.,** Modification, by aspirin and indomethacin, of the haemodynamic and prostaglandin releasing effects of E coli endotoxin in the dog, *Br. J. Pharmacol.,* 61, 175, 1977.
59. **Parratt, J. R. and Sturgess, R. M.,** The protective effect of sodium meclofenamate in experimental endotoxin shock, *Br. J. Pharmacol.,* 53, 466P, 1976.
60. **Cook, J. A., Wise, W. C., Butler, R. R., Reines, H. D., Rambo, W., and Halushka, P. V.,** The potential role of thromboxane and prostacyclin in endotoxin and septic shock, *Am. J. Emerg. Med.,* 2, 28, 1984.
61. **Cook, J. A., Wise, W. C., Knapp, D. R., and Halushka, P. V.,** Sensitization of essential fatty acid-deficient rats to endotoxin by arachidonate pretreatment: role of thromboxane, *Circ. Shock,* 8, 69, 1981.
62. **Ball, H. A., Cook, J. A., Wise, W. C., and Halushka, P. V.,** Role of thromboxane, prostaglandins and leukotrienes in endotoxin and septic shock, *Intensive Care Med.,* 12, 116, 1986.
63. **Reines, H. D., Halushka, P. V., Cook, J. A., Wise, W. C., and Rambo, W.,** Plasma thromboxane concentrations are raised in patients dying with septic shock, *Lancet,* July 24, 174, 1982.

64. **Rie, M., Peterson, M., Kong, D., Quinn, D., and Watkins, D.,** Plasma prostacyclin increases during acute human sepsis, *Circ. Shock,* 10, 232, 1983.
65. **Lefer, A. M.,** Significance of lipid mediators in shock states, *Circ. Shock,* 27, 3, 1989.
66. **Decker, K.,** Eicosanoids, signal molecules of liver cells, *Sem. Liver Dis.,* 5, 175, 1985.
67. **Perez, H. D., Roll, F. J., and Bissell, D. M.,** Production of chemotactic activity for polymorphonuclear leukocytes by cultured rat hepatocyes exposed to ethanol, *J. Clin. Invest.,* 74, 1350, 1984.
68. **Dawson, W., Jessup, S. J., McDonald-Gibson, W., Ramwell, P. W., and Shaw, J. E.,** Prostaglandin uptake and metabolism by the perfused rat liver, *Br. J. Pharmacol.,* 39, 585, 1970.
69. **Anggard, E. and Samuelsson, B.,** Prostaglandins and related factors. Metabolism of prostaglandin E1 in guinea-pig lung: the structures of two metabolites, *J. Biol. Chem.,* 239, 4097, 1964.
70. **Tran-Thi, T. A., Gyufko, K., Henninger, H., Busse, R., and Decker, K.,** Studies on synthesis and degradation of eicosanoids by rat hepatocytes in primary culture, *J. Hepatol.,* 5, 322, 1987.
71. **Sego, T., Nakayama, R., Okumura, T., and Saito, K.,** Metabolism of prostaglandins D2 and F2 alpha in primary cultures of rat hepatocytes, *Biochim. Biophys. Acta,* 879, 330, 1986.
72. **Hagmann, W., Denzlinger, C., and Keppler, D.,** Role of peptide leukotrienes and their hepatobiliary elimination in endotoxin action, *Circ. Shock,* 14, 223, 1984.
73. **Nunes, G., Blaisdell, F. N., and Margaretten, W.,** Mechanism of hepatic dysfunction following shock and trauma, *Arch. Surg.,* 100, 546, 1970.
74. **Holman, J. M. and Saba, T. M.,** Hepatocyte injury during post-operative sepsis: activated neutrophils as potential mediators, *J. Leuk. Biol.,* 43, 193, 1988.
75. **Saba, T., Tanaka, J., Yukihiro, K., Jones, R. T., Cowley, A., and Trump, P. F.,** Hepatic cellular injury following lethal Escherichia coli bacteremia in rats, *Lab. Invest.,* 47, 304, 1982.
76. **Spitzer, J. A., Turco, E. R., Deaciuc, I. V., and Ruth, B. L.,** Perturbation of transmembrane signaling mechanism in acute and chronic endotoxemia, in *Pathophysiological Role of Mediators and Mediator Inhibitors in Shock,* Alan R. Liss, New York, 1987, 401.
77. **Rodriguez de Turco, E. B. and Spitzer, J. A.,** Metabolic fate of arachidonic acid in hepatocytes of continuously endotoxemic rats, *J. Clin. Invest.,* 81, 700, 1988.
78. **Rodriguez de Turco, E. B. and Spitzer, J. A.,** Early effects of Escherichia coli endotoxin infusion on vasopressin-stimulated breakdown and metabolism of inositol lipids in rat hepatocytes, *Biochem. Biophys. Res. Commun.,* 155, 151, 1988.
79. **Rodriguez de Turco, E. B. and Spizer, J. A.,** Impairments in vasopressin-stimulated inositol lipid metabolism in hepatocytes of septic animals, *Circ. Shock,* 25, 299, 1988.
80. **Voelkel, E. F., Levine, L., Alper, C. A., and Tashjian, A. H.,** Acute phase reactants ceruloplasmin and haptoglobin and their relationship to plasma prostaglandins in rabbits bearing the XV2 carcinoma, *J. Exp. Med.,* 78, 1078, 1977.

81. **Pepys, M. B. and Baltz, M. L.,** Acute phase proteins with special reference to C-reactive protein and related proteins (Pentaxins) and serum amyloid A protein, in *Advances in Immunology,* Vol. 34, Academic Press, New York, 1983, 141.
82. **Kunkel, S. L., Wiggin, R. C., Chensue, S. W., and Larrick, J.,** Regulation of macrophage tumor necrosis factor production by prostaglandin E2, *Biochem. Biophys. Res. Commun.,* 137, 404, 1986.
83. **Taffet, S. B. and Russel, S. W.,** Macrophage-mediated tumor cell killing: regulation of expression of cytolytic activity by prostaglandin E, *J. Immunol.,* 126, 424, 1981.
84. **Billiar, T. R., Lysz, T. W., Curran, R. D., Bentz, B. G., Machiedo, G. W., and Simmons, R. L.,** Hepatocyte modulation of Kupffer cell prostaglandin E2 production *in vitro, J. Leuk. Biol.,* 47, 304, 1990.
85. **Lysz, T. W., Billiar, T. R., Curran, R. D., Simmons, R. L., and Machiedo, G. W.,** Kupffer cell-hepatocyte interactions and the changes in 1-[^{14}C]-arachidonate incorporation in response to endotoxin in vitro, *Prostaglandins,* 39, 497, 1990.
86. **Wright, S. D., Tobias, R. S., Ulevitch, R. J., and Ramos, R. A.,** Lipopolysaccharide (LPS) binding protein opsonizes LPS-bearing particles for recognition by a novel receptor on macrophages, *J. Exp. Med.,* 170, 1231, 1989.
87. **Tobias, P. S., Mathison, J. C., and Ulevitch, R. J.,** A family of lipopolysaccharide binding proteins in response to gram-negative sepsis, *J. Biol. Chem.,* 263, 13479, 1988.
88. **Tobias, P. S., Soldau, K., and Ulevitch, R. J.,** Identification of a lipid A binding site in the acute phase reactant lipopolysaccharide binding protein, *J. Biol. Chem.,* 264, 10867, 1989.
89. **Nolan, J. R.,** The role of endotoxin in liver injury, *Gastroenterology,* 69, 1346, 1975.

INDEX

D

E

F

G

H

I

K

L

U–V

W–Z